This graduate/research level text introduces the theory of multi-electron transitions in atomic, molecular and optical physics, emphasizing the emerging topic of dynamic electron correlation.

The book begins with an overview of simple binomial probabilities, classical scattering theory, quantum scattering and correlation, followed by the theory of single electron transition probabilities. Multiple electron transition probabilities are then treated in detail. Various approaches to multiple electron transitions are covered including the independent electron approximation, useful statistical methods and peturbation expansions treating correlation in both weak and strong limits. The important topic of the dynamics of electron correlation is a central theme in this book. The text contains a comprehensive summary of data for few and many electron transitions in atoms and molecules, including transitions on different atomic centers, fast ion-atom and electron-atom interactions, and recent observations using synchrotron radiation. Emphasis is given to methods that may be used by non-specialists.

This text provides a pedagogic introduction to graduate students and researchers new to this developing field, but will also serve as a valuable reference for atomic, chemical and optical scientists interested in correlation and multi-electron transitions.

CAMBRIDGE MONOGRAPHS ON ATOMIC, MOLECULAR AND CHEMICAL PHYSICS

ELECTRON CORRELATION DYNAMICS IN ATOMIC COLLISIONS

CAMBRIDGE MONOGRAPHS ON ATOMIC, MOLECULAR AND CHEMICAL PHYSICS

1. R. Schinke **Photodissociation Dynamics**
2. L. Frommhold **Collision-Induced Absorption in Gases**
3. T. F. Gallagher **Rydberg Atoms**
4. M. Auzinsh and R. Ferber **Optical Polarization of Molecules**
5. I. E. McCarthy and E. Weigold **Electron–Atom Collisions**
6. V. Schmidt **Electron Spectrometry of Atoms using Synchrotron Radiation**
7. Z. Rudzikas **Theoretical Atomic Spectroscopy**
8. J. H. McGuire **Electron Correlation Dynamics in Atomic Collisions**

Electron Correlation Dynamics in Atomic Collisions

J. H. McGUIRE
Tulane University

CAMBRIDGE UNIVERSITY PRESS
Cambridge, New York, Melbourne, Madrid, Cape Town, Singapore, São Paulo

Cambridge University Press
The Edinburgh Building, Cambridge CB2 2RU, UK

Published in the United States of America by Cambridge University Press, New York

www.cambridge.org
Information on this title: www.cambridge.org/9780521480208

First published 1997
This digitally printed first paperback version 2005

A catalogue record for this publication is available from the British Library

Library of Congress Cataloguing in Publication data

McGuire, J. H. (James Horton), 1942–
Electron correlation dynamics in atomic collisions / J.H. McGuire.
p. cm. – (Cambridge monographs on atomic, molecular, and chemical physics ; 8)
Includes bibliographical references.
ISBN 0 521 48020 5
1. Collisions (Nuclear physics) 2. Electron configuration.
3. Atomic transition probabilities. I. Title. II. Series.
QC794.6.C6M46 1997
539.7′2112–dc21 96–44743 CIP

ISBN-13 978-0-521-48020-8 hardback
ISBN-10 0-521-48020-5 hardback

ISBN-13 978-0-521-01859-3 paperback
ISBN-10 0-521-01859-5 paperback

Dedicated to

Meredith Mallory, Jr.

Contents

Preface

One of the central questions of science is: how are complex things made from simple things? In many cases larger systems are more complicated than their smaller subsystems. In biology and chemistry the issue is how to understand large molecules in terms of atoms. In atomic physics one may strive to understand properties of many electron systems in terms of single electron properties. The general theme is interdependency of subsystems, or 'correlation'.

Correlation may be regarded as a conceptual bridge from properties of individuals to properties of groups or families. In atoms and molecules correlation occurs because electrons interact with one another – the electrons are interdependent. This electron correlation determines much of the structure and dynamics of many electron systems, i.e., how complex electronic systems are made from single electrons. Complexity is the more significant idea, but complexity may be seldom, if ever, understood. Correlation is the key to complexity.

Understandably, much has been done on the correlation of static systems. There are many excellent methods and computer codes to evaluate energies and wavefunctions for complex atomic and molecular systems. However, the dynamics of these many electron systems is less well understood. Hence, the dynamics of electron correlation is a central theme in this book.

The dynamics of electron correlation may affect single electron transitions. However, this effect is sometimes difficult to separate from other effects. Correlation is usually dominant in multiple electron transitions for fast collisions since there is not enough time for the collision partners to interact more than once. This means that multiple electron transitions in fast collisions provide an unobstructed view of the dynamics of electron correlation.

Since there are numerous processes that depend on the transition of more than a single electron, multiple electron transitions are of interest in their own right. Also recent advances in the production of highly charged ions and in synchrotron radiation provide new experimental tools for studying multiple electron transitions.

So the motivation for this book is to look at an emerging topic that is of interest in its own right and that impacts on a question of broad scientific interest, namely, the question of correlation (or the many body problem, or the construction of complex atomic systems ...).

The audience for whom this book is intended includes specialists in the field, graduate students and engineers. The first chapter includes a review of basic concepts in simple probability, classical scattering theory and a brief introduction to quantum mechanics. Then a background in one electron transitions in atomic and molecular transitions is developed. The third chapter addresses the formulation of interactions in many electron atomic and molecular systems. Later chapters deal with interactions of many electron targets with charged particles, with projectiles carrying electrons, and with photons.

An effort has been made to include practical methods for analysis and interpretation of data which can be used without extensive theoretical background or large computer codes. While much of this book focuses on relatively simple fast interactions, examples for intermediate and slow interactions related to chemistry, biology and condensed matter are also included. The emphasis in this book is on ideas and techniques that are simple enough for a beginning graduate student to pick up quickly and interesting enough to attract the attention of a person with a little curiosity.

Jim McGuire
Murchison-Mallory Professor of Physics

1
Introduction

This introductory chapter begins with a review of uncorrelated classical probabilities and then extends these concepts to correlated quantum systems. This is done both to establish notation and to provide a basis for those who are not experts to understand material in later chapters.

1.1 Probability of a transition

Seldom does one know with certainty what is going to happen on the atomic scale. What can be determined is the probability P that a particular outcome (i.e. atomic transition) will result when many atoms* interact with photons, electrons or protons. A transition occurs in an atom when one or more electrons jump from their initial state to a different final state in the atom. The outcome of such an atomic transition is specified by the final state of the atom after the interaction occurs. Since there are usually many atoms in most systems of practical interest, we can usually determine with statistical reliability the rate at which various outcomes (or final states) occur. Thus, although one is unable to predict what will happen to any one atom, one may determine what happens to a large number of atoms.

1 Single particle probability

A simple basic analogy is tossing a coin or dice. Tossing of a coin is analogous to interacting with an atom. In the case of a simple coin there

* Kalmer [1995] argues that the concept of a transition probability P for a single atomic event is not mathematically or conceptually well defined. Knowing that the probability for an atomic transition is 50% does not predict what will happen to any one specified atom any more than knowing that the probability for rain is 50% will give an unambiguous prediction of how much rain there will be on a specific day.

are two outcomes: after the toss one side of the coin ('heads') will either occur or it will not occur. Thus, the outcome of this dynamic process is either 'heads' or 'not heads'. Under normal circumstances one cannot predict with certainty whether a particular coin toss will result in 'heads' or 'not heads', but the probability P that 'heads' will occur is usually known (e.g. $P = 1/2$ if the probability of both outcomes is the same). Since either 'heads' or 'not heads' must occur, the probability for 'not heads' is

$$Q = 1 - P . \tag{1.1}$$

The outcome of any one coin toss cannot be predicted with certainty, although the probabilities P and $Q = 1 - P$ may usually be determined either experimentally or theoretically. If there are N events, then one can predict with statistical certainty that NP events will be 'heads' and NQ events will be 'not heads'.

With real coins it is possible to weight the coins so that P and Q differ from 1/2. After many tosses ($N >> 1$) with weighted coins NP of the coins will have the outcome 'heads' and NQ will be 'not heads' with $NP + NQ = N$.

Atomic transitions involving a single electron follow this same pattern. The probability P that a given outcome (called a 'transition to a final state') will occur is usually known (i.e. may be calculated). Methods for calculating P are given in chapter 2. If something (such as a beam of particles or a transient electric field) interacts with (i.e. 'tosses') a system of N_A atoms, one may calculate P and find that $N_A P$ electron transitions will occur as a result of the interaction and $N_A Q$ will not occur. If the interaction reoccurs at a known rate, the rate at which electron transitions occur can be determined. In principle such an analysis may be used to understand processes ranging from the operation of electrical devices (e.g. storage of binary information in a computer) to chemical and biological reactions (e.g. the rate at which food is metabolized).

2 *Two particle probability*

Let us now extend the above example to the case of two weighted coins. If coin 1 has a probability P_1 that the outcome 'heads' will occur, and coin 2 has probability P_2, then the probability that both events occur (i.e. that both coins end as 'heads') is,

$$P = P_1 \cdot P_2 . \tag{1.2}$$

Here the P_j ($j = 1, 2$) are probabilities of the outcome 'heads' for the independent coins 1 and 2. One may use this to model an atomic system with two distinguishable independent electrons (1 and 2) each of which has two states $|f>$ and $|i>$. Here $|f>$ means that a transition has occurred (e.g. 'heads') and $|i>$ means that the transition has not occurred ('not heads'). Since total probability is conserved, then after a collision each electron has either made a transition to a final state $|f>$ or has not and remains in the initial state $|i>$. This means that,

$$P_j + Q_j = 1\,, \tag{1.3}$$

which corresponds to (1.1) above.

After an atomic interaction (corresponding to tossing two coins) there are four possible states:

neither makes a transition	P_{00}^2	$=$	$(1-P_1)(1-P_2)$	$\xrightarrow{P_1=P_2=P}$	$(1-P)^2$
only #1 makes a transition	P_{10}^2	$=$	$P_1(1-P_2)$	$\xrightarrow{P_1=P_2=P}$	$P(1-P)$
only #2 makes a transition	P_{01}^2	$=$	$P_2(1-P_1)$	$\xrightarrow{P_1=P_2=P}$	$P(1-P)$
both make a transition	P_{11}^2	$=$	P_1P_2	$\xrightarrow{P_1=P_2=P}$	P^2
	sum=1		sum=1		sum=1

The right hand column gives the uncorrelated probabilities when $P_1 \to P_2 = P$. All columns sum to 1 reflecting the fact that the total probability for all probabilities sums to one.

It is at this point straightforward to list all uncorrelated probabilities for a system of two distinguishable electrons with three possible states: ionization, capture and elastic scattering (no transition), where P_I is the ionization probability, P_C is the capture probability and $Q = 1 - P_C - P_I$ is the probability for elastic scattering. In this case our analogy to tossing coins corresponds to throwing two coins each of which has three sides (or tossing the same coin twice if the coins [atoms] are identical as in the right hand column in the table above).

In the right hand column of the above table where $P_1 = P_2 = P$, one can write the uncorrelated probability that $n = 0, 1, 2$ of the two electrons made a transition as,

$$P = \binom{2}{n} P^n (1-P)^{2-n}\,, \tag{1.4}$$

where $\binom{2}{n}$ is the binomial coefficient, i.e., $\binom{2}{0} = 1$. $\binom{2}{1} = 2$ and $\binom{2}{2} = 1$. Such binomial distributions are commonly encountered [Eadie *et al.*, 1971] in statistical physics.

3 N particle binomial probability

A simple and useful generalization of (1.4) may be obtained by expanding $(P+Q)^N$ where $Q = 1 - P$, namely,

$$\begin{aligned} 1 = (P+Q)^N &= \sum_{n=0}^{N} \binom{N}{n} P^n Q^{N-n} \\ &= \sum_{n=0}^{N} \binom{N}{n} P^n (1-P)^{N-n} , \end{aligned} \tag{1.5}$$

where the binomial coefficient $\binom{N}{n} = \frac{N!}{(N-n)!n!}$ is the number of ways of arranging n of N electrons. This is the probability of the outcome of n 'heads' after tossing N identical two sided coins (where $P = 1/2$ if the coins are not unequally weighted).

The quantity,

$$P_N^n = \binom{N}{n} P^n (1-P)^{N-n} , \tag{1.6}$$

is called the probability function, corresponding to the transition of exactly n of N electrons each one of which has a single electron transition probability P. This is defined as an exclusive probability in which other possible outcomes (transitions) are excluded.

In the binomial distribution the average number $< n >$ of electrons undergoing a transition is given by $< n >= NP$, and the square of the standard deviation $\Delta n^2 =< n^2 > - < n >^2$ is found from the recursion relation $=< n(n-1) >^2 - < n^2 > - < n >= NP(1-P)$. These and other results are derived in Appendices at the end of this book.

A simple sum rule, corresponding to a useful inclusive probability, is easily developed from (1.6). Let us consider an atom with J shells (or $N = N_1 + N_2 + ... + N_J$ weighted two sided coins of which N_1 coins have a probability P_1 of 'heads', N_2 a probability P_2 of 'heads', ... and N_J with a probability P_J). There are N_1 electrons in shell 1, N_2 electrons in shell 2, ... and N_J electrons in shell J. The exclusive probability for exactly n_1 electrons undergoing a transition in shell 1, n_2 electrons in shell 2,...and n_J in shell J is given by,

$$\begin{aligned} P^{n_1+n_2..+n_J}_{N_1+N_2..+N_J} &= \binom{N_1}{n_1} P_1^{n_1}(1-P_1)^{N_1-n_1}\binom{N_2}{n_2} P_2^{n_2}(1-P_2)^{N_2-n_2}\ldots\ldots \\ &\times \binom{N_J}{n_J} P_J^{n_J}(1-P_J)^{N_J-n_J} . \end{aligned} \tag{1.7}$$

Let us assume that $P_s = P_{N_s} << 1$ in some shell s. The inclusive probability is defined as the probability that a particular outcome (or transition) occurs in any of the possible outcomes (i.e. includes all possibilities with the particular outcome). The inclusive probability for the transition of one electron from the sth shell, summing over all other shells, is,

$$\begin{aligned} P &= \sum_{n_1=0}^{N_1}\binom{N_1}{n_1} P_1^{n_1}(1-P_1)^{N_1-n_1}\ldots\binom{N_s}{1} P_s(1-P_s)^{N_s-1}\ldots \\ &\times \sum_{n_J=0}^{N_j}\binom{N_J}{n_J} P_J^{n_J}(1-P_J)^{N_J-n_J} \simeq \binom{N_s}{1} P_s = N_sP_s , \end{aligned} \tag{1.8}$$

where terms of order P_s^2 and smaller have been neglected and (1.5) is used to do the sums. The result is that the inclusive probability is equal to N_sP_s, i.e. just what one would expect if one had only N_s independent electrons with equal probability P_s.

If P is not small compared to 1 in (1.8), then one may not ignore terms of order P^2 and it may be easily shown that,

$$NP \geq \sum_{n=1}^{N}\binom{N}{n} P^n(1-P)^{N-n} \geq NP(1-P)^{N-1} , \tag{1.9}$$

where the middle term is the inclusive probability for a binomial distribution.

In most cases it is the inclusive probability (not the exclusive probability) that applies to experimental data [McGuire and Macdonald, 1975]. This extends in a useful way the overly simple notion that passive electrons are frozen. If the passive electrons are independent, a sum over all final states (corresponding to many experimental situations) gives a probability of 1 that each passive electron did or did not make a transition. In this case the passive electrons may in effect be ignored and only the active electron need be considered.

The binomial distributions of (1.5) - (1.6) may be extended to a system with three possible outcomes (such as a three sided coin shaped like a beachball [or dice] with three flat sides) with probabilities P_1, P_2 and $Q = 1 - P_1 - P_2$. Let $P = P_1 + P_2$. Then,

$$\begin{aligned} 1 &= (P+Q)^N = \sum_{n=0}^{N} \binom{N}{n} P^n Q^{N-n} \qquad (1.10) \\ &= \sum_{n=0}^{N} \binom{N}{n} (P_1+P_2)^n Q^{N-n} = \sum_{n=0}^{N} \binom{N}{n} \sum_{k=0}^{n} \binom{n}{k} P_1^k P_2^{n-k} Q^{N-n} , \end{aligned}$$

so that the exclusive probability of having 'k' transitions into state 1, 'l' transitions into state 2 and 'm' in the remaining state in a system with N interactions is,

$$P_N^{k\ l\ m} = \binom{N}{n} \sum_{k=0}^{n} \binom{n}{k} P_1^k P_2^l Q^m . \qquad (1.11)$$

where $l = n - k$, $m = N - n$ and $k + l + m = N$. The general case with an arbitrary number of possible outcomes may be generated using multinomial distributions discussed next. Eq(1.11) corresponds to a multinomial distribution with three possible outcomes.

There are numerous applications of binomial distributions in science ranging from atomic scattering [McGuire, 1992] to collisions with solid surfaces [Vana *et al.*, 1995] to electrical stimuli of cells on earthworms [Manivannan *et al.*, 1992].

4 *Multinomial probabilities*

It is not necessary to use the same value of P for all electrons in a given shell. Different electrons may have different values of P (e.g., due to different binding energies). In this case a multinomial distribution [Eadie *et al.*, 1971; Aberg, 1987; Abramowtiz and Stegun, 1964, chapter 24] may be used, corresponding to,

$$1 = (P_1 + P_2 + ...P_m)^N = N! \sum_{n_1, n_2 ... n_m}^{N} \prod_{i=1}^{m} P_i^{n_i} / n_i! , \qquad (1.12)$$

where $\sum_{i=1}^{m} P_i = 1$ and $\sum^N$ is restricted by the condition that $n_1 + n_2 + ...n_m = N$ and $P_1 + P_2 + ... + P_m = 1$. In this distribution m final states for each electron (or sides of a coin or dice) are possible instead of 2 as in the binomial distribution of (1.6).

Both binomial and multinomial distributions are used in various independent electron approximations used to describe multiple electron transitions (Cf. section 4.1.4). Binomial distributions apply if the single electron transition probabilities P are the same (or nearly the same) for a group (or shell) of N electrons.

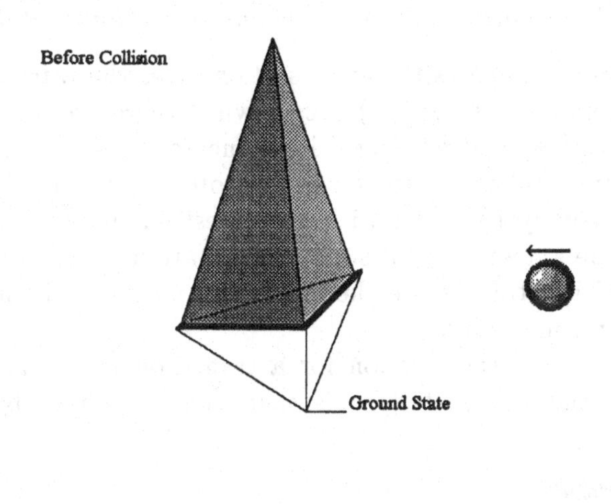

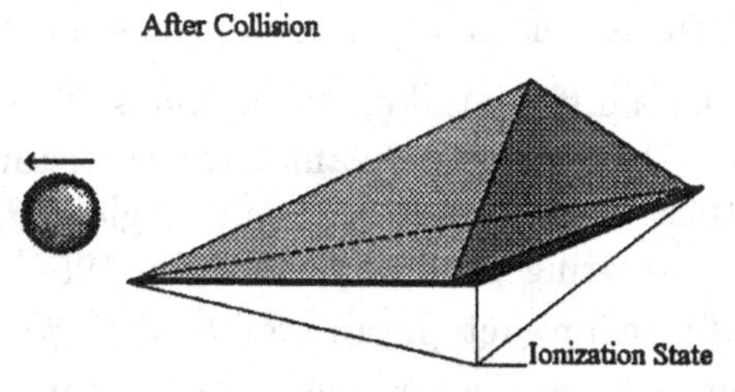

Fig. 1.1. The probability of a particular outcome is proportional to the cross section (i.e. cross sectional area) corresponding to the outcome.

1.2 Cross sections and reaction rates

One would often like to know what can happen to a system under certain conditions, the frequency with which a particular outcome occurs, and the mechanisms responsible for each of the outcomes. The concept of a reaction cross section is useful in addressing each of these three queries. Knowing the cross section for a particular outcome with given initial conditions enables one to determine whether a given outcome is possible, the rate at which it occurs, and how one may model the mechanisms that describe the change from the initial condition of the system to a particular final outcome (figure 1.1). The concept of a cross section connects relatively small atoms (typically 10^{-10} meters) with much larger detectors (typically 10^{-1} meters). This happens because the scattering angle θ does not depend on scale, i.e., $\theta_{atom} = \theta_{detector} = \theta$.

1 Differential cross sections and transition probabilities

For reactions with atoms the cross section gives the rate at which a particular outcome (or transition) occurs when there are a large number of atoms undergoing various transitions due to some external interaction (often a beam of photons, electrons or protons). The initial state $|i>$ of the atoms (usually the state with the lowest energy or 'ground state') is assumed to be known and for simplicity all atoms are usually assumed to be in this initial state before the interaction begins. The final state $|f>$ of these atoms may differ.

The differential cross section for a transition from the initial atomic state $|i>$ to the final state $|f>$ is operationally defined by,

$$d\sigma / sin\theta d\theta d\phi = \frac{\text{number of transitions per time from } |i> \text{ to } |f> \text{ going into } (\theta, \phi)}{\text{number of incident particles per time per area in the incident beam}} . \tag{1.13}$$

A typical picture describing a differential cross section is shown in figure 1.2. All of the projectiles in the incident beam have the same incident velocity $\vec{v}$. These projectiles scatter into different angles (θ, ϕ). In a classical description of the scattering process [Goldstein, 1950] the angles (θ, ϕ) corresponds to a definite impact parameter $\vec{b}$ = (b,ϕ) (where ϕ_b is the same as ϕ). The impact parameter $\vec{b}$ is equal to the distance of closest approach that the incident particle would have if it did not scatter (i.e. if it stayed on a straight line trajectory). Note that $\vec{b} \cdot \vec{v} = 0$ where $\vec{v}$ is the velocity of the projectile before the interaction. In figure 1.2 only scattering from one atom is shown for a definite impact parameter $\vec{b}$. The angle of scattering (θ, ϕ) may be classically determined from the impact parameter $\vec{b}$ and the interaction potential V between the incoming projectile and the target atom.

It is explained in introductory texts in classical mechanics [Goldstein, 1950] that,

$$\begin{aligned} \left(\frac{d\sigma}{sin\theta d\theta d\phi} \right) sin\theta d\theta d\phi &= \left(\frac{d\sigma}{sin\theta d\theta d\phi} \right) \left(\frac{d(cos\theta, \phi)}{d\vec{b}} \right) d\vec{b} \\ &= P(\vec{b}) d\vec{b} \, . \end{aligned} \tag{1.14}$$

Here $d\vec{b} = db_x db_y = bdbd\phi$ is the area of the ring between $\vec{b}$ and $\vec{b}+d\vec{b}$ which corresponds to the solid angle $d\Omega = sin\theta d\theta d\phi = |d(cos\theta)| d\phi$ into which the projectile scatters†. $P(\vec{b})$ is the probability that a transition from $|i>$

† Because θ usually increases as b decreases, $d\theta / db < 0$ normally.

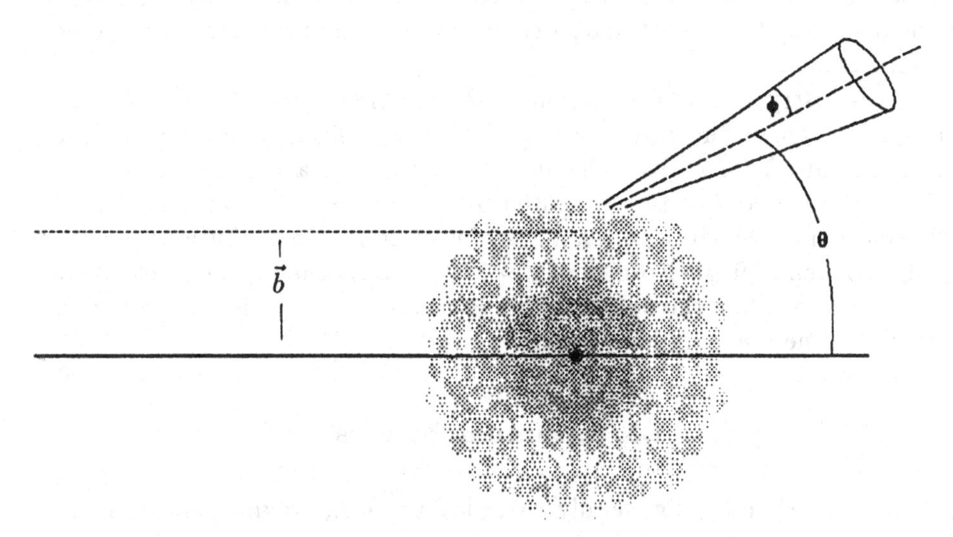

Fig. 1.2. The differential cross section $d\sigma$ is the area between the impact parameter $\vec{b}$ and $\vec{b} + d\vec{b}$ weighted with the probability P of a transition.

to $|f>$ occurs at $\vec{b}$. Thus a cross section corresponds to a geometric area weighted with the probability that a transition occurs. This is analogous to tossing N identical coins [or atoms] shaped like a multi-sided beachball [or dice] with flat sides of different area with a probability P_f of landing on side f. The probability of landing on side f is proportional to the area of that side.

All atoms of the same element are identical. For the same transition (or outcome) all atoms scatter in the same way and give the same area $d\sigma$. This area varies with the outcome (or final state) of the reaction because some outcomes have a higher probability than others.

2 Total cross section

The total cross section is the integral of the differential cross section into (θ, ϕ) over all (θ, ϕ). Using (1.14) one has,

$$\sigma = \int (d\sigma / sin\theta d\theta d\phi))\ sin\theta d\theta d\phi = \int P(\vec{b}) d\vec{b}\,. \tag{1.15}$$

The total cross section is the probability of a particular transition integrated over all impact parameters.

Both σ and $P(\vec{b})$ are experimentally observable. $P(\vec{b})$ is found by observing only those transitions that occur at a particular impact parameter $\vec{b}$ (i.e. a particular scattering angle, (θ, ϕ)). The total cross section σ is integrated over $\vec{b}$. The relation between (θ, ϕ) and the impact parameter $\vec{b}$ provides a way to connect macroscopic outcomes and rates to microscopic atomic transitions.

In chemistry and biology atomic cross sections are necessary for the determination of the rates for chemical and molecular reactions. The rate constant k for a given chemical reaction is an average over velocity of the atomic cross section weighted by an appropriate over a statistical distribution of velocities (often a Maxwell-Boltzman distribution) [Levine and Bernstein, 1974]. Note that $\sigma \sim 1/\mathrm{v} \rightarrow \infty$ as the incident velocity v tends to zero, when the outgoing flux is greater than the incoming flux in reactions where energy is released.

1.3 Quantum features

Classical mechanics is generally regarded as an approximation to quantum mechanics which is necessary to provide an adequate description of atoms and their interactions.‡ Classical mechanics (excluding classical wave phenomena) denotes Newtonian physics of non-interfering localized particles (or distributions of particles). On the atomic scale interference occurs. Locality of any finite object is unrealistic: no object is confined to a single point in space or time. That is, non-locality is required by finiteness§. An overly simple wave description which uses a single frequency (or wavelength) is also unrealistic since an infinite universe is required for such a perfectly non-local wave. Quantum mechanics is a model which describes objects of finite size in both space and time using wavepackets¶. Fourier analysis is a central mathematical tool in quantum

‡ Gryzinski [1987a,b] adopts the unconventional point of view that quantum mechanics may be understood in terms of classical mechanics. He argues, for example, that the de Broglie wavelength λ may be obtained by requiring that the phase of the spin of the electron increase by 2π over a distance of length λ and that the Schrödinger equation follows from the stability of mechanical systems in the presence of perturbations.

§ While locality of an object may be unrealistic, locality of space and time is usually taken for granted. The Uncertainty Principle states that objects (not space or time) are non-local. Experimental tests of Bell's theorem indicate that physical observables may not be described in terms of locally defined variables. However, quantum mechanics, which may be regarded as a non-local, statistical description, is apparently adequate to describe all observables in physics. Thus, it appears impossible to formulate a physically verifiable, non-statistical, local description of quantum phenomena.

¶ The localized trajectories of Newtonian particles may be regarded as the geometric ray limit of classical optics (although there is no locality along the trajectory). In most cases the primary difference between quantum mechanics and classical optics is the inclusion of particle identity in quantum mechanics, e.g. antisymmetry of the wave amplitude for mechanics. The validity of classical methods and the relation of quantum to classical terms is discussed by Beigmann [1995].

Quantum mechanics gives three complementary ways to picture atoms and their constituents: a particle picture, a wave picture and a probability picture. In the particle picture protons, electrons and photons are regarded as indivisible point particles. Although the particle picture is useful in explaining how particle detectors operate, it does not provide a description of interference‖, nor the non-locality of electrons on the atomic scale. In the wave picture objects have a de Broglie wavelength, $\lambda = h/p$, and interference is well described, but the indivisibility of the electron (or photon) is not explained. The probabilistic picture describes the shapes of electron clouds by using $|\psi(\vec{r})|^2$ to determine the density of the electron cloud where $\psi(\vec{r})$ is the wave amplitude for the object (or system of objects) being described. All of these pictures are used in this book. However, the incoming projectile is treated as a wavepacket that is well localized about its classical trajectory whenever possible, although in principle the wave description is usually better justified. Thus, in this semi-classical approximation charged incoming projectiles are treated as classical particles and the electrons are treated quantum mechanically. It is the activity of the electrons that is of primary interest.

1 The Hamiltonian

The wave amplitude or wavefunction Ψ gives the most complete description possible of a quantum system: usually the goal of the quantum scientist is to find Ψ exactly, or in a useful approximation if necessary. One usually approximates Ψ in multi-electron systems. In principle one may find Ψ by finding the eigenfunctions of the basic Schrödinger equation,

$$H\Psi = E\Psi \,, \tag{1.16}$$

where H is the Hamiltonian of the system.

For systems with more than one particle it is both useful and conventional to separate the system into a target (usually an atom initially in its ground state) and a projectile. The total Hamiltonian H then becomes,

$$\begin{aligned} H &= H_{projectile} + H_{target} + V \\ &= K + H_T + V \,. \end{aligned} \tag{1.17}$$

more than one electron.

‖ Quantum teleportation corresponding to long range non-local effects as in the Einstein-Rosen-Podolsky paradox provides a non-classical example of quantum interference.

Here K is the kinetic energy of the incoming projectile, H_T is the Hamiltonian for the target atom and V is the interaction connecting the projectile and the target. It is usually assumed that the eigenfunctions $|s>$ for each of the atomic states s are known, but that the wavefunction for the full system is not known.

In the semi-classical approximation one treats the incoming projectile as a particle with a well defined trajectory $\vec{R}(t)$. If the incident speed v is large enough, then the straight line trajectory $\vec{R}(t) = \vec{b} + \vec{v}t$ may be used. Now, $V = V(\vec{R}(t)) = V(t)$ is a known time dependent interaction that can cause electron transitions between different states $|s>$ in the atom.

2 The electronic wave amplitude

Since the physical and chemical properties of materials (including biological materials) are determined by the state of the electrons, it is the electronic wavefunction ψ which is useful to understand. In principle ψ may be found by solving,

$$H_{el}\psi = (H_T + V)\psi = \epsilon\psi \ , \tag{1.18}$$

where H_{el} is the Hamiltonian governing the electrons. Initially at $t = -\infty$, $\psi = |i>$ and $V = 0$. During the collision V is non zero. Then ψ changes according to,

$$\psi(t) = U(t, -\infty)\ \psi(-\infty) = Te^{-i\int_{-\infty}^{t} V(t)dt}\ \psi(-\infty) \ . \tag{1.19}$$

Here $U(t_2, t_1) = Te^{-i\int_{t_2}^{t_1} V(t)dt}$ is called the evolution operator because it changes $\psi(t_1)$ into $\psi(t_2)$. T is a time ordering operator which preserves causality (Cf. sections 3.2.3 and 7.1.2). Note that if $V = 0$, then $U = 1$ and nothing changes (i.e. there are no transitions). It is V that causes transitions in the atoms from their initial state $|i>$ to a variety of final states $|f>$.

3 The transition probability

After the interaction, as $t \to +\infty$, many outcomes are possible corresponding to different atomic states $|s>$ (including no transition, $|s> = |i>$). Mathematically this means that after the interaction the wavefunction is in a linear superposition of eigenstates, namely,

$$\psi(t \to +\infty) = \sum_s a_{si}|s> \ . \tag{1.20}$$

One may regard ψ as a simple vector in a basis of orthonormal eigenstates $|s>$ which act as mutually perpendicular unit vectors. This means that $< f|U = 1|s> = < f|s> = \delta_{fs}$, meaning that an eigenstate $|s>$ does not change if $V = 0$. Initially $\psi(-\infty) = |i>$. Here $< i|i> = \int |\psi(-\infty)|^2 d\vec{r}$, and $\int |\psi(\vec{r}, -\infty)|^2 d\vec{r} = 1$ because there is a 100% probability of finding the object described by ψ somewhere. Hence, at $t = -\infty, \psi(-\infty) = |i>$ is a unit vector because $< i|i> = 1$. Since matter is neither created nor destroyed in atomic interactions, $\psi(t)$ remains a unit vector which rotates during the time of interaction. The evolution operator $U(t, t_0)$ rotates ψ without changing its length. Such an operator is called a unitary operator. It is conventional to use orthonormal eigenstates $|s>$.

Using orthonormality, the state vector $\psi(+\infty)$ from (1.20) may be projected onto a particular final state $|f>$ yielding,

$$\begin{aligned} < f|\psi(+\infty)> &= < f|\sum_s a_{si}|s> = \sum_s a_{si} < f|s> \\ &= \sum_s a_{si}\delta_{fs} = a_{fi} \,. \end{aligned} \tag{1.21}$$

Since ψ is a unit vector,

$$\begin{aligned} 1 &= \int |\psi|^2 = \sum_f \sum_s a_{fi} < f|s> a_{si} \\ &= \sum_f |a_{fi}|^2 \,. \end{aligned} \tag{1.22}$$

Now $|a_{fi}|^2$ may be identified as the probability for a transition from an initial atomic state $|i>$ to a particular final state $|f>$ because $|a_{fi}|^2$ corresponds to the intensity of a particular outcome divided by the intensity of all possible outcomes.

Thus the probability P for a transition from an initial state $|i>$ to a final state $|f>$ is the square of the probability amplitude a_{fi} namely,

$$P = |a_{fi}|^2 = |< f|U(+\infty, -\infty)|i>|^2 = |< f|Te^{-i\int_{-\infty}^{+\infty} V(t)dt}|i>|^2 \,. \tag{1.23}$$

This definition of the transition probability is general. It applies to both single particle and multi-particle systems; it applies to systems of particles that are independent and systems of particles that are correlated.

1.4 Correlation

Correlation, broadly defined, is interaction between elements of a system. Correlated means interdependent; that is, correlated particles influence one another. Electron-electron correlation occurs because electrons interact with each other. It is correlation that causes DNA to act as more than a collection of independent atoms. Correlation may be defined in both space and time. In general correlation is the interdependency of individual members in a group: how individuals affect one another. A system of correlated objects may have properties beyond those which are possible from a collection of independent members. Correlation is a broad concept reaching beyond science. A more detailed description of correlation is given in section 6.1.

1 Definition of correlation

Physical definition: Correlation usually refers to correlation between electrons, although time correlation effects are briefly considered in the analysis of second order effects in section 7.1.1-2. Electrons are correlated because they interact with one another. Because electrons carry a negative charge$-e$ they repel each other with an interaction potential e^2/r_{12} where r_{12} is the distance between the electrons. This interaction potential is called the 'correlation interaction' (Cf. (4.7) for the precise definition). Electron correlation is an example of spatial correlation because this correlation depends on the position of the electrons. Because the electrons push each other around, the shape of the electron cloud of a system of electrons is influenced by the constituent electrons. When transitions occur the electrons in the system can also influence the transition rate due to their correlation interaction. If the electrons are not correlated, they are said to be uncorrelated or independent. The concept of correlation is discussed further in section 6.1.

Mathematical definition: Correlation is mathematically defined as something other than a product of independent functions. Thus, correlation means non-separable. If $F(x_1, x_2, ...x_N)$ describes the properties of a system of N objects (e.g. N electrons), and $f_j(x_j)$ describes the property of a single object, then F is correlated if,

$$F(x_1, x_2, .., x_j, ...x_N) \neq f_1(x_1)f_2(x_2)...f_N(x_N) = \prod_j f_j(x_j) \,. \quad (1.24)$$

If $F(x_1, x_2, .., x_j, ...x_N) = \prod_j f_j(x_j)$, then the properties, f_j, are said to be uncorrelated or independent.

The solution ψ to (1.18) is a product of single electron terms $\prod_j \psi_j(x_j)$ if the Hamiltonian is a sum of single electron terms $H_{el} = \sum_j H_{el,j}$. It follows that the probability P in (1.23) is then a product of single electron probabilities, i.e., $P = \prod_j P_j$. This follows from the method of separation of variables commonly used to solve partial differential equations.

2 *Correlated probabilities*

The probability P for a specified outcome of N objects is correlated if,

$$P \neq P_1 P_2 \ldots P_N = \prod_j P_j \,, \tag{1.25}$$

where P_j is the probability of an outcome for a single independent object. If $P = \prod_j P_j$, the probability is uncorrelated and the particles involved are independent. All the probabilities considered in the first part of this chapter (e.g. for outcomes in tossing N independent coins) are uncorrelated. A central part of the challenge of understanding the N body problem is to determine the correlated probability P for many body systems.

3 *Other applications*

Correlation is a widely used concept. In statistical mechanics correlation is a generalization of the standard deviation $\varsigma = \sqrt{(x - <x>)^2}$ [Sokolnikoff and R.M.Redheffer, 1958, pp. 663-671]. Correlation is used in pattern recognition. If ς is small, then the correlation is large and two patterns are similar (e.g. '1' and 'l', but not '1' and '2'). There is a correlation coefficient τ which varies between 0 and 1 given in Eq(6.2) that may be used in both pattern recognition and atomic interactions. But τ has not yet been widely used to describe atomic interactions. There is also a cluster expansion for many particle systems which defines an N particle correlation function requiring correlation of all of the N particles. This $N + 1$ particle correlation function may be generated from the N particle correlation function using the correlation interaction between two particles (e.g. the correlation potential between two electrons). This progression is referred to as the BBGKY hierarchy (Cf. section 6.1). Time correlation is used to describe properties of fluids. Most studies have been confined to two particle or pair correlation because experimental observations of higher order correlation effects are usually difficult. Examples of time correlation include time ordering effects (Cf. section 7.1.2) and memory.

2
Single electron transition probabilities

There are, in general, many possible transitions of electrons in atoms. In some processes of practical interest more than one electron may undergo a transition. Such multiple electron transitions are the topic of the next and subsequent chapters. In this chapter the simpler topic of single electron transitions is considered, where the activity of a single electron in an atom is the focus of attention. Even this relatively simple case may be impossible to fully understand if the electron of interest is influenced by other electrons in the atom. So in this chapter the interdependency of electrons in the system is ignored. That is, the electrons are treated independently. Typically, such an independent electron is regarded as beginning in an initial state characterized by some effective nuclear charge Z_T and a set of quantum numbers n, ℓ, m, s, m_s from which all possible properties (e.g., energy, shape, magnetic properties, etc.) may be determined. Interaction with something else, (usually a particle of charge Z and velocity, v), may cause a transition to a different final state of the atom.

The simplest transition occurs in interaction of atomic hydrogen with a structureless projectile. There are various ways to evaluate the transition probability for such a system. Exact calculations usually require use of a computer. Approximate calculations may be done more easily. Calculations for many electron systems are often done approximately using single electron transition probabilities. For example, these single electron transition probabilities are used for both single and multiple transitions of independent electrons as discussed in both the previous and following chapters. Single electron probabilities are also widely used when the 'active' electron picture is used to describe the activity of a single independent electron in an atom with many other electrons [Cf. section 3.3.6].

In this and subsequent chapters atomic units are used, i.e., $m_e = \hbar =$

$e^2 = 1$. In these atomic units the unit of mass is the mass of an electron $m_e = 9.11 \times 10^{-31}$ kg, the unit of distance is the distance of an electron from the nucleus of a hydrogen atom in its ground state $a_0 = 5.29 \times 10^{-11}$ m, the unit of time $t_0 = v_0/a_0 = 2.42 \times 10^{-17}$ s. is the time it takes an electron in the ground state of hydrogen at a speed $\mathrm{v}_0 = 2.19 \times 10^6$ m/s to travel the distance a_0, and the unit of charge is the charge of a proton $|e| = 1.60 \times 10^{-19}$ Coulombs. Cross sections may be sensibly measured either in units of a_0^2 or in units of πa_0^2, the cross sectional area of a sphere of radius a_0. Here units of a_0^2 are used.

2.1 Formulation

The electron transition probabilities P may be employed to find many useful properties of a system. For example, the average energy transfer is simply the sum of the individual energy transfers for various transitions weighted by the probability of such a transition. Cross sections are also often useful. As discussed in the previous chapter, the total cross section (or effective cross sectional area), σ, for a particular transition is found by summing the transition probability over all impact parameters, $\vec{b}$, of the interaction, i.e., $\sigma = \int P(\vec{b})d\vec{b}$. The stopping power of a charged particle moving through matter (which is needed to find the distance the charged particle penetrates) is found by summing the energy transfer of the projectile over the cross section for various transitions at various projectile energies [Knudsen and Reading, 1992].

1 Complete atomic system

The transition probability P and subsequent cross sections and other properties may be evaluated quantum mechanically* from the probability amplitude a defined by,

$$\begin{aligned} P &= |a|^2 \\ a &= < f \mid \psi_i > . \end{aligned} \tag{2.1}$$

As discussed below, $< f|$ is the wavefunction of the final atomic state of the active electron, and $|\psi_i >$ is the wavefunction of the fully interacting electron which evolves from an initial atomic state $|i >$ to $|\psi_i >$ when the interaction is finished. In general the total wavefunction Ψ gives all

* Classical probabilities and cross sections evaluated using Newton's Laws are considered in the appendix of this chapter.

information possible about the system†. It is the evaluation of the slightly simpler dynamic electronic wavefunction $\psi_i(t)$ that is of primary interest.

Total wavefunction. The total wavefunction is determined by the total Hamiltonian H. In particular the total wavefunction Ψ is found from the time dependent Schrödinger equation,

$$H\Psi = i\frac{\partial}{\partial t}\Psi \,. \tag{2.2}$$

The solution to this Schrödinger equation is conventionally described in terms of eigenstates $|\Psi_\alpha>$ where each of the eigenstates has a definite energy E_α. The solution Ψ is expressed as a linear superposition of these eigenstates,

$$\Psi = \sum_\alpha a_\alpha \Psi_\alpha \,. \tag{2.3}$$

The initial conditions determine the coefficients α at $t \to -\infty$, i.e. before the interaction begins. Usually the system starts in an eigenstate which corresponds to a well defined state of the atom (e.g. the ground state). The eigenstates $|\Psi_\alpha>$ form an orthogonal set of unit vectors such that $| < \Psi_\beta|\Psi_\alpha > |^2 = \int \Psi^*_\beta(\vec{r})\Psi_\alpha(\vec{r})d\vec{r} = \delta_{\beta\alpha}$. The projection of Ψ onto a particular eigenstate $|\Psi_\alpha>$ is,

$$\begin{aligned} < \Psi_\alpha|\Psi > &= \int \Psi^*_\alpha(\vec{r})\Psi(\vec{r})d\vec{r} = \sum_\gamma a_\gamma < \Psi_\alpha|\Psi_\gamma > \\ &= \sum_\gamma a_\gamma \delta_{\alpha\gamma} = a_\alpha \,. \end{aligned} \tag{2.4}$$

The probability that the state $|\Psi>$ will be found in the eigenstate $|\Psi_\alpha>$ is,

$$P_\alpha = |a_\alpha|^2 \,. \tag{2.5}$$

Total Hamiltonian. The Hamiltonian for a particle of charge Z interacting with an electron on a target of nuclear charge Z_T is,

† In a Newtonian system the trajectory, $\vec{r}(t)$, indicates where a particle is at all times in the past and in the future. The grand vision of Newtonian physics is to predict what will happen to everything by finding $\vec{r}(t)$ for all particles. It is largely the difficulty of the classical many body problem that prevents this from being done. In quantum mechanics the goal is to find $\Psi(t)$ which gives all possible properties (including $\vec{r}(t)$). In large part it is the difficulty of the many body problem which limits the extent to which people can do this.

$$H = \left(-\nabla_R^2/2M + ZZ_T/R\right) - \left(\nabla_r^2/2 - Z_T/r\right) - \frac{Z}{|\vec{R} - \vec{r}|} \quad (2.6)$$
$$= H_P + H_T + H_{int} ,$$

where $-i\nabla_R$ represents the momentum of the projectile of mass M, $-i\nabla_r$ the momentum of the target electron, H_P is the Hamiltonian for the projectile, H_T the Hamiltonian for the target, and $H_{int} = V$ is the interaction between the target and the projectile. It is this interaction V that causes the electronic transitions in the atomic target‡.

Semi-classical approximation. In the semi-classical approximation the projectile is treated as a classical point particle with a well defined trajectory $\vec{R}(t)$ and the electrons are treated quantum mechanically. This approximation is generally used to describe fast heavy small charged particles. The conditions under which this approximation is valid are discussed in detail in sections 3.2.1 and 3.2.3 and the appendix of chapter 3. It is usually valid for projectiles whose mass M is greater than that of an electron. The semi-classical approximation is usually not a good approximation for describing incident photons or incident electrons that are not fast. As shown in section 3.2.1, the problem of finding the total wavefunction for all particles is now replaced by the simpler problem of finding the wavefunction for the electron ψ_{el} satisfying,

$$H_{el}\ \psi_{el} = i\frac{\partial}{\partial t}\psi_{el} , \quad (2.7)$$

generated from the electronic Hamiltonian,

$$H_{el} = H_T + H_{int} = -\nabla_r^2/2 - Z_T/r - Z/|\vec{R} - \vec{r}| . \quad (2.8)$$

2 Atomic electrons

Total electronic wavefunction. The total electronic wavefunction for a dynamic system is seldom known analytically. However, the total wavefunction may always be expressed as a linear superposition of any complete set of functions. It is conventional to choose the functions used to be

‡ The total Hamiltonian H may be partitioned such that the instantaneous Coulomb interaction V is replaced by the difference between the instantaneous interaction V and its average $V - \bar{V} = V - \langle i|V|i \rangle$ removing the long range $1/R$ Coulomb tail [Cf. section 3.3.4].

the set of orthogonal unit vectors $|s>$ that are the eigenfunctions of the target atom. Specifically,

$$\psi_{el} = \sum_s a_s e^{-iE_s t} |s> \ . \tag{2.9}$$

It is also conventional to apply the boundary condition that $|\psi_{el}>$ is in the initial state $|i>$ of the target atom before the interaction at $t \to -\infty$. The electronic wavefunction satisfying this initial condition is denoted by $|\psi_i>$ or sometimes $\psi_i(\vec{r})$.

During the time when the interaction is non-zero, the electronic wavefunction changes from $|i>$ to $|\psi_i(t)>$ which is expressed by,

$$|\psi_i(t)> = \sum_s a_{si}(t)\ |s> \ . \tag{2.10}$$

Here it is assumed that the atomic eigenfunctions $|s>$ are known and that the problem of evaluating ψ_i is equivalent to the problem of evaluating the probability amplitudes a_{si}. Conventional methods used to find the a_{si} are discussed in sections 2.2 - 2.6 below. Note that the probability amplitudes a_{si} for an electronic transition from $|i>$ to $|s>$ change during the time that the interaction is non-zero. It is a_{si} after the interaction has occurred at $t \to +\infty$ that one normally wishes to find.

Target electron eigenfunctions. The Hamiltonian of a hydrogen-like atom consisting of a single electron and a nucleus of charge Z_T is,

$$H_T = -\nabla^2/2 - Z_T/r \ , \tag{2.11}$$

where ∇_r is now more simply denoted by ∇.

The eigenstates $|s>$ are solutions of the differential equation[§],

$$H_T|s> = \left(-\nabla^2/2 - Z_T/r\right)|s> = E_s|s> \ , \tag{2.12}$$

where a time dependence of $e^{-iE_s t}$ has been used, but is not included in $|s>$ which depends only on $\vec{r}$.

The time independent Schrödinger equation of (2.12) may be solved by the conventional method of separation of variables in spherical coordinates. The eigensolutions obtained [Hill, 1996; Messiah, 1961] are,

---

[§] With approximate classical methods one faces the differential equation $-\nabla V = m\frac{d^2\vec{r}}{dt^2}$ which is arguably more complicated than the correct Schrödinger equation (2.12).

$$
\begin{aligned}
|s>=|n\ell m> &= (Z_T)^{-\frac{3}{2}} N_{n\ell} F_{n\ell}(\frac{2r}{nZ_T}) Y_\ell^m(\theta,\phi) \qquad (2.13)\\
F_{n\ell}(x) &= x^\ell e^{-x/2} L_{n-\ell-1}^{2\ell+1}(x) ,
\end{aligned}
$$

where

$$
N_{n\ell} = \frac{2}{n^2}\sqrt{\frac{(n-\ell-1)!}{[(n+\ell)!]^3}}
$$

$Y_\ell^m(\theta,\phi)$ is the spherical harmonic and

$$
L_p^k(x) = \sum_{s=0}^{p}(-1)^s \frac{(p+k)!]^2}{(p-s)!(k+s)!s!} x^s ,
$$

is the Laguerre Polynomial [Abramowitz and Stegun, 1964]. The hydrogenic wavefunctions are discussed in more detail in the appendices at the end of this book.

Classification of final states. It is conventional to classify the final states of the projectile-target system of the target and projectile $\psi_{el}(t \to +\infty)$ into four groups.

1. Elastic scattering. Here $|f>=|i>=|n_i, \ell_i, m_i>$ so that the electronic state is unchanged and there is no exchange of energy.
2. Excitation. In this case $|f>=|n\ell m>$ is a bound excited state of the target (not equal to $|i>$).
3. Ionization. Now $|f>$ is not a bound state, but is in the continuum of unbound states.
4. Transfer Then $|f>$ is a bound state of the projectile.

This classification of final states is widely used to describe atomic collisions. However, there are difficulties with this classification. The boundary between bound and unbound states is blurred by stray electromagnetic fields so that excitation and ionization are not always well separated. Electrons in the conduction band of a solid are not easily described with this atomic classification. Electron capture to the continuum of the projectile denotes an unbound electron under the influence of the nuclear charge of the projectile and thus belongs to ionization. However, it is not unreasonable to classify this final state under transfer since the states are described as continuum states of the projectile. Also, whenever a basis set containing both target and projectile target eigenstates is used there is a potential mathematical difficulty in overcounting since such a basis set is over complete.

Coupled channel equations. The total electronic wavefunction ψ_{el} may be found by determining the probability amplitudes $e^{iE_s t} a_s(t)$ which yield the probability of finding ψ_{el} in the state $|s>$ at any time t, as denoted by (2.9). The equations that must be solved for the probability amplitudes a_s given below are expressed in terms of matrix elements $< f|V|s >$ of the interaction term $V = H_{int}$ of (2.7) which couples the states (or channels) of the system. These equations are called coupled channel equations [Bartschat, 1993; Henry and Kingston, 1988].

Imposing the initial condition of (2.10) and using (2.9) in (2.7), one has,

$$
\begin{aligned}
H_{el}\ \psi_{el} &= (H_T + V)\left(\sum_s a_{si}(t) e^{-iE_s t} |s>\right) \qquad (2.14)\\
&= i\frac{\partial}{\partial t}\left(\sum_s a_{si}(t) e^{-iE_s t} |s>\right)\\
&= i\sum_s \dot{a}_{si}(t) e^{-iE_s t} |s> + E_s \sum_s a_{si}(t) e^{-iE_s t} |s> \ .
\end{aligned}
$$

Projecting $< f|e^{iE_f t}$ onto the above equation and using (2.12) yields,

$$
\sum_s e^{-i\omega_{fs} t} < f|V|s > a_{si} = i\sum_s e^{-i\omega_{fs} t} \dot{a}_{si} \delta_{fs} = i\dot{a}_{fi} \ , \qquad (2.15)
$$

where $\omega_{fs} = E_s - E_f$, so that,

$$
\dot{a}_{fi} = -i\sum_s e^{-i\omega_{fs} t} < f|V|s > a_{si} \ . \qquad (2.16)
$$

Integrating and applying the initial condition that $a_{fi}(-\infty) = \delta_{fi}$, one has,

$$
a_{fi}(t) = \delta_{fi} - i\int_{-\infty}^{t} \sum_s e^{-i\omega_{fs} t} < f|V(t')|s > a_{si}(t') dt' \ , \qquad (2.17)
$$

where $V = V(t)$ because it contains $\vec{R}(t)$ in (2.7). Eq(2.16) and equivalently (2.17) are the coupled channel equations in the particle picture. A similar set of equations is easily derived in the wave picture. A more complete development of coupled channel equations may be found in McDowell and Coleman [1970]. Many authors have solved the coupled channel equations numerically [Fritsch and Lin, 1991].

The expansion used above for the coupled channel equations is referred to as the atomic orbital (AO) expansion because the basis set is composed of atomic orbitals (or channels). In principle any complete set of states may be used. At low collision velocities the projectile and target form a quasi-molecule whose internuclear axis changes with time. A complete set of these quasi-molecular (MO) states is commonly used for slow collisions. Other choices, including a mixture of AO and MO states, are also used. Expansions in Sturmian functions and Gaussian functions are also commonly employed. The Sturmian basis is similar to the hydrogenic eigenfunctions, except that unbound states (and sometimes higher bound states) are represented by functions that are mathematically similar to bound state hydrogenic wavefunctions. A range of states in the continuum state is then represented by a single function. This is an example of a pseudostate where an eigenstate is treated approximately. The accuracy of such coupled channel methods depends on the numerical accuracy of the calculation, e.g. the number of states included in the calculation. If only two states are used, the equations may often be solved analytically [McDowell and Coleman, 1970].

Sometimes it is useful to use the evolution operator $U(t, t_0)$ instead of the probability amplitude a_{fi}.

3 Evolution of an electron

Evolution operator. The evolution operator $U(t, t_0)$ is the Green function operator that takes $\psi(t_0)$ to $\psi(t)$, namely,

$$\psi(t) = U(t, t_0)\psi(t_0) , \tag{2.18}$$

with the causality condition that $U(t, t_0) = 0$ if $t < t_0$.

Using (2.18) and (2.2) in (2.18), one has,

$$\begin{aligned} \dot{a}_{fi} &= < f|\dot{\psi} > \; = \; < f|\dot{U}|i > \\ &= -i \sum_s e^{iE_f t} < f|V|s > e^{-iE_s t} e^{iE_s t} < s|\psi_i > e^{-iE_i t} \\ &= -i < f|V_I U|i > , \end{aligned} \tag{2.19}$$

where $V_I = e^{iH_T t} V e^{-iH_T t}$. Thus,

$$\frac{dU}{U} = -iV_I(t)dt . \tag{2.20}$$

This can be easily integrated under the constraint of causality¶, i.e. that

¶ The quantum evolution operator satisfies the same causality condition as does the corresponding classical Green function, namely that cause is constrained to occur before effect.

$U(t, t_0) = 0$ if $t < t_0$, and the boundary condition that $U(t_0, t_0) = 1$, to give [Goldberger and Watson, 1964],

$$
\begin{aligned}
U(t,t_0) &= 1 - i\int_{t_0}^{t} V_I(t')U(t',t_0)dt' && (2.21)\\
&= 1 \ -i\int_{-\infty}^{t} V_I(t_1)dt_1 + \frac{(-i)^2}{2}T\int_{-\infty}^{t} V_I(t_2)dt_2 \int_{-\infty}^{t} V_I(t_1)dt_1 + ...\\
&\quad +\frac{(-i)^n}{n!}T\int_{-\infty}^{t} V_I(t_n)dt_n .... \int_{-\infty}^{t} V_I(t_2)dt_2 \int_{-\infty}^{t} V_I(t_1)dt_1 + ...\\
&= 1 \ -i\int_{-\infty}^{t} V_I(t_1)dt_1 + \frac{(-i)^2}{2}\int_{-\infty}^{t} V_I(t_2)dt_2 \int_{-\infty}^{t_2} V_I(t_1)dt_1 + ...\\
&\quad + \frac{(-i)^n}{n!}\int_{-\infty}^{t} V_I(t_n)dt_n \int_{-\infty}^{t_n} V_I(t_{n-1})dt_{n-1} ... \int_{-\infty}^{t_2} V_I(t_1)dt_1 + ...\\
&= \sum_{n=0}^{\infty}\left(\frac{1}{n!}\right) T\left(-i\int_{-\infty}^{\infty} V_I(t')dt'\right)^n\\
&\equiv Te^{-i\int_{t_0}^{t} V_I(t')dt'} \ ,
\end{aligned}
$$

where T is the time ordering operator which is required to enforce causality.

The probability amplitude a_{fi} is now given by,

$$a_{fi}(t) = < f|U(t, -\infty)|i > \ . \tag{2.22}$$

This is equivalent to (2.17).

Perturbation theory. Here a simple version of first order perturbation theory (or Born expansion) is considered. A more complete development of perturbation theory is given in chapter 7.

In first order perturbation theory it is assumed that the initial state is not significantly perturbed. This corresponds to setting $a_{si} = \delta_{si}$ in (2.17). Then for $f \neq i$,

$$a_{fi}(t) = -i\int_{-\infty}^{t} e^{-i\omega_{fi}t} < f|V(t')|i > dt' \ . \tag{2.23}$$

Nevertheless, time may not be localized in the correct quantum description within the limit of the uncertainty principle $\Delta E\Delta t \geq 1$ which follows from the classical band width theorem $\Delta\nu\Delta t \geq 1$ which is a general mathematical property of Fourier transforms. In classical waves this band width theorem limits the number of radio or television stations that may be carried over a given range of transmission frequencies.

Usually the $\lim_{t\to+\infty}$ is taken. This result also follows immediately from (2.22) by taking only the first term in an expansion in the interaction V in (2.21).

This first order approximation is valid if higher order terms in $-i\int V(t)dt$ are small, i.e $(\int V(t)dt)^2 << \int V(t)dt$. For the Coulomb interaction V defined by (2.6), using $\vec{R}(t) = \vec{b} + \vec{v}t$ with $dz = \mathrm{v}dt$ one has‖,

$$\int V(t)dt = \int \frac{Z}{|\vec{R}(t) - \vec{r}|}dt = Z/\mathrm{v}\int \frac{dz}{|\vec{R}(t) - \vec{r}|} \simeq Z/\mathrm{v} \,. \tag{2.24}$$

The parameter Z/v is referred to as the Massey parameter. If $Z/\mathrm{v} << 1$, then first order perturbation theory is usually valid.

Unitarity. In a correct calculation the probability of a transition may not exceed unity. This condition follows from the normalization condition that $\int |\Psi|^2 = 1$ which may be easily shown from (2.3) and orthonormality to be equivalent to,

$$\sum_s |a_{si}|^2 = 1 \,. \tag{2.25}$$

This result is referred to as unitarity. It means that the total probability of all possible outcomes is unity. Unitarity holds for all exact calculations and is used to test the numerical accuracy of coupled channel computer codes for the a_{si}.

First order perturbation theory is not exact and in this approximation the transition probability $P = |a_{si}|^2$ may exceed unity when the Massey parameter Z/v is not small. An easy way to avoid this unphysical result when P is not small compared to one is to use the replacement [Sidorovitch *et al.*, 1985],

$$P \to \left(1 - e^{-P}\right) \,. \tag{2.26}$$

While this remedy is easy, it is not well justified. Nevertheless, it is convenient and it corrects an obvious difficulty.

An alternative method may be used to force unitarity in a system with several different probabilities P_j, where $\sum_j P_j > 1$. One may replace P_j by

‖ This argument is more sensible if V is replaced by $V - \bar{V}$ since $\int dz/|\vec{R}(t) - \vec{r}| \to \infty$.

$$\tilde{P}_j = P_j / \sum_j P_j$$

The new $\tilde{P}_j$ satisfy $\sum_j \tilde{P}_j = 1$, and also satisfy the differential equation (with $\dot{P} = \partial P/\partial t$),

$$(\sum_k P_k)\dot{\tilde{P}}_j = \dot{P}_j - \tilde{P}_j(\sum_k \dot{P}_j)$$

which enforces unitarity on each as well as all of the P_j.

Scaling laws. While first order perturbation results are not always valid, they are simple. One of their convenient features is the scaling of matrix elements, amplitudes, probabilities and cross sections with charge and velocity. These scaling relations hold whenever eigenstates $|s>$ of the target are used to describe both the initial state $|i>$ and the final state $|f>$. Thus, these scaling laws hold for elastic scattering, excitation of the target and ionization described by a Coulomb wave of the target. These rules do not apply to electron transfer where the initial state is a target state but the final state is a state of the projectile. Under these conditions, one has the following scaling.

1. Matrix elements:

$$\begin{aligned} V_{fi}(Z, Z_T, \mathrm{v}, \vec{b}) &= <f|V|i> \\ &= \frac{Z}{Z_T} V_{fi}(Z=1, Z_T=1, \frac{\mathrm{v}}{Z_T}, Z_T\vec{b}) \,. \end{aligned} \tag{2.27}$$

2. Probability amplitude:

$$\begin{aligned} a_{fi}(Z, Z_T, \mathrm{v}, \vec{b}) &= -i \int e^{i\omega t} <f|V|i> dt \\ &= \frac{Z}{Z_T} a(Z=1, Z_T=1, \frac{\mathrm{v}}{Z_T}, Z_T\vec{b}) \,. \end{aligned} \tag{2.28}$$

3. Transition probability:

$$P(Z, Z_T, \mathrm{v}, \vec{b}) = |a|^2 = (\frac{Z}{Z_T})^2 P(Z=1, Z_T=1, \frac{\mathrm{v}}{Z_T}, Z_T\vec{b}) \,. \tag{2.29}$$

4. Total cross section:

$$\sigma(Z, Z_T, \mathrm{v}) = \int P d\vec{b} = \frac{Z^2}{Z_T^4} \sigma(Z=1, Z_T=1, \frac{\mathrm{v}}{Z_T}) \,. \tag{2.30}$$

These scaling laws may be used to express any of the above quantities in terms of the corresponding quantity for $p+H$.

These relations may be derived by noting that: i) the first order matrix element is linear in Z_T, ii) $\vec{r}$ scales as $1/Z_T$, and iii) $< i|i >= 1$ for all charges and velocities. Then,

$$\begin{aligned} < f|V|i > &= < f|\frac{Z}{|\vec{R}-\vec{r}|}|i > \\ &= \frac{Z}{Z_T} < f\left|\frac{1}{\frac{\vec{b}}{Z_T}+\frac{\vec{v}}{Z_T}t-\frac{\vec{r}}{Z_T}}\right|i > = \frac{Z}{Z_T}\tilde{V}_{fi} \end{aligned} \tag{2.31}$$

where $\tilde{V}_{fi}$ has the same numerical value for all Z and Z_T at a fixed value of v/Z_T. Using $z' = \mathrm{v}t' = Z_T\tilde{z}'$ and $\omega = \Delta E = Z_T^2\Delta\tilde{E}$, where $\tilde{z}'$ and $\Delta\tilde{E}$ are scale invariant, it follows that,

$$\begin{aligned} a_{fi}(Z, Z_T, \mathrm{v}, \vec{b}; t) &= -i\int_{-\infty}^{t} e^{i\omega t'} < f|V|i > dt' \\ &= -i\int_{-\infty}^{t} e^{i(Z_T^2\Delta\tilde{E}t)}\frac{Z}{Z_T}\tilde{V}_{fi}dt' \\ &= -i\frac{Z}{Z_T}\int_{-\infty}^{t} e^{i(\Delta\tilde{E}\tilde{z}'Z_T/\mathrm{v})}\,\tilde{V}_{fi}\,dt' \\ &= -i\frac{Z}{Z_T}\int_{-\infty}^{\frac{Z_T}{\mathrm{v}}\tilde{z}} e^{i(\Delta\tilde{E}\tilde{z}'Z_T/\mathrm{v})}\,\tilde{V}_{fi}\,\frac{Z_T}{\mathrm{v}}\,d\tilde{z}' \\ &= \frac{Z}{Z_T}a_{fi}(Z=1, Z_T=1, \mathrm{v}/Z_T, Z_T\vec{b}; \frac{Z_T}{\mathrm{v}}\tilde{z}')\,. \end{aligned} \tag{2.32}$$

The remaining scaling laws in first order perturbation theory, (2.29) and (2.30), follow immediately.

In general the first order probability scales as Z^2 according to (2.29). Thus, the first order probability is the same for protons as for anti-protons at the same velocity. Moreover, at projectile high velocities where the impact parameter of an incident electron is reasonably well localized, the total cross sections are independent of the mass of the projectile. This follows from,

$$\sigma = \int |a_{fi}|^2 d^2\vec{b} = \int_{q_{min}=q_{\|}=\Delta E/\mathrm{v}}^{q_{max}} |f((\vec{q})|^2 d^2 d\vec{q}_{\perp}\ ,$$

in the limit that $q_{max} \to \infty$. In this limit the total cross section for incident protons, electrons, anti-protons and positrons of equal velocity v are the same in first order perturbation theory. When $Z/v << 1$, these first order results are usually valid.

In the sections 2.2 - 2.4 simple first order transition probabilities are considered in more detail for excitation, electron transfer and ionization.

4 Partial wave expansion

In general projectiles are described by wavepackets quantum mechanically. If the wavepacket is localized in a relatively small region, then the projectile may be regarded as a point particle. On the other hand, if the wavepacket is much larger than the target, the projectile may be regarded as a wave. The total cross section may be equivalently expressed in terms of either the probability amplitude a_{fi} in the particle picture, or the scattering amplitude $f(\vec{q})$ in the wave picture, i.e., as expressed by (11.32),

$$\sigma = \int |a_{fi}|^2 d\vec{b} = \frac{1}{k_i^2} \int |f(\vec{q})|^2 d\vec{q}_\perp \ , \tag{2.33}$$

where $q^2 = (\vec{k}_f - \vec{k}_i)^2 \simeq k^2 + k^2 - 2k^2 cos(\theta)$ with $k_f \simeq k_i \simeq k$ at high collision velocities. Then, $d\vec{q}_\perp \simeq -2k^2 sin(\theta) d\theta d\phi$, where θ is the scattering angle of the projectile, i.e. $cos(\theta) = \hat{k}_f \cdot \hat{k}_i$, where $\vec{k}_i$ and $\vec{k}_f$ are the initial and final momenta of the projectile. The scattering amplitude f and the commonly used transition (or T) matrix element T_{fi} are related by $f = -\mu/2\pi T_{fi}$ where μ is the reduced mass of the system.

An insight into the nature of the probability amplitude a may be found by expanding the scattering amplitude factors $f(\theta)$ in partial waves, namely,

$$f(\theta) = \frac{1}{2k} \sum_{\ell=0}^{\infty} (2\ell + 1) a_\ell P_\ell(cos\theta) \ , \tag{2.34}$$

where $P_\ell(cos\theta)$ is the Legendre Polynomial. After integrating isotropically over ϕ and using the the orthogonality relation for P_ℓ following (11.5), the total cross section (in units of a_0^2) is,

$$\begin{aligned} \sigma &= \frac{2\pi}{k^2} \int_{q_{min}}^{q_{max} \simeq \infty} |f(q)|^2 q dq = 2\pi \int_0^\pi |f(\theta)|^2 sin\theta d\theta \\ &= \frac{\pi}{k^2} \sum_{\ell=kb}^{\infty} (2\ell + 1)|a_\ell|^2 \simeq \frac{\pi}{k^2} \int_0^\infty (2\ell + 1)|a_{\ell=kb}|^2 d\ell \\ &= 2\pi \int_0^\infty |a_{\ell=kb}|^2 b db \ . \end{aligned} \tag{2.35}$$

That is, the probability amplitude a may be regarded as a semi-classical approximation to the amplitude a_ℓ of the ℓth partial wave of the scattering amplitude. Inverting (2.34) yields,

$$a_\ell = k \int_0^\pi f(\theta) P_\ell(cos\theta) sin\theta d\theta \ . \tag{2.36}$$

This is equivalent to (11.31). The above equation provides a useful way to evaluate a from f, which is usually easier to evaluate mathematically than a. On the other hand, the transition probability $|a|^2$ has a familiar conceptual interpretation in the limit of classical point particles. Also, the constraint of unitarity that everything must go somewhere is easily expressed in the particle picture as $\sum_f |a_{fi}|^2 = 1$. This constrains the individual probability amplitude a_{fi} for transition from state $|i>$ to state $|f>$ by the useful and obvious condition $|a_{fi}|^2 \leq 1$.

2.2 Excitation probabilities

To evaluate transition probabilities exactly it is in general necessary to calculate the probability amplitudes numerically. However, it is often convenient to use simple analytic results to estimate single electron transition probabilities. The first order results may often be evaluated analytically and are reliable when the probabilities are small. A simple approximate form that is sometimes used for probabilities that are not small is given in (2.26). When multiple ionization occurs, factors of $(1-P)^n$ constrain P to values that are not especially large. For these reasons it is sensible to have the simple first order results available for the transition probabilities. Excitation of the atomic target is the simplest case and is considered in this section.

The transition probability for an electron initially in the target state, $|i>$, to a final state, $|f>$, is expressed in first order perturbation theory as**,

$$P = |a_{fi}|^2 = |\int_{-\infty}^{\infty} e^{i\Delta E_{fi}t} < f|V|i > dt|^2 \ . \tag{2.37}$$

The target wavefunctions for hydrogenic targets are given in the appendices at the end of the book. Because of the scaling laws discussed in the section above, however, it is only necessary to consider the case of $p+H$,

** Except in the case of elastic scattering the orthogonality of $|f>$ and $|i>$ removes the problem of whether to use V or $V - \bar{V}$, since terms that depend only on R (and not r), such as $\bar{V}$, do not contribute.

since transition probabilities for other systems may be generated using (2.29).

1 1s – 2s, 2p transitions

For the case of excitation from the ground state to the first excited state, namely, $|i> = |1s>$ and $|f> = |2s>$, using the wavefunctions in the appendices of this book one has,

$$<2s|V|1s> = \int d^3r \frac{2}{\sqrt{4\pi}} e^{-r} | \frac{1}{|\vec{R}-\vec{r}|} \frac{1}{2\sqrt{8\pi}} (2-r)e^{-r/2} \,. \tag{2.38}$$

Using (11.16) and integrating over angles one has,

$$\begin{aligned} <2s|V|1s> &= -\frac{1}{\sqrt{2}} \int_0^R r^2 dr (2-r) e^{-3r/2}/R \\ &- \frac{1}{\sqrt{2}} \int_R^\infty r^2 dr (2-r) e^{-3r/2}/r \\ &= -\frac{2^{3/2}}{3^3} (2+3R) e^{-3R/2} \,, \end{aligned} \tag{2.39}$$

where from Gradshteyn and Ryzhik [1980] (2-321-2),

$$\int^r r^n e^{ar} dr = e^{ar} \left(\frac{r^n}{a} + \sum_{k=1}^{n} (-1)^k \frac{n(n-1)...(n-k+1)}{a^{k+1}} r^{n-k} \right) . \tag{2.40}$$

Now,

$$a_{2s,1s} = -i \int_{-\infty}^{\infty} <2s|V|1s> e^{-i\frac{3}{8}t} dt \,, \tag{2.41}$$

where $\Delta E_{2s,1s} = -(\frac{1}{2} - \frac{1}{8}) = -\frac{3}{8}$. Using $R^2 = b^2 + v^2t^2 = b^2 + z^2$, and noting that V is even in t,

$$a_{2s,1s} = -i \frac{2^{5/2}}{3^3} \int_0^\infty (2+3R) e^{-3R/2} cos(\frac{3}{8}\frac{z}{v}) dz/v \,. \tag{2.42}$$

From Gradshteyn and Ryzhik (3.914),

$$\int_0^\infty e^{-\beta\sqrt{b^2+z^2}} cos(\omega t) dz = \frac{\beta b}{\omega^2+\beta^2} K_1(b\sqrt{\omega^2+\beta^2}) \,, \tag{2.43}$$

where K_1 is a modified Bessel function of the second kind. The second term in (2.42) may be expressed in terms of the derivative of the first term, $-\frac{d}{d\beta}\left[\frac{\beta b}{\sqrt{\omega^2+\beta^2}}K_1(b\sqrt{\omega^2+\beta^2}\right]$. Using $\frac{dK_1(x)}{dx} = \frac{1}{x}K_1(x) - K_2(x)$, it follows that,

$$a_{2s,1s} = i\frac{2^{5/2}}{3^3}\left[\frac{2\beta^3 b^2}{\beta^2+\omega^2}\right] K_1(b\sqrt{\omega^2+\beta^2} \quad , \tag{2.44}$$

with $\beta = 3/2$ and $\omega = 3\text{v}/8$. Then the first order probability for a $1s-2s$ transition in atomic hydrogen caused by an interaction with a proton is,

$$P(Z=1, Z_T=1, \text{v}, b) = \frac{2^{13}\text{v}^6}{3^4(1+16\text{v}^2)}K_2\left(\frac{3\beta}{8v}\sqrt{1+16\text{v}^2}\right) \quad . \tag{2.45}$$

The first order probability amplitude for an electron initially in the $1s$ state of an atom with charge Z_T interacting with a projectile of charge Z_T and going to a final state with $n = 2, \ell = 1$ and $m = +1$ may be evaluated in a similar fashion giving,

$$\begin{aligned} &a_{2p+,1s}(Z, Z_T, \text{v}, \vec{b}; t=\infty) = \\ &-\frac{i12Z_T^5}{\text{v}\alpha^6}e^{i\varphi}\left[|q_{\|}|K_1(|q_{\|}|b) - \beta K_1(\beta b) - \frac{\alpha^2 b}{2}K_0(\beta b) - \frac{\alpha^4 b^2}{8\beta}K_1(\beta b)\right] \\ &= \frac{Z}{Z_T}a_{2p+,1s}(Z=1, Z_T=1, \text{v}/Z_T, Z_T\vec{b}; t=\infty) \ , \end{aligned} \tag{2.46}$$

where φ is the azimuthal angle of $\vec{b}$, K_i the modified Bessel function of the second kind, $\alpha = 3Z_T/2$, $\beta = \sqrt{\alpha^2 + q_{\|}^2}$ and $q_{\|} = \Delta E_{2p+,1s}/\text{v}$.

2 $n, \ell, m - n', \ell', m'$ transitions

The first order transition probability amplitudes, a_{fi}, may be expressed in closed form for an arbitrary transition from an arbitrary initial bound state, $|i\rangle = |n, \ell, m\rangle$, to an arbitrary final bound state, $|f\rangle = |n', \ell', m'\rangle$. This has been done by Straton [1991]. Here the interaction potential V is given by,

$$V = -\frac{Z_p}{|\vec{R} - \nu\vec{r}|} + \frac{Z_p}{R} \ , \tag{2.47}$$

where $\nu = M_t/(M_t + m_e) \neq 1$ is now included. Here the long range Coulomb tail does not contribute. This is the same as $V - \bar{V}$ at large R.

The possibility of different effective charges Z_T for the target nucleus initially and finally is incorporated by using $\lambda = /ZT$ in $|i\rangle = |n, \ell, m\rangle$ and $\lambda' = Z_T'$ in $|f\rangle = |n', \ell', m'\rangle$. To include cases where the electron is replaced by a particle of mass m and charge Z, the parameters $\nu = m/m_e$ and Z, which normally assume the values of $\nu = 1$ and $Z = -1$. As above, $\omega = \Delta E_{fi}$. The parameter γ is defined as $\gamma = \frac{\lambda'}{n'} + \frac{\lambda}{n} + \eta_1$, where $\eta_1 > 0$ imposes a factor of $e^{-\eta R}$ onto V, changing the Coulomb potential to a Yukawa potential. These terms are used to define,

$$A = \sqrt{\frac{\gamma^2}{\nu^2} + \frac{\omega^2}{v^2}} . \tag{2.48}$$

The general state-to-state one electron transition probability amplitude may now be expressed in terms of derivatives of an analytic function:

$$D = \left(-i\frac{\partial}{\partial\omega}\right)^p A^j K_j(bA) . \tag{2.49}$$

Then, using Gradshteyn and Ryzhik (8.486.14) and (0.432.1) one has,

$$D = \sum_{r=0}^{p} \frac{p!(-b)^r(2\omega)^{p-2r}}{(p-2r)!2^r v^{2r}} A^{j-r} K_{j-r}(bA) . \tag{2.50}$$

Then the general form for state-to-state transitions in hydrogen becomes[††],

$$\begin{aligned} ia_{fi}(b) \;&\equiv\; \mathrm{A}^{\lambda' \lambda}_{n'\ell'-m',n\ell m}(\omega, b, \nu, Z_e) \;= \\ &(-1)^{-m'} \frac{Z_p Z_e}{\mathrm{v}} N_{n'\ell'} N_{n\ell} \, (\lambda'\lambda)^{3/2} \\ &\times \sum_{s'=0}^{n'-\ell'-1} \frac{(-1)^{s'}(\lambda'/n')^{s'+\ell'}}{(n'-\ell'-1-s')!(2\ell'+1+s')!s'!} \end{aligned}$$

[††] Five errors appear in the Straton [1991]: The first sum in the second line should have had $s'!$ rather than $s!$ in the denominator. The third line was missing a factor of $2(-1)^{m'}$ and the $\delta_{m\,-m'}$ was missing the '$-$' sign. The exponent of b in the q sum, $2q + m' + m$, was missing the final '+.' The absolute value signs for ω were missing in the third and eighth lines [the ω on the seventh line should be a signed quantity since it comes from differentiating $exp(-i\omega t)$ with respect to ω, not $|\omega|$]. The factors of 4π have been canceled. There are also two errors in the a_j^h. The correct version is:

$$\begin{aligned} a_i^i &\equiv 1 \\ a_i^{i-1} &= -\binom{i+1}{2} \\ a_i^{i-k} &= -\sum_{m=0}^{k-1} a_i^{i-m} \binom{i+k-2m}{2k-2m} (2k-2m-1)!! \;, 1 < k < (i+1)/2 \\ a_i^k &\equiv 0 \;, k < i/2 . \end{aligned}$$

$$\times \sum_{s=0}^{n-\ell-1} \frac{(-1)^s(\lambda/n)^{s+\ell}}{(n-\ell-1-s)!(2\ell+1+s)!s!} \Big\{ (-1)^{1+m'} \delta_{\ell\ell'}\, \delta_{m\,-m'}$$

$$\times\, 2^{s'+\ell'+s+\ell+1}\, \frac{(s'+\ell'+s+\ell+2)!}{\gamma^{s'+\ell'+s+\ell+3}}\; K_0(\frac{b|\omega|}{\mathrm{v}})$$

$$+ \sum_{L=L_{min}(2)}^{\ell'+\ell} \; 2^L \left[(2\ell'+1)(2\ell+1)(2L+1)\right]^{1/2}$$

$$\times \begin{pmatrix} \ell' & \ell & L \\ 0 & 0 & 0 \end{pmatrix} \begin{pmatrix} \ell' & \ell & L \\ m' & m & -m'-m \end{pmatrix}$$

$$\times \sum_{I=0}^{\left\{ {M/2 \atop (M-1)/2} \right\}} \frac{(-1)^I(2I-1)!!M!}{(M-2I)!(2I)!}\; \frac{1}{\gamma^{M+1}}\; 2^{M-1}\; \frac{1}{\nu^{L+1}}$$

$$\times \left[(2L+1)\,(L+m'+m)!(L-m'-m)!\right]^{1/2}$$

$$\times \sum_q \frac{(-1)^q\, b^{2q+m'+m}\, \mathrm{v}^p}{2^{2q+m'+m}(q+m'+m)!q!p!}\; 2(-i)^p$$

$$\times \sum_{r=0}^{p} \frac{(p)(p-1)(p-2)\cdots(p-2r+1)\;\; (-b)^{p-r}(2\omega)^{p-2r}}{r!2^{p-r}\mathrm{v}^{2(p-r)}}$$

$$\times \sum_{k=0}^{N} \binom{N+k}{2k} (2k-1)!!\,(2L+N-k)! \left\{ (2b)^{-L}\, \frac{\nu^{2L+1}}{\gamma^{2L+1}} \right.$$

$$\times \frac{\sqrt{\pi}}{\Gamma(\frac{1}{2}+L)} \left(\frac{|\omega|}{\mathrm{v}}\right)^{L-p+r} \quad K_{L-p+r}(\frac{b|\omega|}{\mathrm{v}})$$

$$-\frac{1}{(2L-1)!!} \sum_{j=1}^{L} \frac{1}{[2(L-j)]!!}\; \frac{\nu^{2j+1}}{\gamma^{2j+1}}\; b^{-j}\; A^{j-p+r} K_{j-p+r}(bA)$$

$$-\frac{1}{(2L)!}\; \frac{\nu}{\gamma}\; A^{-p+r}\; K_{-p+r}(bA)$$

$$- \sum_{h=0}^{N-k-1} \sum_{j=h}^{N-k-1} \frac{a_j^h}{(2L+1+j)!}\; \frac{\gamma^{2h+1}}{\nu^{2h+1}} b^{h+1}\; A^{-(h+1+p-r)}$$

$$\times K_{h+1+p-r}(bA) \Big\} . \tag{2.51}$$

Except for $p = L - m' - m - 2q$ and Z_e (denoting the charge of the electron), the various constants are the same as those defined by Straton [1991]. The index '(2)' on the summation over L indicates that one is to sum in steps of two. In practice this '(2)' may be ignored because the appropriate $3-j$ coefficients are zero. Examples of the first order probability amplitudes of (2.51) for specific one electron transitions are given by Straton [1991]. The transition probability is found from $P = |a_{fi}|^2$.

It is possible to produce large cross sections with small transition probabilities that may be correctly evaluated using the first Born approximation. An example is the $2s \rightarrow 2p$ transition probability in the limit where the $2s - 2p$ energy splitting becomes small. In this case even weak perturbations at large distances can induce a transition [McGuire et al., 1979].

Application to pseudostates (e.g. using correlated configuration interaction initial and final state asymptotic wavefunctions) is straightforward and follows the description in section 7.2.3.

2.3 Mass transfer probabilities

In this section a transition probability for the transfer of an electron from the target atom to the projectile is developed in first order perturbation theory. A more thorough review of the theory of mass transfer in fast collisions is given by Bransden and Dewangan [1988]. In the relatively simple case of mass transfer in first order perturbation theory, the initial state $|i>$ is a bound state of the target, while the final state $|f>$ is a bound state of the projectile. Since the projectile is moving, these states are non-orthogonal. The first order amplitude for mass transfer was first properly derived by Bates [1958] from coupled channel equations using an over complete set of basis states that includes states on both the target and the projectile. The resulting expression for the first order amplitude is,

$$a_{fi} = \int_{-\infty}^{\infty} e^{i\omega t} \frac{< f|V - \bar{V}|i >}{(1 - |< f|i >|^2)} dt \,. \tag{2.52}$$

Here $\bar{V}$ may be chosen as either $< i|V|i >$ or $< f|V|f >$. As noted in a footnote earlier in this chapter, this is equivalent to a Gramm-Schmidt orthogonalization of the final state unit vector $|f>$ with respect to the initial state vector $|i>$ except for the factor of $1 - |< f|i >|^2$, which is usually close to 1. This same correct first order result was derived from the distorted wave Born approximation by Bassel and Gerjouy [1960].

However, the above correct first order expression is often replaced by the earlier first order approximation of Oppenheimer [1928] and Brinkman and Kramers [1930], which corresponds to,

$$a_{fi} = \int_{-\infty}^{\infty} < f|V|i > e^{i\omega t} dt \ . \tag{2.53}$$

This OBK amplitude is relatively easy to evaluate.

1 $n, \ell, m - n', \ell', m'$ transitions

Evaluation of electron capture is relatively straightforward in the wave picture in the OBK approximation. Since the final state, $|f>$, is moving with a translational velocity $\vec{v}$ with respect to the initial state, $|i>$, it is conventional to use two related momentum transfers [McDowell and Coleman, 1970, chapter 8],

$$\vec{p} = a\vec{k}_k - \vec{k}_i \quad \text{and} \quad \vec{q} = b\vec{k}_i - \vec{k}_f \ , \tag{2.54}$$

where

$$a = \frac{M_P}{M_P + 1}, \qquad b = \frac{M_T}{M_T + 1} \ ,$$

and

$$\Delta E_{fi} = \frac{p^2}{a} - \frac{q^2}{b} \ .$$

Then it is convenient to express the scattering amplitude in q-space as,

$$\begin{aligned} f &= -\frac{\mu}{2\pi} < f|V|i > = -\frac{\mu}{2\pi} < n'\ell'm'|\frac{-Z}{|\vec{R} - \vec{r}|}|n\ell m > \\ &= \left(\frac{\mu}{2\pi}\right) (2\pi)^3 \, g^*_{n'\ell'm'}(-\vec{p})\psi_{n\ell m}(\vec{q}) \ , \end{aligned} \tag{2.55}$$

where μ is the reduced mass and $\psi_{n\ell m}(\vec{q})$ is the momentum space wave-function defined by (11.13) and (11.16), and,

$$g_{n'\ell'm'}(\vec{p}) = \int d\vec{r'} e^{i\vec{p}\cdot\vec{r}} \frac{-Z}{r'} \psi_{n'\ell'm'}(\vec{r'}) \ , \tag{2.56}$$

with $\vec{r}\,' = \vec{R} - \vec{r}$, which is the coordinate of the electron relative to the projectile nucleus. Now, $g_{n'\ell'm'}(\vec{p})$ may be related to $\psi_{n'\ell'm'}(\vec{p})$ by using the Schrödinger equation for the projectile bound state,

$$\left(-\nabla^2/2 + Z^2/2n'^2\right) \psi_{n'\ell'm'}(\vec{r'}) = \left(\frac{Z}{r'}\right) \psi_{n'\ell'm'}(\vec{r'}) \,. \tag{2.57}$$

Since,

$$\nabla^2 e^{i\vec{p}\cdot\vec{r'}} = -p^2 e^{i\vec{p}\cdot\vec{r'}} \,, \tag{2.58}$$

after multiplying (2.57) by $e^{i\vec{p}\cdot\vec{r'}}$ and integrating one obtains,

$$g_{n'\ell'm'}(\vec{p}) = \frac{1}{2}(p^2 + Z^2/n'^2)\psi_{n'\ell'm'}(\vec{p}) \,, \tag{2.59}$$

which conveniently relates $g_{n'\ell'm'}(\vec{p})$ to $\psi_{n'\ell'm'}(\vec{p})$.

As a consequence,

$$f = \frac{\mu}{2\pi}\,(2\pi)^3\,\frac{1}{2}(p^2 + Z^2/n'^2)\;\psi_{n'\ell'm'}(\vec{p})\;\psi_{n\ell m}(\vec{q}) \,, \tag{2.60}$$

so that the OBK scattering amplitude for electron capture may be expressed in terms of a product of wavefunctions in momentum space and not in coordinate space. An expression for the more general Born Oppenheimer (or Jackson and Schiff, 1953) amplitude has been given by Deb and Sil [1996].

The differential cross section equal to $|f|^2$, may also be expressed in terms of a product (but not an overlap) of wavefunctions in momentum space. This leads to the convenient result that sums over the final state quantum numbers m' and ℓ' may be done readily using the Fock sum rule of (11.14). This leads to the same functional form of the cross section for all final n' summed over m' and ℓ'. Unfortunately, this simplification occurs only in momentum space.

In the case of $1s-1s$ electron capture, the scattering amplitude of (2.60) may be expressed explicitly using the wavefunctions given after (11.16) of the appendices, namely,

$$\begin{aligned} f_{1s,1s} &= \left(\frac{\mu}{2\pi}\right) 32\pi(p^2 + Z^2/n'^2)\frac{Z^{5/2}}{(p^2+Z^2)^2}\frac{Z_T^{5/2}}{(q^2+Z_T^2)^2} \\ &= \left(\frac{\mu}{2\pi}\right) 2^5\pi\frac{ZZ_T)^{5/2}}{(q^2+Z_T^2)^3} \,, \end{aligned} \tag{2.61}$$

where $q^2 + Z_T^2 = p^2 + Z_T^2$ has been used which follows from (2.54).

2 1s − 1s transitions

The probability amplitude for a transition for electron transfer from $|i> |1s>_{target}$ to $|f>=|1s>_{projectile}$ may be evaluated by using (2.61) in (2.36). For fast heavy projectile the scattering occurs mostly at $\theta << 1$ so that,

$$a_{1s,1s} = k\int_0^{\pi} f(\theta) sin\theta d\theta \simeq \frac{\mu k}{2\pi}\int_0^{\pi} 2^5\pi(ZZ_T)^{5/2}(Z_T^2+q^2)^{-3}\theta d\theta \,. \quad (2.62)$$

It follows from (2.54) that,

$$q^2 = \frac{\mathrm{v}^2}{4}\left[\left(1-\frac{Z^2-Z_T^2}{\mathrm{v}^2}\right)^2 + 4\mu^2\theta^2\right], \quad (2.63)$$

where $\mathrm{v} = k/\mu$ is the collision velocity. Defining,

$$S = \frac{\mathrm{v}}{Z_T}\left[\left(1-\frac{Z^2-Z_T^2}{\mathrm{v}^2}\right)^2\right],$$

and $y = \mu \mathrm{v}\theta/Z_T$, then,

$$f = \left(\frac{\mu}{2\pi}\right)\frac{1}{(2\pi)^3}2^5\pi\frac{1}{\left[Z_T^2\left(1+\frac{S^2}{4}+y^2\right)^3\right]}\,. \quad (2.64)$$

When $\ell >> 1$ in (2.36), one has

$$P_\ell(cos\theta) \simeq J_0(\ell\theta) = J_0(kb\theta) = J_0(Z_T by)\,,$$

and

$$\begin{aligned} a_{1s,1s} &= \frac{\mu k}{2\pi}\int_0^{\pi} f(\theta)\, J_0(\ell\theta)\theta d\theta \quad (2.65)\\ &= 2^5\pi(ZZ_T)^{5/2}k\int_0^{\infty}\frac{J_0(\tilde{b}y)}{\left[Z_T^2\left(1+\frac{S^2}{4}+y^2\right)\right]^3}\left(\frac{Z_T}{k}\right)^2 ydy\,. \end{aligned}$$

From Gradshteyn and Ryzhik (6.565-4),

$$\int_0^{\infty}\frac{yJ_0(by)}{[1+\beta^2+y^2]^{n+1}}dy = \frac{X^nK_n(X)}{2^n n!(1+\beta^2)^n}\,, \quad (2.66)$$

with $X = \tilde{b}(1+\beta^2)^{1/2}, \tilde{b} = Z_T b, \beta^2 = S^2/4$, and K_n is a modified Bessel function of order n. Then,

$$a_{1s,1s} = \frac{2(ZZ_T)^{5/2}}{\mathrm{v}} \frac{X^2 K_2(X)}{Z_T^4(1+S^2/4)^2}, \tag{2.67}$$

so that the probability for the $1s - 1s$ transfer is,

$$P_{1s,1s} = \frac{4}{Z_T^8 \mathrm{v}^2}(ZZ_T)^5 \frac{X^2 K_2(X)}{\left(1+\frac{S^2}{4}\right)^4}. \tag{2.68}$$

Here X acts as a scaled impact parameter and S acts as a scaled velocity.

The total cross section is found by integrating over impact parameters. Noting that $\int_0^\infty [X^2 K_2(X)]^2 dX = 32/5$, one has (in units of a_0^2),

$$\sigma = \frac{2^8\pi}{5v^2} \frac{(ZZ_T)^5}{Z_T^{10}(1+S^2/4)^5}. \tag{2.69}$$

This OBK first order cross section gives the correct scaling with Z_T, ZT and v at high velocities. However, the absolute cross sections and probabilities are usually larger than those observed even at high velocities.

3 Special features of electron capture

In the process of electron capture the electron is transferred from one atomic nucleus to another nucleus in relative motion. This has a number of interesting consequences [McGuire *et al.*, 1996a; Briggs and Macek, 1990].

1. The set of basis states used is over complete when it includes states on both the target and the projectile.

2. The initial and final states of the electron are not orthogonal, i.e., $< f|i > \neq 0$. Formulas based on a single set of orthonormal functions may not be valid.

3. The scaling laws discussed in section 2.1.3 that are based on a single basis set above do not apply.

4. The electron is moving in the final state with respect to the rest frame of the initial state. Consequently electron translation factors must be included. These translation factors complicate the calculations.

5. Unlike excitation and ionization, the minimum momentum transfer, q_{min} in (2.35), depends on the mass of the projectile. Consequently the cross section for capture by a positron is higher than that of a proton of equal velocity at high velocities.

6. Classically the transfer of mass to a fast projectile violates conservation of energy and momentum in a single collision, and is therefore forbidden. The simplest process that is kinematically allowed is a two step (or Thomas) process [Shakeshaft and Spruch, 1979].

7. Since the simplest process that is classically allowed is a two step process, the second order perturbation amplitude dominates over the first order amplitude at high energies [Miraglia *et al.*, 1981; Rivarola, 1981; Simony, 1981; Briggs and Macek, 1990; McGuire *et al.*, 1996a]. In most cases second order effects are small and may be ignored for multiple electron transitions. An exception is noted in section 7.4.

2.4 Ionization probabilities

1 Continuum of the target

When an atom is ionized, the final state of the electron is a continuum state, i.e., $|f>=|\vec{k}>$. The continuum states may be regarded as an analytic continuation of bound states, $|n,\ell,m>$ to unbound states $|\vec{k}>$ with the angular parts described by the Legendre $Y_{\ell m}(\theta,\phi)$ and $n\rightarrow -iZ_T/k$. Thus, in a sense, ionization may be regarded as excitation to continuum states. Here it is assumed that the projectile is sufficiently fast that the charges of the final target ion and outgoing electron screen each other from the fast projectile. Thus the fast projectile is treated as a plane wave and the ion and outgoing electron as a two body Coulomb wavefunction.

In principle it is straightforward to evaluate the first order probability, namely,

$$\frac{dP}{d\vec{k}} = |a_{\vec{k}i}|^2 = |-i\int dt' e^{i\omega_{if}t'} <\vec{k}|V(t')|i>|^2 , \tag{2.70}$$

with a total cross section for ionization defined by,

$$\sigma = \int d\vec{k}\frac{d\sigma}{d\vec{k}} = \int d\vec{b}\int d\vec{k}\frac{dP}{d\vec{k}}(\vec{b}) . \tag{2.71}$$

The normalization condition on the Coulomb waves is $<\vec{k}'|\vec{k}>=C\delta(\vec{k}'-\vec{k})$ where usually $C=(2\pi)^3$ [McDowell and Coleman, 1970]. That is, ionization usually refers to a sum over all continuum states and the transition to a particular $\vec{k}$ is regarded as a differential probability and cross section.

The two body continuum wavefunction that is orthogonal to the bound state hydrogenic wavefunctions is the Coulomb wavefunction [McDowell and Coleman, section 5.4] defined by,

$$|\vec{k}>= \psi_{2C}^{+} = e^{-\eta\pi/2}\Gamma(1+i\eta)e_1^{i\vec{k}\cdot\vec{r}}F_1[-i\eta, 1; i(kr - \vec{k}\cdot\vec{r})] , \qquad (2.72)$$

where $\eta = Z_T/k$ and ${}_1F_1$ is the confluent hypergeometric function. For large k, the Coulomb wavefunction reduces quite slowly to a plane wave [Joachain, 1983]. The same nuclear charge, Z_T, is conventionally used for the final state as for the initial state, even though this may not correspond to the net charge of the target ion at large distances after the collision.

Calculation of the total ionization cross section using the two body Coulomb wavefunction is straightforward in momentum space, and is given in units of a_0^2 by [McDowell and Coleman, section 7.2],

$$\begin{aligned} \sigma &= \frac{2^{11}\pi Z_p^2}{v^2}\int_0^{k_{max}} \frac{kdk}{(1-e^{-2\pi Z_T/k})} \\ &\times \int_{q_{min}}^{q_{max}} dq \frac{q^2 + \frac{1}{3}(k^2 + Z_T^2)}{q[Z_T^4 + 2Z_T(k^2+q^2) + (q^2-k^2)^2]^3} \\ &\times\ exp\left[-\frac{2Z_T}{k} tan^{-1}\frac{2Z_T k}{Z_T^2 - k^2 + q^2}\right] . \end{aligned} \qquad (2.73)$$

Here k_{max}, q_{max} and q_{min} may be found from the constraints of conservation of energy and momentum. For fast projectiles, $k_{max} \simeq \infty, q_{max} \simeq \infty$ and $q_{min} \simeq \Delta E/2\mathrm{v}$, where $\Delta E = I + k^2/2$ is the energy transferred to the electron in the collision.

Cross sections in first order perturbation theory for ionization of atomic hydrogen by a fast projectile of unit charge are given in Table 2.1 in the appendix of this chapter. Cross sections for ionization evaluated in first order perturbation obey the scaling relations of (2.30), namely $\sigma(Z, Z_T, \mathrm{v}) = \frac{Z^2}{Z_T^4}\sigma(Z=1, Z_T=1, \mathrm{v}/Z_T)$, where $\sigma(Z=1, Z_T=1, \mathrm{v}/Z_T)$ is listed in Table 2.1 for $Z_T = 1$.

Evaluation of the probability amplitude, $a_{\vec{k}i}$, is tedious. Calculations have been carried out by Hansteen, Johansen and Kocbach [1975], and the resulting probabilities have been tabulated for ionization from initial hydrogenic states with principal quantum numbers n = 1, 2 and 3. The probabilities for ionization from n = 1 are given in Table 2.2 of the appendix to this chapter.

2 Continuum of the projectile

In the usual first order development for ionization (given in the section above) the final state wavefunction is taken to be a two body Coulomb wavefunction for the outgoing electron in the presence of the nuclear charge of the target with the fast projectile regarded as a plane wave. This description is not accurate if the outgoing electron remains closer to the projectile than the residual target ion. When the velocity of the ejected electron matches the velocity of the outgoing positively charged projectile, a cusp can arise in the energy spectra of the ionized electrons. This effect is known as electron capture to the continuum and has been observed experimentally in collisions of fast ions with atoms [Crooks and Rudd, 1970; Dettmann *et al.*, 1974; Rodbro and Andersen, 1979; Stolterfoht *et al.*, 1987b]. From a theoretical point of view the electron capture to the continuum occurs when the final state is described by a two body Coulomb wavefunction for the electron in the field of the projectile rather than the field of the target [Salin, 1969; Macek, 1970; Dettmann *et al.*, 1974; Burgdörfer, 1984; Briggs and Macek, 1990; Moiseiwitsch, 1994]. These states are a continuation of the bound states of the projectile into the continuum. At high projectile velocities, electron capture to the continuum contributes little to the total cross section for all transitions from bound to unbound states because the capture cross section decreases much faster with velocity (v^{-11}) than ionization to two body Coulomb states of the target ($ln\ \mathrm{v}^2/\mathrm{v}^2$).

2.5 Long range Coulomb effects

The $1/R$ Coulomb potential is well established in physics. Like gravity, the Coulomb force diminishes so that the lines of electric force which cross any sphere at large R remain the same as R increases. This remarkable dependence on the distance R however, causes some mathematical difficulty. Because the potential falls off so slowly with R, the integral equations used to generate the quantum wavefunctions do not smooth any roughness in the source terms (i.e. they are not mathematically compact). One consequence is that it is not possible to guarantee that solutions are unique. Another problem is that the on-shell limit of the off-energy-shell Coulomb T-matrix does not exist due to a phase divergence [Briggs and Macek, 1990; Chen and Chen, 1972]. This can cause difficulty when integrating over off-shell Coulomb scattering amplitudes. A third problem is numerical. One may never be certain that one has integrated out far enough in R to guarantee convergence of the numerical results. Fortunately, the problem of two interacting charged particles is exactly solvable. Unfortunately, the wavefunction for three charged particles is not well known. Consequently, approximate wavefunctions such as those described above are used.

The problem of the long range Coulomb problem has been addressed in various ways [Taylor, 1972, chapter 12]. One simple approach is to remove the difficulty in principle by noting that charged particles tend to be screened by particles of opposite charge which are attracted. For most electrically charged ions in chemistry and biology screening occurs at distances of several Bohr radii, unless something prevents such screening to occur. In outer space screening distances are of the order of meters. In principle one may eliminate troublesome $1/R$ terms in the electronic wavefunctions by pairing off terms in the Hamiltonian so that one uses the difference between the instantaneous and the average Coulomb potentials. The remaining long range terms (if any) are then transferred to the projectile-nucleus scattering. This is described in section 3.3.4.

Alternatively it is possible to develop a formalism that does not depend on the long range nature of the interaction potential [Taylor, 1972; Dollard, 1964; Amrein *et al.*, 1970].

1 Continuum of target and projectile

Another approach is to use an approximate three body Coulomb wavefunction [Redmond, unpublished; Cheshire, 1964] used by a number of authors who have employed the continuum distorted wave (CDW) approximation [Crothers, 1996; Crothers and Dube, 1992; Belkic *et al.*, 1979; McCarroll and Salin, 1968; Cheshire, 1964]. This three body Coulomb wavefunction is often referred to as the BBK wavefunction following a widely quoted paper by Brauner, Briggs and Klar [1989]. There were similar approaches including an approximate solution of the three body Faddeev-Mercuriev equations [Godunov *et al.*, 1983] and a multiple scattering approach [Miraglia, 1983]. This three body Coulomb wavefunction has the symmetric form of a product of three two body Coulomb wavefunctions, one for each pair of interacting particles, namely [Briggs, 1993],

$$\begin{aligned}\psi_{3C} &= (2\pi)^{-\frac{3}{2}} \exp(-\pi\eta_a/2)\Gamma(1-i\eta_a)e_1^{i\vec{k}_a\cdot\vec{r}_a} F_1(i\eta_a, 1; -i[k_a r_a + \vec{k}_a\cdot\vec{r}_a]) \\ &\times (2\pi)^{-\frac{3}{2}} \exp(-\pi\eta_b/2)\Gamma(1-i\eta_b)e_1^{i\vec{k}_b\cdot\vec{r}_b} F_1(i\eta_b, 1; -i[k_b r_b + \vec{k}_b\cdot\vec{r}_b]) \\ &\times \exp(-\pi\eta_{ab}/2)\Gamma(1-i\eta_{ab})e_1^{i\vec{k}_{ab}\cdot\vec{r}_{ab}} F_1(i\eta_{ab}, 1; -i[k_{ab} r_{ab} + \vec{k}_{ab}\cdot\vec{r}_{ab}] ,\end{aligned} \tag{2.74}$$

with

$$\eta_a = -Z_T/k_a \quad \eta_b = Z_T Z_T/k_b \quad \eta_{ab} = Z/2k_{ab} \quad k_{ab} = \frac{|\vec{k}_b - \vec{k}_a|}{2} , \tag{2.75}$$

where k_a is the momentum of the outgoing electron, k_b the momentum of the outgoing projectile, and k_{ab} the momentum of the electron relative to the outgoing projectile. Calculated results at low collision energies using this three body Coulomb wavefunction are not well normalized in comparison to observed results, especially at low collision velocities [Briggs, 1993].

A somewhat more elaborate wavefunction has been introduced by Alt and Mukhamedzhanov [1993] which is asymptotically correct if any interparticle separation is large. Numerical close-coupling methods have been developed by Bray *et al.* [1994]. An overview is given by Jones and Madison [1995].

An approximate four body Coulomb wavefunction has been expressed as a product of two body Coulomb wavefunctions by Mercuriev and Faddeev [1985]. This four body Coulomb wavefunction has been used in numerical calculations by Godunov and Schipakov [1995].

Threshold. At energies close to the ionization threshold the long range Coulomb interaction changes the energy dependence of the cross section for ionization from E^1 valid for short range interactions‡‡ to E^m, where $m > 1$ is an irrational number [Wannier, 1953]. This unusual exponent depends on the nature of the asymptotic three body Coulomb wavefunction. For electron impact on a hydrogenic one electron target of nuclear charge Z_T, the Wannier threshold dependence is,

$$\sigma \sim E^m \qquad m = \frac{\sqrt{(100Z_T - 9)/(4Z_T - 1)} - 1}{4} . \tag{2.76}$$

Experiments in which shielding from stray charges is carefully eliminated have confirmed this result [Read,1992; Hall *et al.*, 1991, 1993; Cvejanovic and Read, 1974]. However, the range of energies E above the ionization threshold for which this dependence applies has been under discussion since Wannier's original paper in 1953.

The nature of the exact wavefunction for three (or more) particles interacting via a deceptively simple $1/R$ interaction has been an intriguing problem in atomic physics for over forty years.

‡‡ The Wigner threshold behavior for short ranged interactions gives $E^{2\ell+1}$ where ℓ is the angular momentum of the scattering amplitude, e.g., E^1 dominates for s waves for e + H near threshold.

2.6 Other methods

Here a brief description is given for some of the theoretical methods referenced in this book. A more general overview of scattering theory has been given by Burke [1996] for electron-atom and electron molecule collisions and by Ford and Reading [1996] for ion-atom and atom-atom collisions.

Coupled channel methods. Some of the most accurate calculations of transition probabilities for single electron transitions are done using the method of coupled channels (Cf. section 2.1.2).

R-matrix method. Another method that is in principle exact and leads to accurate numerical results is the reaction-matrix or R-matrix method [Burke, 1996; Greene, 1995; Greene and Aymar, 1991; Burke and Robb, 1975]. In this method one matches a numerically accurate wavefunction with an asymptotic wavefunction at a boundary.

Other numerical methods. There are a variety of methods for integrating the Schrödinger equation directly including lattice representations, method of splines, continued fractions, Hylleras method and Monte Carlo integration [Schultz *et al.*, 1994; Bottcher *et al.*, 1994; Wang and Callaway, 1994; Pindzola, 1995; Feagin, 1995]. Bottcher [1988, 1985] and Burke [1996] have surveyed some of these.

Distorted wave method. In the distorted wave Born approximation (DWBA) [Jakubassa-Amundsen, 1989; Madison, 1996], the full Hamiltonian is partitioned into,

$$H = H_T + W + V' \,, \tag{2.77}$$

such the eigenfunctions of $H_T + W$ are known and V' is as small as possible. Then the perturbation expansion is made in the residual interaction, V'. Examples of DWBA-like methods include the continuum distorted wave (CDW) method [Crothers and Dube, 1992; Cheshire, 1964] in which the analytic properties of two body Coulomb continuum waves are utilized, and the strong potential Born (SPB) approximation [Jakubassa-Amundsen, 1989; McGuire and Sil, 1986; Briggs *et al.*, 1982; Macek and Alston, 1982] in which an expansion in the weaker of Coulomb potentials in the initial and final channels is used while the strong potential is retained to all orders.

Eikonal methods. In the eikonal approximation [Maidagan and Rivarola, 1984; Eichler and Chan, 1979], the Schrödinger equation,

$$\left(\frac{\nabla^2}{2M} + V\right)\psi = E\psi\,, \tag{2.78}$$

is rewritten by setting $\psi = e^{i\vec{k}\cdot\vec{r}}\chi$ such $k^2/2M = E$ whence the exact equation becomes,

$$\left(\frac{\nabla^2}{2M} + i\frac{\vec{k}}{M}\cdot\nabla + V\right)\chi = 0\,. \tag{2.79}$$

If χ is slowly varying, the first term is relatively small and is neglected, leaving the eikonal equation,

$$i\mathrm{v}\frac{d\chi}{dz} = -V\chi\,, \tag{2.80}$$

where $\frac{\vec{k}}{M}\cdot\nabla = \mathrm{v}\frac{d}{dz}$ where $\hat{z} = \hat{k}$. Then, integrating (2.80),

$$\chi \approx e^{\frac{i}{\mathrm{v}}\int V dz}\,. \tag{2.81}$$

This eikonal wavefunction is related to the semi-classical WKB method [Messiah, 1961, chapter VI]. The Glauber approximation [Glauber, 1959; Golden, 1975] may be obtained by subsequently neglecting the component of the momentum transfer parallel to the incident momentum $\vec{k}$. It has been shown that, for heavy projectiles, the Glauber approximation is valid over a wider range of collision velocities v than the corresponding Born approximation [McGuire, 1983].

1 Strong electric fields

Consider a one electron atom of nuclear charge Z_T in a strong electric field, $\vec{E}(t) = -\frac{\partial\vec{A}}{\partial t}$, where $\vec{A}$ is the vector potential in the Coulomb gauge. The Hamiltonian for this system is,

$$H = -\frac{1}{2}\nabla^2 - Z_T/r - \vec{r}\cdot\vec{E}(t)\,. \tag{2.82}$$

If one neglects the Z_T/r interaction of the electron with the atomic nucleus, then H becomes,

$$H = -\frac{1}{2}\nabla^2 - \vec{r}\cdot\vec{E}(t)\,, \tag{2.83}$$

which corresponds to a free electron in an electric field. The solution to (2.83) corresponding to an asymptotic momentum, $\vec{k}$, is,

$$\phi_k(\vec{r},t) = \frac{1}{(2\pi)^{3/2}} e^{i[\vec{k}\cdot\vec{r} - \frac{1}{2}k^2 t - \vec{r}\cdot\vec{P}(t) - J(t)]} , \quad (2.84)$$

where $\vec{P}(t)$ is a momentum transfer given by,

$$\vec{P}(t) = \int_0^t d\tau \vec{E}(\tau) , \quad (2.85)$$

and $J(t)$ is an energy spread induced by the momentum transfer,

$$J(t) = \int_0^t d\tau [\frac{1}{2}P^2(\tau) - \vec{k}\cdot\vec{P}(\tau)] . \quad (2.86)$$

It may be verified that ϕ_k is an eigenfunction of (2.83), thus satisfying,

$$\left(i\frac{\partial}{\partial t} + \frac{1}{2}\nabla^2 + \vec{r}\cdot\vec{E}(t)\right)\phi_k = 0 . \quad (2.87)$$

These continuum states ϕ_k are called Volkov Keldysh states.

2 Strong projectile fields

Presnyakov *et al.* [1995] have extended this method to find an approximate solution for the impact of highly charged projectiles on atomic targets. The Schrödinger equation for a projectile of charge Z and trajectory $\vec{R}(t)$ incident on a one electron target of nuclear charge Z_T is,

$$\left(i\frac{\partial}{\partial t} + \frac{1}{2}\nabla^2 + Z_T/r + \frac{Z}{|\vec{R}(t) - r|} - \frac{ZZ_T}{R(t)}\right)\psi_k = 0 . \quad (2.88)$$

Neglecting the $ZZ_T/R(t)$ term and retaining only the dipole contribution to the projectile electron interaction, one obtains,

$$\left(i\frac{\partial}{\partial t} + \frac{1}{2}\nabla^2 + Z/r + \vec{r}\cdot\vec{E}(t)\right)\psi_k = 0 . \quad (2.89)$$

where $\vec{E}(t) = Z\vec{R}(t)/R^3(t) = -\frac{\partial\vec{A}}{\partial t}$. The solution to this equation given by Presnyakov *et al.* [1995] is written as,

$$\psi_k = \phi_k G . \quad (2.90)$$

When $\frac{\partial G}{\partial t}$ is neglected, then $G = \exp(\pi\nu/2)\Gamma(1+i\nu)F_1(-i\nu, 1; i(pr+\vec{p}\cdot\vec{r}))$ with $\vec{p} = \vec{k} - \vec{A}$ and $\nu = Z_T/p$. The Volkov-Keldysh states ϕ_k are recovered if $Z_T = 0$ so that G = 1.

3 Over the barrier model

A simple and surprisingly useful model for estimating estimating probabilities and cross sections for electron transfer in slow ion-atom collisions is the over the barrier model introduced by Ryufuku, Sasaki and Watanabe [1980]. This is a classical model in which transfer occurs when the Coulomb potential between the two nuclei is low enough so the potential barrier between the initial and final states lies below the energy of the electron. When this occurs, the probability for electron transfer from the target to the projectile is the same as from the projectile to the target, namely $P = 1/2$. The cross section for single electron transfer is then,

$$\sigma = \frac{1}{2}\pi R_p^2 , \tag{2.91}$$

where R_p is the maximum internuclear distance for which the electron may go over the barrier between the nuclei.

The value of R_p is easily evaluated in this simple model as illustrated by Ryufuku, Sasaki and Watanabe [1980] for a bare ion of charge incident on a hydrogen atom. In this case the height of the barrier is given by the sum of the Coulomb potentials due to the two nuclei. The potential for this barrier at a distance x from the projectile is,

$$V(x) = -Z/x - 1/(R-x) , \tag{2.92}$$

where R is the internuclear distance. The maximum height of the barrier, $V(x)$, for $0 < x < R$ is,

$$V_{max} = -(Z^{1/2}+1)^2/R. \tag{2.93}$$

The electron is transferred if the following relations are satisfied,

$$-\frac{1}{2} - Z/R = -Z^2/2n^2 - 1/R , \tag{2.94}$$

$$-\frac{1}{2} - Z/R > -(Z^{1/2}+1)^2/R , \tag{2.95}$$

where the left hand side of (2.94) is the potential of the initial $1s$ state of hydrogen perturbed by the Coulomb potential of the projectile and the right hand side is the energy of the final state of the electron in a principal quantum number n of the projectile of charge Z perturbed by the remaining bare proton. The second condition (2.95) insures that the electron may go over the maximum barrier height.

The value of n satisfying both (2.94) and (2.95) may be shown to be,

$$n \leq n_p \,, \tag{2.96}$$

with

$$n_p = \left[\left\{ (2Z^{1/2}+1)/(Z+2Z^{1/2}) \right\}^{1/2} \right] , \tag{2.97}$$

where $[x]$ denotes the largest integer value not exceeding x. The corresponding value of R satisfying both (2.94) and (2.95) is,

$$R_n = \frac{2(Z-1)}{Z^2/n^2 - 1} \,. \tag{2.98}$$

Electrons may transfer over the potential barrier for all internuclear distances $R \leq R_p$ where,

$$R_p = \frac{2(Z-1)}{Z^2/n_p^2 - 1} \,. \tag{2.99}$$

This is the value of R_p conventionally used in (2.91) in the over the barrier model for bare ions of charge Z incident on atomic hydrogen.

It is clear from (2.97) that $n_p \leq Z$, and from (2.98) that $R_{n-1} < R_n$ and $R_{n-1} - R_{n-2} < R_n - R_{n-1}$ for $n \leq n_p$, i.e. the crossing distance $R-n$ becomes progressively smaller with decreasing n and the separation between two adjacent crossing points also decreases with decreasing n. From these facts one can identify n_p of the principal quantum number of the final state most likely to be occupied by the transferred electron, since probabilities of electron capture into states of principal quantum number n results mainly from impact parameters b between R_{n-1} and R_n.

The over the barrier model has been extended to multiple capture by Niehaus [1986].

2.7 Observations

Total cross sections for excitation, ionization and transfer of a single electron in various atomic targets in collisions with charged particles have been observed in many experiments [Trajmar and McConkey, 1996; Phaneuf, 1996; Cocke, 1996]. A survey of benchmark measurements for proton

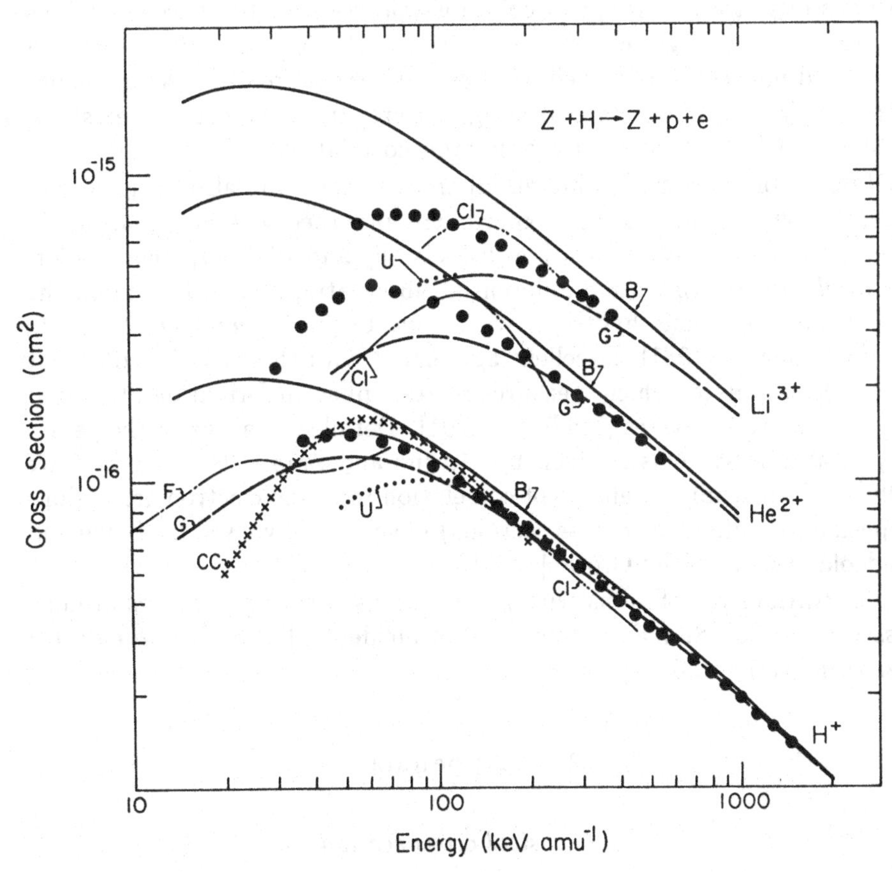

Fig. 2.1. Cross sections for single ionization of atomic hydrogen by impact of H^+, He^{2+} and Li^{3+} from McGuire *et al.* [1982]. Data (black dots) shown are from Shah and Gilbody [1982]. More recent data may be found from Gilbody [1995] and McDaniel and Mansky [1994]. The experimental errors of $\pm3\%$ are the size of the black dots. Theoretical results shown are the first Born approximation (B), classical CTMC (Cl), coupled channels (CC), finite element method (F), Glauber approximation (G) and unitarized distorted wave Born approximation (U).

and higher charged ion impact has been given by Gilbody [1994, 1995] while observed cross sections for electron atom scattering have been summarized by McCarthy and Weigold [1990; 1995]. An useful guide to data compilations is given by McDaniel and Mansky [1994]. Data for ionization of atomic hydrogen by impact of bare heavy charged particles are shown in figure 2.1. Excitation data for atomic hydrogen have been taken by Detleffsen *et al.* [1995]. In the case of atomic hydrogen good agreement of experimental observation now exists with a variety of theoretical methods over a broad range of collision velocities for total cross sections for most strong single electron transitions. In few and many electron atoms and molecules, observed total cross sections are often in agreement with theory [Amusia, 1996], although discrepancies sometimes exist for systems in which there is strong electron correlation.

Observations of various differential cross sections are also plentiful and generally agreement between experiment and theory is also reasonably good for single electron transitions [McCarthy and Weigold, 1995; Andersen *et al.*, 1988]. Extensive differential observations of excitation to the $n = 2$ level of atomic hydrogen were done by by Weigold *et al.* [1980] and by Williams [1981]. Excellent agreement with theory is found except for the R parameter which measures relative phase information of excited singlet and triplet states [Madison, 1991]. Detailed analysis of the geometry of atomic systems is given by Greene and Zare [1982] and by Zare [1988]. Observations of effects of correlation in single electron transitions (especially in differential cross sections) have been reviewed by Boyle and Pindzola [1995] and by Crowe [1987].

Observations of collisions with incident ions carrying electrons are discussed in section 8.5. Experiments with incident photons are briefly discussed in section 9.5.

2.8 Appendix

1 *Classical cross sections*

Cross sections for single electron transitions may be calculated classically. A simple form for these classical single electron cross sections is provided by the binary encounter approximation (BEA). Here a cross section for two body Coulomb scattering between a projectile of velocity $\vec{v}$ and an electron of velocity $\vec{v}_2$ may be found by integrating over the known two body cross section per energy transfer $d\sigma(v_2, v, \Delta E)/d\Delta E$, namely,

$$\sigma(v_2, v) = \int_{E_{min}}^{E_{max}} \frac{d\sigma(v_2, v, \Delta E)}{d\Delta E} d\Delta E \,. \tag{2.100}$$

In BEA, an average over the direction of $\vec{v}_2$ is made. Excitation, ionization and charge transfer differ only in the range of energy transfers over which the integration is done. However, there is a difficulty with charge transfer in that the range of energy transfers overlaps with excitation and ionization so that double counting may occur.

Next the cross section for a given electron velocity, v_2, is averaged over a velocity distribution, $\rho(v_2, v_0)$ appropriate to the atomic electron. For a closed atomic shell a common choice is $\rho(v_2, v_0) = \frac{8}{\pi^2}\frac{v_0^5}{(v_0^2+v_2^2)^4}$ corresponding to the square of the momentum space wavefunction for a hydrogen like electron, where $v_0 = Z_T/n$ is the Bohr velocity for the orbit in question (Cf. (11.14) of the appendices to this book). The cross section then is,

$$\begin{aligned} \sigma(v) &= N\int d\vec{v}_2\rho(v_2, v_0)\sigma(v_2, v) = \sigma(v, v_0) = \sigma(v/v_0) \quad (2.101) \\ &= \sigma(V) = \frac{NZ^2\sigma_0}{U^2}G(V) \, . \end{aligned}$$

Here $\sigma_0 = 6.56 \times 10^{-10}\mathrm{m}^2\mathrm{eV}^2$, U is the binding energy of the electron in eV and N is the number of electrons in the atomic shell. Since $V = v/Z_T$ and $U \sim Z_T^2$, this BEA cross section obeys the same scaling rules for excitation and ionization as does the quantum mechanical first Born approximation, namely, (2.30). Unlike the more complete quantum first Born approximation, the function $G(V)$ is the same for all atomic shells for the BEA cross sections described here. Thus, this BEA $G(V)$ gives a single universal curve for all projectiles and targets.

For ionization the universal BEA $G(V)$ has been expressed analytically [Gryzinski, 1965; Gerjuoy, 1966; Vriens, 1966]. The result of Vriens is expressed for $V > 1/2$ as,

$$\begin{aligned} G(V) &= \frac{1}{4V^2}\left\{\frac{35}{6} + \frac{35}{3\pi}tan^{-1}c + \frac{128}{9\pi}\left(V^3b^3 - b^{\frac{3}{2}}\right)\right. \quad (2.102) \\ &+\frac{cb}{3\pi}\left(35 - \frac{58}{3}b - \frac{8}{3}b^2\right) \\ &+\frac{2}{3\pi}Vab\left[(5-4v^2)\left(3a^2 + \frac{3}{2}ab + b^2\right) - cV\left(\frac{15}{2} + 9a + 5b\right)\right] \\ &-\frac{16}{\pi}Va^4ln(4V^2+1) - V^2a\left(1 + \frac{2}{\pi}tan^{-1}c\right) \\ &\left.\times\left(\frac{5}{2} + 3a + 4a^2 + 8a^3\right)\right\} , \end{aligned}$$

where,

$$c = V - \frac{1}{4}V, \qquad b = (1+c^2)^{-1}, \qquad a = (1+V^2)^{-1} \, . \qquad (2.103)$$

$G(V)$ has been tabulated by McGuire and Richard [1973], who also give a model for calculating transition probabilities. At large V the BEA cross section for ionization varies as V^{-2} in contrast to $V^{-2}ln\ V$ given correctly by the first Born term in quantum mechanics (tabulated below for ionization from the $n = 1$ level).

2 Tables of ionization cross sections and probabilities

Table 2.1 *Cross section (in units of $a_0^2 = 2.80 \times 10^{-13} m^2$) in the first Born approximation for ionization of atomic hydrogen initially in its ground state by a fast projectile of unit charge and velocity v (in a.u. $= 2.19 \times 10^6 m/s$).*

Velocity	Cross section	Velocity	Cross section
0.1	0.00522	3.0	2.48
0.2	0.0407	4.0	1.54
0.3	0.395	5.0	1.06
0.4	1.43	6.0	0.772
0.5	3.00	7.0	0.590
0.6	4.66	8.0	0.467
0.7	6.05	9.0	0.379
0.8	7.00	10.0	0.315
0.9	7.53	20.0	0.0910
1.0	7.72	30.0	0.0435
1.2	7.43	40.0	0.0256
1.4	6.73	50.0	0.0169
1.6	5.94	100.0	0.00463
1.8	5.20	200.0	0.00124
2.0	4.54		

Cross sections for ionization evaluated in first order perturbation obey the scaling relations of (2.30), namely,

$$\sigma(Z, Z_T, \mathrm{v}) = \frac{Z^2}{Z_T^4}\sigma(Z = 1, Z_T = 1, \frac{\mathrm{v}}{Z_T})$$

where $\sigma(Z = 1, Z_T = 1, \frac{\mathrm{v}}{Z_T})$ is listed in Table 2.1 for $Z_T = 1$. For 25.6 MeV $O^{+7}(1s) + He^{+2} \rightarrow O^{+8} + He^{+2} + e^-$ the relative velocity v is 8 a.u. and the scaled velocity $\mathrm{v}/Z_T = 1$ so that $\sigma = 2^2/8^4(7.72) = 0.00754a_0^2 = 2.11 \times 10^{-19}\mathrm{cm}^2$. For systems with two electrons in the $n = 1$ state, the cross sections are doubled.

Table 2.2. *Ionization probability* $P(Z, Z_T, X = 6.33Z_T/\mathrm{v}, Z_T b)$ *in the first Born approximation for p + H(1s) → p + p + e as a function of collision velocity v and impact parameter b. These results are from Hansteen, Johansen and Kocbach [1975] who define* $X \equiv Z_T/\sqrt{E} = 6.33Z_T/v$, *where E is the energy of the projectile in MeV/amu so that X is a scaled inverse velocity in units of the velocity of the initial state,* Z_T. *Ionization probabilities for other systems may be found from the scaling relation (2.29).*

b/X	1.	2.	3.	4.	5.	6.
.2	3.52-2	1.45-1	3.26-1	5.34-1	7.24-1	8.68-1
.4	3.37-2	1.37-1	3.06-1	4.98-1	6.70-1	7.89-1
.6	3.16-2	1.27-1	2.79-1	4.48-1	5.96-1	6.89-1
.8	2.92-2	1.16-1	1.49-1	3.93-1	5.16-1	5.87-1
1.0	2.69-2	1.04-1	2.20-1	3.41-1	4.39-1	4.94-1
1.2	2.43-2	9.35-2	1.93-1	2.93-1	3.71-1	4.13-1
1.4	2.18-2	8.25-2	1.66-1	2.51-1	3.12-1	3.41-1
1.6	1.93-2	7.21-2	1.45-1	2.13-1	2.61-1	2.81-1
1.8	1.68-2	6.25-2	1.24-1	1.80-1	2.17-1	2.29-1
2.2	1.25-2	4.53-2	8.78-2	1.25-1	1.46-1	1.49-1
2.6	8.93-3	3.19-2	6.00-2	8.32-2	9.51-2	9.44-2
3.0	6.33-3	2.18-2	3.99-2	5.41-2	6.02-2	5.83-2
4.0	2.78-3	8.71-3	1.42-2	1.77-2	1.83-2	1.64-2
5.0	1.42-3	3.75-3	5.34-3	5.80-3	5.35-3	4.33-3
6.0	8.15-4	1.83-3	2.20-3	2.03-3	1.61-3	.00+0
7.0	5.15-4	9.78-4	9.92-4	00+0	.00+0	.00+0
8.0	3.44-4	5.53-4	.00+0	00+0	.00+0	.00+0

b/X	7.	8.	9.	10.	11.	12.
.2	7.51-1	9.69-1	9.38-1	8.75-1	7.96-1	7.08-1
.4	8.59-1	8.69-1	8.35-1	7.72-1	6.96-1	6.13-1
.6	7.41-1	7.42-1	7.05-1	6.44-1	5.73-1	4.99.1
.8	6.23-1	6.15-1	5.76-1	5.20-1	4.56-1	3.90-1
1.0	5.15-1	5.01-1	4.63-1	4.10-1	3.54-1	2.98-1
1.2	4.22-1	4.03-1	3.66-1	3.19-1	2.70-1	2.23-1
1.4	3.43-1	3.22-1	2.87-1	2.45-1	2.03-1	1.65-1
1.6	2.77-1	2.55-1	2.23-1	1.87-1	1.52-1	1.20-1
1.8	2.22-1	2.00-1	1.71-1	1.42-1	1.13-1	8.73-2
2.2	1.40-1	1.21-1	1.00-1	7.92-2	6.05-2	4.51-2
2.6	8.57-2	7.16-2	5.72-2	4.33-2	3.20-2	2.27-2
3.0	5.14-2	4.13-2	3.18-2	2.33-2	1.64-2	1.12-2
4.0	1.31-2	9.66-3	6.78-3	4.51-3	2.87-3	0.00
5.0	3.20-3	0.00	0.00	0.00	0.00	0.00

b/X	13.	14.	15.	16.	17.	18.
.2	6.19-1	5.34-1	4.58-1	3.91-1	3.31-1	2.79-1
.4	5.31-1	4.54-1	3.85-1	3.24-1	2.72-1	2.26-1
.6	4.26-1	3.59-1	3.00-1	2.49-1	2.05-1	1.68-1
.8	3.28-1	2.72-1	2.23-1	1.82-1	1.47-1	1.18-1
1.0	2.46-1	2.00-1	1.61-1	1.28-1	1.02-1	7.99-2
1.2	1.81-1	1.44-1	1.13-1	8.85-2	6.86-2	5.27-2
1.4	1.31-1	1.02-1	7.85-2	5.99-2	4.53-2	3.42-2
1.6	9.37-2	7.16-2	5.39-2	4.01-2	2.96-2	2.18-2
1.8	6.64-2	4.98-2	3.67-2	2.67-2	1.92-2	1.37-2
2.2	3.29-2	2.34-2	1.64-2	1.15-2	7.93-2	5.38-3
2.6	1.58-2	1.09-2	7.24-3	4.78-3	3.13-3	2.05-3
3.0	7.49-3	4.85-3	3.09-3	1.97-3	1.22-3	7.52-4
4.0	0.00	0.00	0.00	0.00	0.00	0.00

b/X	19.	20.	21.	22.	23.	24.
.2	2.35-1	1.98-1	1.66-1	1.39-1	1.17-1	9.81-2
.4	1.88-1	1.56-1	1.29-1	1.07-1	8.84-2	7.31-2
.6	1.37-1	1.11-1	9.05-2	7.35-2	5.96-2	4.83-2
.8	9.43-2	7.52-2	5.99-2	4.76-2	3.78-2	3.00-2
1.0	6.26-2	4.88-2	3.79-2	2.95-2	2.29-2	1.78-2
1.2	4.03-2	3.07-2	2.33-2	1.76-2	4.34-2	1.02-2
1.4	2.56-2	1.89-2	1.40-2	1.03-2	7.63-3	5.68-3
1.6	1.60-2	1.16-2	8.34-3	5.99-3	4.31-3	3.14-3
1.8	9.81-3	7.00-3	4.91-3	3.43-3	2.40-3	1.71-3
2.2	3.62-3	2.45-3	1.65-3	1.09-3	7.22-4	4.89-4
2.6	1.31-3	8.35-4	5.37-4	3.39-4	2.11-4	1.36-4
3.0	4.68-4	2.80-4	1.69-4	.00+0	.00+0	.00+0

b/X	25.	26.	27.	28.
.2	8.29-2	7.02-2	5.95-2	5.02-2
.4	6.08-2	5.06-2	4.22-2	3.52-2
.6	3.93-2	2.61-2	2.11-2	2.13-2
.8	2.38-2	1.89-2	1.51-2	1.20-2
1.0	1.38-2	1.07-2	8.31-3	6.45-3
1.2	7.68-3	5.81-3	4.39-3	3.32-3
1.4	4.19-3	3.08-3	2.27-3	1.67-3
1.6	2.25-3	1.62-3	1.16-3	8.31-4
1.8	1.19-3	8.33-4	5.82-4	4.06-4
2.2	3.24-4	2.14-4	1.41-4	9.32-5
2.6	8.51-5	5.32-5	3.33-5	2.08-5

b/X	29.	31.	33.	35.
.2	4.30-2	3.13-2	2.30-2	1.70-2
.4	2.95-2	2.07-2	1.47-2	1.04-2
.6	1.74-2	1.16-2	7.84-3	5.31-3
.8	9.56-3	6.09-3	8.90-3	2.51-3
1.0	5.02-3	3.04-3	1.85-3	1.13-3
1.2	2.51-3	1.44-3	8.27-4	4.77-4
1.4	1.23-3	6.65-4	3.61-4	1.96-4
1.6	5.96-4	3.07-4	1.58-4	8.18-5
1.8	2.83-4	1.38-4	6.71-5	3.28-5
2.2	6.15-5	2.67-5	1.16-5	5.03-6
2.6	1.29-5	5.01-6	1.93-6	7.45-6

Ionization probabilities for projectiles of charge Z on targets with nuclear charge, Z_T, may be found from the scaling relation (2.29), namely,

$$P(Z, Z_T, \mathrm{v}, \vec{b}) = (\frac{Z}{Z_T})^2 P(1, 1, \frac{\mathrm{v}}{Z_T}, Z_T\vec{b}) = (\frac{Z}{Z_T})^2 P(1, 1, X, Z_T b)\,.$$

For 1.6 MeV $He^{+2} + O^{+7}(1s) \rightarrow He^{+2} + O^{+8} + e^-$, $X = 6.33$ (corresponding to $\mathrm{v}/Z_T = 1$). At an impact parameter of 0.125 a_0, $Z_T b = 1$, and interpolating from the table, $P = (2/8)^2(0.499) = 0.0312$. Hansteen, Johansen and Kocbach [1975] tabulate probabilities for initial states with principle quantum numbers $n = 1, 2$ and $n = 3$ with various effective charges for Z_T.

Note on the classical limit for Coulomb scattering.

In general cross sections are regarded as the square of the amplitude of an outgoing spherical wave,

$$d\sigma/d\Omega = |f(\theta)|^2\,. \tag{2.104}$$

As r tends to infinity, the full quantum wavefunction is the sum of an unscattered part and an outgoing spherical wave with amplitude, $f(\theta)$,

$$\psi \rightarrow e^{i\frac{\vec{k}\cdot\vec{r}}{\hbar}} + f(\theta)\frac{e^{ikr}}{r}\,. \tag{2.105}$$

Here factors of $\hbar$ are retained so the classical limit where $\hbar \rightarrow 0$ can be considered.

It is straightforward to show [Messiah, 1961; Goldberger and Watson, 1964] that for simple elastic two body scattering via a potential $V(\vec{r})$,

$$f(\theta) = \frac{M\hbar}{2\pi} \int e^{i\frac{\vec{k}\cdot\vec{r}}{\hbar}} V(\vec{r}) d\vec{r} \,. \tag{2.106}$$

Now, consider $V(r) = C/r^n$. Then,

$$f(\theta) = \frac{M\hbar}{2\pi} \int e^{i\frac{\vec{k}\cdot\vec{r}}{\hbar}} \frac{C}{r^n} d\vec{r} \approx \frac{M\hbar}{2\pi} C a^{-n} \,, \tag{2.107}$$

where $ka/\hbar \approx 1$, i.e. $a =\simeq \hbar/k$. Thus,

$$f(\theta) \simeq \frac{M\hbar}{2\pi} C \left(\frac{k}{\hbar}\right)^n \,, \tag{2.108}$$

so that,

$$d\sigma/d\Omega \sim \hbar^{2(n-1)} \,. \tag{2.109}$$

For $n = 1$ the quantum cross section is independent of n, so that the classical limit where $\hbar \to 0$ is the same as the quantum result. It is well known that for two body Coulomb scattering the quantum and the classical scattering amplitudes are the same except for an overall phase. This appears to be an accident of nature for the r^{-1} Coulomb potential.

3

Formulation of multi-electron transition probabilities

The interaction of a single isolated electron in an atomic target with an incident projectile is, for the most part, relatively well understood (Cf. chapters 2 and 9). However, if the electrons are not independent, then it is more difficult to determine the properties of these many body systems*. Dealing with transitions of many electrons may also depend on the correlation interaction between the electrons. In this chapter the problem of dealing with multiple electron transitions is considered in detail and a general formulation is given for interactions of atoms with incoming charged particles. Interaction with light and with projectiles carrying electrons is given in chapters 8-10. The many body question of correlation is discussed in sections 1.4 and 6.1.

3.1 Terms in the Hamiltonian

The Hamiltonian for scattering of an atom of nuclear charge Z_T with N fully correlated electrons by a particle of charge Z and mass M using atomic units is,

$$\begin{aligned} H &= -\frac{\nabla^2}{2M_r} + \frac{ZZ_T}{R} - \sum_{j=1}^{N} \frac{Z}{|\vec{R} - \vec{r}_j|} \\ &\quad + \sum_{j=1}^{N} \left[\frac{-\nabla_j^2}{2} - \frac{Z_T}{r_j} + \sum_{\substack{k \\ (k>j)}} \frac{1}{|\vec{r}_k - \vec{r}_j|} \right] \end{aligned} \tag{3.1}$$

* It is conventionally assumed that all interactions in the N electron (or more generally the N body) problem may be expressed as sums of known two body interactions. The necessity for such simplicity is not clear.

$$= K + V + H_T \,,$$

where M_r is the reduced mass of the projectile, and

$$K = -\frac{\nabla^2}{2M_r} \,, \tag{3.2}$$

and

$$\begin{aligned} V &= \frac{ZZ_T}{R} - \sum_{j=1}^{N} \frac{Z}{|\vec{R} - \vec{r}_j|} \\ &= Z\sum_{j=1}^{N}\left[\frac{1}{R} - \frac{1}{|\vec{R} - \vec{r}_j|}\right] + \frac{Z(Z_T - N)}{R} \\ &\equiv \sum_{j=1}^{N} V_j + \frac{Z(Z_T - N)}{R} \,. \end{aligned} \tag{3.3}$$

Here the long range Coulomb tail of the interaction V between the incoming projectile and the atomic target vanishes when $Z_T = N$. The Hamiltonian of the target atom is defined by,

$$H_T = \sum_{j=1}^{N}\left[-\frac{\nabla_j^2}{2} - \frac{Z_T}{r_j} + \sum_{\substack{k \\ (k>j)}} |\vec{r}_k - \vec{r}_j|^{-1}\right] , \tag{3.4}$$

where $\vec{R}$ is the position of the projectile and $\vec{r}_j$ the coordinate to the jth electron. The origin is located at the nucleus of the target atom. The $|\vec{r}_k - \vec{r}_j|^{-1}$ Coulomb (electron correlation) interaction in the unperturbed (static) atomic Hamiltonian H_T gives rise to electron correlation.

3.2 The many electron wavefunction

Total knowledge of a system of many electrons is contained in the exact wavefunction Ψ for the entire system. This wavefunction may in principle be found by solving the wave equation of Schrödinger, namely,

$$i\frac{\partial \Psi}{\partial t} = i\dot{\Psi} = H\Psi \,. \tag{3.5}$$

The exact wavefunction is seldom fully known, especially if there are many electrons.

1 *Separation of the electronic from the nuclear wavefunction*

It is now assumed that the internuclear motion may be separated from the electron motion and that the internuclear motion may be treated classically, so that the internuclear trajectory $\vec{R}(t)$ is well defined, e.g., $\vec{R}(t) = \vec{b} + \vec{v}t$ where $\vec{b}$ is the impact parameter of the projectile and $\vec{v}$ is the projectile velocity. The resulting equation derived in the appendix of this chapter for the electron wavefunction is now time dependent, namely,

$$i\frac{\partial \psi_{el}}{\partial t} = H_{el}\psi_{el} \ . \tag{3.6}$$

The Hamiltonian for the electrons H_{el} defined as,

$$H_{el} = H_T + \sum_j V_j = H_T + V - \frac{Z(Z_T - N)}{R} \ , \tag{3.7}$$

which is now explicitly time dependent since the V_j depend on $\vec{R}$ which is explicitly time dependent. Consequently, the evolution operator U is not simply given by $\exp[-iH_{el}(t - t_0)]$, but rather by a more complicated expression containing time ordering, namely, (2.21) above or (3.17) below, which leads to use of the interaction representation. Other representations are more difficult to use because they do not take advantage of the fact that both $\vec{R}(t)$ and H_0 are known.

2 *Interaction representation*

In dealing with probability amplitudes it is sensible to use the interaction (or intermediate) representation in which $V(t)$ is a known function of time t. In the particle picture (i.e., the semi-classical approximation) the trajectory $\vec{R}(t)$ of the projectile is localized and from the end of section 2.1.1 the interaction potential V becomes $V(\vec{R}(t)) = V(t)$ which is a known function of t. In principle one could work in another representation such as the Heisenberg representation. In practice this is difficult because of the coupling between the projectile and the target expressed [Magnus, 1954] as an infinite series of commutator terms in an exponential. Without these exponential commutator terms, the $exp\{i\omega t\}$ term (e.g., in (2.17) is missing in the first order amplitude in an expansion in Z and the resulting cross sections become infinite as with the Magnus results of Eichler [1977]. Thus, the coupling between the electronic and nuclear motion may be more difficult to deal with in representations other than the interaction representation.

The interaction (or intermediate) representation is intermediate between the Schrödinger representation in which time dependence is carried by the wavefunction and the Heisenberg representation where the operators carry the time dependence [Messiah, 1961, p.321]. In the Schrödinger representation one considers,

$$i\frac{\partial|\psi>}{\partial t} = H|\psi>= (H_T + V)|\psi> \ . \tag{3.8}$$

It is often the case that the eigenstates, $|s>$ of H_T are known such that

$$H_T|s>= \epsilon_s|s> \ . \tag{3.9}$$

One may take advantage of this knowledge by defining the interaction state such that,

$$\psi_I(t) = e^{iH_Tt}\psi(t) \ , \tag{3.10}$$

and

$$V_I(t) = e^{iH_Tt}Ve^{-iH_Tt} \ . \tag{3.11}$$

Direct substitution into (3.8) yields,

$$i\frac{\partial\psi_I}{\partial t} = V_I(t)\psi_I(t) \ . \tag{3.12}$$

Here the wavefunction is determined by the interaction and hence the name interaction representation.

It is now sensible to define the evolution operator $U_I(t, t_0)$ by,

$$\psi_I(t) = U_I(t, t_0)\psi_I(t_0) \qquad t > t_0 \ . \tag{3.13}$$

It is straightforward to show that,

$$i\frac{\partial U_I}{\partial t} = V_I U_I \ . \tag{3.14}$$

Messiah [p.321] points out that in transforming to the intermediate representation, one may find that the evolution operator is more slowly varying in time and therefore easier to evaluate. Since the evolution operator is used only in the interaction representation in this book, U_I is simply replaced by U.

3 The evolution operator

In the simplest cases one has $U(t, t_0) = e^{-iH_T(t-t_0)}$. In the interaction representation, however, the evolution operator, or Green function $U(t, t_0)$ is governed by (3.14), [Goldberger and Watson, 1964, p. 45],

$$i\frac{\partial U(t,t_0)}{\partial t} = V_I(t)U(t,t_0) \,, \tag{3.15}$$

where,

$$V_I(t) \equiv e^{iH_T t}V(t)e^{-iH_T t} \,. \tag{3.16}$$

$V_I(t)$ is not a sum of single electron operators because H_T in (3.4) is not a sum of single electron terms due to the $|\vec{r}_k - \vec{r}_j|^{-1}$ electron-electron interaction. Eq(3.15) may be formally solved for $U(t,t_0)$ (with T as the time ordering operator),

$$\begin{aligned} U(t,t_0) &= T\exp\left[-i\int_{t_0}^{t} V_I(t')dt'\right] \qquad (3.17)\\ &= T\sum_n 1/n!\left(-i\int_{t_0}^{t} V_I(t')dt'\right)^n \\ &= T\Big[1+(-i)\int_{t_0}^{t} dt_1 V_I(t_1) + (-i)^2/2!\int_{t_0}^{t} dt_2 V_I(t_2)\int_{t_0}^{t} dt_1 V_I(t_1) \\ &\quad +(-i)^3/3!\int_{t_0}^{t} dt_3 V_I(t_3)\int_{t_0}^{t} dt_2 V_I(t_2)\int_{t_0}^{t} dt_1 V_2(t_1) + \ldots\Big] \\ &\equiv 1+(-i)\int_{t_0}^{t} dt_1 V_I(t_1) + (-i)^2\int_{t_0}^{t} dt_2 V_I(t_2)\int_{t_0}^{t_2} dt_1 V_I(t_1) \\ &\quad +(-i)^3\int_{t_0}^{t} dt_3 V_I(t_3)\int_{t_0}^{t_3} dt_2 V_I(t_2)\int_{t_0}^{t_2} dt_1 V_2(t_1) + \ldots, \end{aligned}$$

namely, Dyson's equation [Goldberger and Watson, 1964, p. 49] including the definition of the time ordering operator T. The time ordering operator T [Cf. section 7.1.2] preserves cause and effect in the propagation of the wavefunction by restricting $t_n \leq t_{n+1}$ in (3.17). This means that $U(t_2,t_1) = 1$ if $t_2 < t_1$, i.e. cause follows effect. This property is the same as in classical physics.

3.3 Transition probability and cross section

The probability amplitude $a = a_{fi}$ for scattering from the asymptotic initial atomic state $|i>$ to the asymptotic final state $|f>$ is given by,

$$a = <f|\psi_i(t=+\infty)> = <f|U(+\infty,-\infty)|i> \,. \tag{3.18}$$

As discussed in section 1.3.3, this corresponds to the projection onto the state vector $|f>$ of the wavefunction $|\psi_i(t)>$ that has evolved from $|i>$.

The probability $P(\vec{b})$ for a transition from $|i>$ to $|f>$ is given by the absolute square of a, and the corresponding cross section by,

$$\sigma_{if} = \int d\vec{b}\, P(\vec{b}) = \int d\vec{b}\ |a|^2 \,. \tag{3.19}$$

This result holds for an arbitrary number N of electrons, including the fundamental case of $N = 1$.

1 Multi-electron effects and electron correlation

Electron correlation is a multi-electron effect. That is, electron correlation arises because there is more than one electron in our system and the electrons affect one another via the residual correlation potential v defined by (4.7). Correlation [discussed in sections 1.4 and 6.1] is usually defined as the difference between exact and uncorrelated quantities. A quantity is uncorrelated if that quantity is determined from a sum of single electron Hamiltonians, even though that quantity obeys appropriate symmetry requirements such as rotational symmetry (conserving angular momentum), parity and the exclusion principle (antisymmetrization of the electron wavefunction). For example, the total wavefunction, the probability amplitude and the probability are uncorrelated if (ignoring symmetries) $\psi(\vec{r}_1, \vec{r}_2) = \psi(\vec{r}_1)\psi(\vec{r}_2)$, $a(1,2) = a(1)a(2)$ and $P(1,2) = P(1)P(2)$ respectively. Note, however, that even in the uncorrelated limit $\sigma(1,2) \neq \sigma(1)\sigma(2)$ due to the integration over impact parameters in (3.19).

The Hartree Fock approximation [Froese Fischer, 1996; 1977] is usually defined to be the uncorrelated non-relativistic static wavefunction. Accordingly, the Time Dependent Hartree Fock (TDHF) approximation [Kuerpick *et al.*, 1995; Gramlich *et al.*, 1986; Stich *et al.*, 1985; Devi and Garcia, 1984] is often used to define the uncorrelated dynamic wavefunction. In practice this requires defining a new basis set of wavefunctions using the variational principle every time the system changes significantly during the collision (which can be difficult). Also, for highly correlated systems an uncorrelated limit is sometimes not easily defined (e.g. there may be no sensible single electron configuration).

2 Scattering, relaxation and asymptotic regions

In a scattering event the precise boundary between the scattering region and the asymptotic region is not easily localized. In most atomic and molecular collisions the interaction occurs within a few Bohr radii of the target nucleus. At high velocities the collision time is often shorter than the relaxation time required for the final state to decay, usually by either

X-ray decay or Auger emission. So the excitation and de-excitation processes can be decoupled in many (but not all) cases. The correlation that occurs outside the scattering and relaxation regions (in both space and time) is referred to as atomic, static or asymptotic correlation (Cf. section 6.2) [Amusia, 1996]. The Auger effect corresponds to a de-excitation process in which the electron-electron interaction causes an excited state to decay with one electron going to a lower state and the other to a higher state. This is sometimes referred to as dynamic atomic correlation [Stolterfoht, 1989]. Correlation occurring during the collision is called scattering correlation [McGuire, 1987] or intermediate correlation [Reading, *et al.*, 1996]. This scattering correlation is a deviation from a 'frozen correlation' approximation used by Martin and Salin [1995].

3 Waves and particles

Sometimes it is convenient to regard the projectile as a well localized particle with a well defined impact parameter $\vec{b}$. This particle picture or impact parameter representation is the semi-classical approximation in which it is assumed that the relation between the scattering angle θ of the projectile and its impact parameter b is one to one. For processes in which the projectile is detected at a well defined scattering angle θ it is usually appropriate to consider the projectile as a wave. This is the wave picture.

Quantum amplitudes may be formulated in either the wave picture using the scattering amplitude f or in the particle picture using the probability amplitude a. The amplitudes f and a are related by a Fourier transform, namely,

$$f(\vec{q}) = \frac{M_r \mathrm{v}}{2\pi} \int d\vec{b}\, e^{i\vec{q}\cdot\vec{b}}\, a(\vec{b}) \,, \tag{3.20}$$

where $\vec{q}$ is the momentum transfer of the projectile, M_r is the reduced mass, v is the velocity of the projectile and $\vec{b}$ is the impact parameter of the projectile. The momentum transfer is defined by $\vec{q} = \vec{k}_f - \vec{k}_i$ where $\vec{k}_f$ and $\vec{k}_i$ are the initial and final momenta of the projectile. Here $f = 0$ if there is no scattering (e.g., if the target does not interact with the projectile) so that U is replaced by $U - 1$ in (3.18). In general both f and a are used according to convenience. The cross section is the same in both the wave and particle pictures [McCarroll and Salin, 1966].

In most experiments the wavepacket of the projectile is much larger than the size of the target. If $\vec{q}$ is well localized (as when both the initial and final momenta of the projectile are experimentally determined with precision), then $\vec{b}$ cannot be localized in accord with the uncertainty principle. Then $\vec{b}$ is regarded as a mathematical rather than a physical variable. There are two conditions to be satisfied if $\vec{b}$ is to be well localized: i) the de Broglie wavelength must be small compared to the distance over which the wavefunction of the target electron varies, and ii) the angle of dispersion, ϕ, must be smaller than the scattering angle, θ. The first condition is easily met for heavy incoming projectiles, is not always valid for incoming electrons, and is usually not valid for incoming photons.

The second condition for the validity of the semi-classical approximation may be applied as follows,

$$\begin{aligned} \Delta y \Delta p &\simeq 1 \\ \Delta p &\simeq p\ sin\phi \simeq p\phi \\ \Delta y &= b\ , \end{aligned} \tag{3.21}$$

so that, with ℓ as the angular momentum,

$$\begin{aligned} bp\phi &= 1\ , \\ \ell\phi &= 1\ , \end{aligned} \tag{3.22}$$

one has the general condition,

$$\theta > \phi = 1/\ell\ . \tag{3.23}$$

For Coulomb scattering in the forward direction, $\theta = d_0/b$, where $d_0 = Z_1 Z_2/\frac{1}{2}M\text{v}^2$ is the distance of closest approach of the projectile and the target nucleus. Then, from $\theta > \phi$, one has $2Z_1Z_2/M\text{v}^2 b > \hbar/M\text{v}b$, which yields,

$$\text{v} < 2Z_1Z_2\ . \tag{3.24}$$

This condition restricts application of the semi-classical approximation to collision velocities that are not too large [Bohr, 1948]. It is more difficult to properly treat the projectile as a finite packet of waves than as a point particles.

4 Long range Coulomb terms

The long range nature of the Coulomb interaction causes physical and mathematical difficulties. The total two body Coulomb cross section is infinite. The kernel of the Lippmann-Schwinger equation is not compact so that there is no guarantee of uniqueness of the solution and corresponding numerical difficulties arise [Joachain, 1983; Merkuriev, 1980; Chen and Chen, 1972]. The on-shell limit of the off-energy-shell Coulomb

T matrix does not exist due to a phase divergence [Gau and Macek, 1975; Roberts, 1985]. Fortunately, the long range Coulomb terms are always shielded in nature at sufficiently large distances.

For scattering from neutral atoms the long range problems may be avoided by formulating atomic scattering in terms of the difference between the instantaneous Coulomb interaction and an average of this Coulomb interaction over target electron densities, i.e. using $V - \bar{V} = \sum_j (V_j - \bar{V_j})$ in place of V. There is some freedom in the partition of the total Hamiltonian H of (3.1), namely,

$$\begin{aligned} H &= K + H_T + V \\ &= K + \bar{V} + H_T + (V - \bar{V}) \\ &\equiv H_N + H_T + (V - \bar{V}) \,. \end{aligned} \tag{3.25}$$

There are some noteworthy features.

1. This gives the correct first Born result for systems in which $< f|i > \neq 0$ (Cf. section 2.3).

2. Either $< i|V|i >$ or $< f|V|f >$ (or possibly some combination) may be used for $\bar{V}$.

3. Using $V - \bar{V}$ for the electronic transitions eliminates the long-range Coulomb tail in all interactions of charged projectiles with electrons of a target atom.

4. In the case of neutral targets the Coulomb interaction between the projectile and the target nucleus is replaced by the short ranged static potential, $\bar{V}$, between the projectile and the entire atomic target with a frozen density of target electrons.

5. Both the electronic interaction and the projectile nucleus interaction are weaker than the corresponding full Coulomb interactions.

6. Unnecessary asymptotic Coulomb terms are removed, e.g., in the case of $p + H \rightarrow H + p$ where asymptotic Coulomb terms are not required.

7. For scattering of charged particles with positive ions in this formulation, the electronic interactions remain non-Coulombic (i.e., short range) and scattering from the nucleus involves (if necessary) only two body Coulomb terms that are well understood.

8. Using $V - \bar{V}$ in place of V also removes to first order non physical terms that arise if non-orthogonal initial and final states are used†.

9. The long-standing problem of the asymptotic wavefunction for three charged particles is not addressed by this formulation.

10. The freedom in partitioning the Hamiltonian is like a gauge transformation. If one changes only V or H_N in (3.25), then the total wavefunction is simply altered by an overall phase term $e^{i\bar{V}t}$.

In many cases, however, there is advantage in retaining long-range Coulomb terms since mathematical techniques have been developed which are useful in analytic evaluation of matrix elements. A useful formulation of scattering amplitudes in terms of Coulomb wavefunctions, the continuum distorted wave (CDW) method, has been developed by Belkic *et al.* [1979].

5 *Exclusive and inclusive cross sections*

In a rigorous quantum mechanical calculation a theorist may evaluate a probability amplitude a_{if} for a transition from a known ϕ_i (e.g., the ground state) to some ϕ_f. Here the final state ϕ_f is completely specified. A quantity such as a probability amplitude, a probability, or a cross section in which the final state of a system of N electrons is fully specified is called exclusive [Lüdde and Dreizler, 1985; Reading, 1973; Briggs and Roberts, 1974; McGuire and Macdonald, 1975; Ben-Itzhak *et al.*, 1988].

In the laboratory it is often not possible to determine what happens to each of the electrons in a collision, especially if the number of electrons in the target is large. Experiments are often designed to detect all those events which include a particular transition. For example, single K-shell ionization data usually include any event that occurs together with ionization of a K-shell electron. Quantities which include a sum over all possible final states of electrons are called inclusive . As a basic example of inclusive cross sections, consider what happens to the full wave function, ψ [McGuire and Macdonald, 1975]. As the system evolves, the initial state ψ_i develops into a linear combination of final states, i.e.,

† Consider the matrix element, $< f|V|i >$, where $< f|i > \neq 0$. This matrix element depends on the zero point energy of the potential, V, and changes if V is replaced by $V + C$ (where C is any constant). This usually unwanted feature can be removed by making a Gramm-Schmidt orthogonalization of $|f >$ with respect to $|i >$, i.e. by replacing $< f|$ by $< f_\perp| = < f| - < f|i >< i|$. Then $< f|V|i >$ is replaced by $< f_\perp|V|i > = < f|V|i > - < f|i >< i|V|i > = < f|V - \bar{V}|i >$. This first order matrix element is independent of the zero point energy C. (Here the difference between $1 - < f|i >$ and 1 is neglected.)

$$\psi_i \underset{H}{\longrightarrow} \psi(t) = \sum_{\{s\}} a_{si}(t)\psi_s \,, \tag{3.26}$$

where ψ_s represents a particular state of ionization and $a_{si}(t)$ represents the probability amplitude for evolving from ψ_i to ψ_s at time t. Choosing $\{\psi_s\}$ as an orthonormal set,

$$\int |\psi_f|^2 d\tau = \int |\sum_{\{s\}} a_{si}(t)\psi_s|^2 \, d\tau = \sum_{\{s\}} |a_{si}(t)|^2 = 1 \,, \tag{3.27}$$

corresponding to conservation of probability (unitarity). This result holds for all t including $t \to \infty$ when $\psi(t)$ converges to its asymptotic limit. Let us now represent the wavefunction of the total system as partially separable, namely,

$$\psi(\vec{r}_p; \vec{r}_1, \vec{r}_2, ..., \vec{r}_{N-1}; \{s\}) = \phi_K(\vec{r}_p)\psi_R \,, \tag{3.28}$$

then each part evolves according to,

$$\psi_i \underset{H}{\longrightarrow} \psi(t = -\infty) = \sum_{\{s\}} a_{si}\psi_s = \sum_{\{k\}} a_{ki}\phi_{Kk} \sum_{\{r\}} a_{ri}\psi_{Rr} \,. \tag{3.29}$$

Using the orthonormality of the $\{\phi_{Kk}\}$ and $\{\psi_{Rr}\}$ the probability amplitude a_{is} for a transition from a particular initial state to a particular final state is a product of two factors corresponding to,

$$a_{si} = a_{ki}a_{ri} \,, \tag{3.30}$$

so that the probability of producing such a state is also a simple product, namely,

$$|a_{si}|^2 = |a_{ki}|^2 |a_{ri}|^2 \equiv P_k P_r \,. \tag{3.31}$$

The corresponding cross section for producing excitation to state k for the participating K electron and state r for the remaining target is,

$$\sigma_{kr} = \int P_k P_r d\vec{b} \,, \tag{3.32}$$

where $\vec{b}$ is the impact parameter of the projectile.

6 The active electron approximation

Summing over all final states of the remaining target then using unitarity,

$$\sum_{\{r\}} P_r = 1 \,, \tag{3.33}$$

one has that,

$$\sum_{\{r\}} \sigma_{kr} = \int P_k \sum_{\{r\}} P_r d\vec{b} = \int P_k d\vec{b} = \sigma_k \,, \tag{3.34}$$

where σ_k corresponds to a standard single electron calculation for exciting the K electron. The K electron is often referred to as the active electron and the remaining R electrons as passive electrons. The active electron approximation is valid if the active electron is weakly correlated with the remaining electrons.

In practice neither the experimentalist nor the theorist usually keeps account of everything that happens in a collision involving N electrons, especially when N is large. Usually one chooses to examine just one (or sometimes a few) active electrons and ignores the remaining passive or spectator electrons. Typically calculations for single electron probabilities or cross sections, which simply omit the passive electrons, are compared to data where the active electron undergoes the specified transition regardless of what the other electrons do. And to the extent that the separability condition of (3.28) holds, this convenient picture is valid.

The simple case above does not incorporate the effects of electron exchange (i.e., the Pauli exclusion principle) corresponding to the asymmetric nature of the electron wavefunction. The effects of electron exchange may be incorporated as done by Lüdde and Dreizler [1985] and by Reading and Ford [1980] described in section 4.1.1.

7 Electron exchange

In principle the consequences of electron exchange are significant. In effect electron exchange replaces the notion of a very large number of classically distinguishable electrons with a single electron complex that must be antisymmetric with respect to the interchange of all occupied states. Because the one universal electron is so complicated‡, it is safe to say that its wavefunction will not be calculated in the near future.

‡ In principle any two people are part of this whole electron. They can be conscious of each other. Thus, this complex electron is apparantly conscious (self-conscious, actually). This complexity arises from correlation, i.e. interdependency of its parts.

In practice it is possible to separate the effects of electron exchange in some cases (Cf. section 4.1.2). Hence, the terms 'Pauli correlation', 'statistical correlation' or 'exchange correlation' [Knudsen and Reading, 1992] are sometimes used to distinguish this exchange effect from other electron correlation. In this book the term 'electron correlation' includes electron exchange in accord with the definition of correlation used in static cases [Löwdin, 1959, 1995].

It is conventional to ignore exchange effects to varying degrees (Cf. section 4.3). One case in which this exchange symmetry may be sensibly ignored is the case of an isolated helium target. In general observables correspond to the square of the matrix element of some operator, namely, $| < f|\mathcal{O}|i > |^2$, where the electronic states $< f|$ and $< i|$ should be properly antisymmetrized. The antisymmetry operator satisfies $\mathcal{A}^2 = \mathcal{A}$. Assuming that $\mathcal{O} = \mathcal{O}(1) + \mathcal{O}(2)$, as is often the case, then the matrix element for helium becomes,

$$\begin{aligned} | < f|\mathcal{O}|i > |^2 &= | < f|\mathcal{A}\,(\mathcal{O}(1) + \mathcal{O}(2))\,|i > |^2 \qquad (3.35) \\ &= | < \frac{\phi_1\phi_2 + \phi_2\phi_1}{\sqrt{2}}|\mathcal{O}(1) + \mathcal{O}(2)|\phi_1'\phi_2 i' > |^2 \\ &= \frac{1}{2}| < \phi_1|\mathcal{O}(1)|\phi_1' > |^2 + | < \phi_2|\mathcal{O}(2)|\phi_2' > |^2 \\ &= \frac{1}{2}|2 < \phi|\mathcal{O}|\phi' > |^2 = 2| < \phi|\mathcal{O}|\phi' > |^2 \,. \end{aligned}$$

While the separation $< f| =< \phi_1\phi_2|$ is not justified, ignoring the effects of the electron exchange is justified in this case since $\phi_1 = \phi_2 = \phi$ in helium. The factor of two simply means that the result is twice as big for two electrons as it would be for a single electron. This convenient result does not hold for most few and many electron systems.

3.4 Methods of computation

Transitions of a few electrons are generally more difficult to evaluate than for single transitions and less difficult than for many electron transitions. Methods to evaluate many electron transition probabilities are either based on a form of the independent electron approximation discussed in chapter 4 or on statistical methods such as those discussed in chapter 5. Simple methods for the transition of a single electron are discussed in chapter 2. Here methods for dealing for few (mostly two) electron transitions are briefly summarized.

Independent electron approximation. The independent electron approximation ignores correlation between electrons so that whatever happens to one electron is independent of the other electrons. Thus, when electron correlation is small, this approximation is valid. If electron exchange is also neglected, then the simple binomial distribution of single electron probabilities is obtained (Cf. sections 4.1.4 and 1.1.3). This reduces the N electron probability to a much simpler product of N one electron probabilities.

1st and 2nd order perturbation theory. Calculations based on perturbation expansions in the scattering potential V and the correlation potential v are often relatively simple both mathematically and conceptually [Briggs and Macek, 1990]. Even so, as of 1995 no calculation exact through third order had yet been done (although the coupled channel method should be noted). Calculations exact through first order in Z/v and all orders in v have been done for double excitation by McGuire and Straton [1990b] using a closure approximation. Subsequently, the second order amplitude for two electron excitation was calculated with inclusion of electron correlation in atomic wavefunctions and with direct summation over the intermediate states without invoking the closure approximation [Godunov and Schipakov, 1993]. Calculations including ionization are more difficult because correlated continuum wavefunctions are not easily determined and electron capture is difficult because of translation factors. Double excitation calculations through second order in Z/v have been done [McGuire and Straton, 1990b] using closure in the intermediate states and confining intermediate states to real $\{\phi\}$. Mechanisms corresponding to specific many body perturbation terms have been considered by Sorensen [Andersen *et al.*, 1987] and by Hino *et al.* [1993]. Some interesting models including some second and third order contributions have been proposed by Vegh [1988]. Calculations in the independent distinguishable electron approximation have been done for double ionization and for double capture. In addition some progress has been made in the case of simultaneous electron transfer and excitation where a combination of amplitudes involving both V and v have been evaluated. Further discussion of these calculations may be found in sections 7.2 and 7.3.

Classical calculations. Olson [1987] has extensively applied fully classical calculations to few electron transitions. In these calculations the electron which classically orbits the target nucleus interacts with an incoming projectile. The orbit of the electron can then be evaluated from coupled equations and followed in time. These calculations are detailed in their description and do not depend on use of perturbation theory in V. Including effects of the electron correlation potential v is more difficult since the atom becomes classically unstable when v is introduced. However, Olson has incorporated some correlation by simply keeping the electrons on opposite sides of the atom [Zajfman and Maor, 1986]. Other classical calculations [Peach *et al.*, 1985; Reinhold and Falcon, 1986; Horbatsch, 1986] include some multi-electron effects using model potentials.

Time dependent Hartree Fock. The independent electron approximation including electron exchange is defined in section 4.3 by the time dependent Dirac Fock (TDDF) approximation. In the non-relativistic limit this reduces to the time dependent Hartree Fock (TDHF) limit. Various authors [Kuerpick *et al.*, 1995; Bottcher *et al.*, 1991; Devi and Garcia, 1984; Gramlich *et al.*, 1986; Stich *et al.*, 1985] have done calculations using TDHF methods. The trajectory of the projectile is considered to be classical and known. The TDHF wavefunction is found self-consistently at each time step in the collision by optimizing the effective central one electron potentials so that the effect of the remaining correlation potential v on the energy is minimized. Thus, a new set of TDHF basis functions is determined at each time step of the collision. The basis functions used to determine the TDHF wavefunctions include atomic expansions, molecular orbital expansions, pseudostate expansions and three-center expansions. Translation factors and continuum states are included.

Forced impulse method. In the forced impulse method developed by Reading and Ford [1987a,b], the system evolves without correlation (i.e. $v = 0$) for short time steps (as in TDHF). The new set of states is then used as a basis set together with the full Hamiltonian with V fixed and $v \neq 0$ to define a correlated wavefunction at the new time. This correlated wavefunction then evolves without correlation to define the next set of basis states which are again used to evaluate a new wavefunction with correlation at the new time step. Thus, dynamic correlation is ignored only for short time steps. As the duration of the time step is reduced to zero the result becomes exact. At present Reading and Ford have used two time steps so that their present results are valid through second order in Z/v. However, in principle, the method may include more time steps.

Coupled channel calculations. In the method of coupled channels the full scattering wavefunction is expanded in a particular set of basis states. The full Hamiltonian including both V and v is then used to define a set of coupled differential equations for the coefficients of the expansion of the exact wavefunction. Choice of the set of basis states can be important since the complete expansion must in practice be truncated. The numerical difficulty of these calculations increases rapidly as the number of basis states is increased. Nonetheless, the coupled channel method is the most complete practical method now available. Coupled channel calculations for few electron transitions have been carried out [e.g., Shingal *et al.*, 1986; Fritsch and Lin, 1983, 1986, 1990; Reading *et al.*, 1984; Bottcher, 1982; and Kimura *et al.*, 1985]. A review is given by Fritsch and Lin [1991].

Other velocity regimes. For the most part both low collision velocities and relativistic velocities [Barat, 1988; Rocin *et al.*, 1989; Barany, 1990] are not included in this book. At intermediate to low collision velocities where the molecular orbital methods may apply, scattering correlation becomes more important and has been examined. Also, the question of successive versus simultaneous interactions during a collision has been considered.

3.5 Classification of multiple electron transitions

Multiple electron transitions are usually grouped in terms of the type of interaction associated with the transition. For transitions between atomic energy levels, the interactions are either internal interactions between the electrons themselves or external interactions with fields (usually electromagnetic fields), particles (e.g. incident electrons or protons) or system of particles (e.g. other atoms). Interactions of electrons with other electrons[§] (either internal or external) lead to electron correlation [see sections 1.4, 6.1, 6.2.4 and 8.1]. Interaction with electromagnetic fields is often associated with the emission (creation) or absorption (annihilation) of a photon. Atomic systems that are in the state of lowest total energy (the ground state) are stable. Transitions of electrons to states of higher energy do not occur unless there is an interaction with an external particle or field.

Transitions without an incident projectile. It is often assumed that the de-excitation transition of an electron may be separated from the excitation process that created an excited state. This useful simplifying assumption is often valid when excitation occurs much faster than de-excitation.

Photoemission. When an electron in an excited atom makes a transition to a state of lower energy, the energy lost by the electron is often carried off by a photon which is created in the process. This is photoemision.

[§] Electrons have the remarkable property that they are identical –i.e indistinguishable. Consequently it is sensible to regard all electrons as a single electron. This universal electron is complicated. Much of the challenge of understanding the physical properties of chemical, biological and macroscopic matter lies in determining the nature of a part of this 'electron of the universe'. This is not easy. So it is conventional to think in terms of many, many simple electrons rather than one complex electron.

It may be described either classically or quantum mechanically [Heitler, 1954] by the interaction of the electron with the external electromagnetic field of the emitted photon. This interaction is the $-q\vec{p}\cdot\vec{A}/c + q^2A^2/c^2$ term in the Hamiltonian [see sections 9.1-2]. If the fields are weak, then the q^2A^2/c^2 term may be neglected. The quantum matrix element which determines the rate of photoemission is $< f|q\vec{p}\cdot\vec{A}/c|i >$. If the momentum transfer q of this interaction is small, i.e. $qr << 1$ (where r is the radius of the transition electron), then one may expand $e^{i\vec{q}\cdot\vec{r}} \simeq 1 + i\vec{q}\cdot\vec{r} + ...$ which describes the oscillating nature of the vector potential, $\vec{A}$, of the photon. This gives an expansion of the matrix element in spherical harmonics, i.e., a partial wave expansion. The first term gives the amplitude for the dipole transition, the second for the quadropole, and so on. Since $1/c = 1/137 = \alpha$, the fine structure constant, this series normally converges rapidly so that the strongest transitions are usually the dipole transitions. Then one may invoke the dipole selection rule which eliminates all non-dipole transitions. Multiple photoemission is usually small.

Auger transition. When an electron in an excited state drops into a lower energy state, the energy lost may be transferred to another electron. This may happen if the two electrons interact with one another via the electron-electron interaction, $1/r_{12}$. This two electron transition is called an Auger transition [Crasemann, 1996; Temkin, 1985; Aberg and Howat, 1982; McGuire, 1975; Auger, 1926]. In the absence of some external interactions, transitions from a particular excited state occur either by photoemission (a one electron transition) or by the Auger process (a two electron transition). Thus the total transition rate for a particular energetically downward transition is the sum of the rates for photoemission and Auger emission.

The Auger transition rate may be evaluated from the matrix element of the interaction which causes the transition, namely, $< f|\frac{1}{r_{12}}|i >$. The energy gained by the second electron is equal to the energy lost by the first. Thus, if the second electron is excited into the continuum (or ionized), its energy is well defined. At this ejected electron energy the combined excitation de-excitation process with the transition of an Auger electron is indistinguishable from simple direct ionization of a single electron since $|i >$ and $|f >$ are the same for both amplitudes. Consequently the two amplitudes may not be decoupled and must be treated coherently. Thus Auger electrons form discrete resonances superimposed on a continuous background of single electron ionization. Two different pathways can produce the same ionized electron state, namely direct excitation of the single electron continuum and direct excitation of a bound state followed by Auger decay into the single ionization continuum. Since these processes are indistinguishable, they interfere. The nature of these interferences has been studied and appropriate parameterization has been developed by Fano [1961] and by Shore [1967] for ionization by photons, by Balashov *et al.* [1973] for ionization by electrons, and by Arcuni and Schneider [1987] and by Godunov *et al.* [1989a] for ionization by ions. The resonance behavior of the Auger process may be in principle described by the use of Green functions (Cf. appendices to this book). However, obtaining the width Γ in the standard expression $\frac{1}{E-E_{res}+i\Gamma}$ from first principles is often not straightforward¶. A theory of post collision effects in photoionization is given by Tulkki *et al.* [1987]. An analysis of autoionizing states distorted by strong fields created by fast highly charged ions has been attempted by Godunov *et al.* [1996].

The inverse of the Auger process occurs when an incident electron falls into a bound state of an atom by giving some of its energy to excite one of the electrons in the atom. This is called dielectronic recombination. It is a resonant process.

The examples above illustrate a useful difference between observably different transitions and explanations of how transitions occur. Transitions are different if either the initial state $|i>$ or the final state $|f>$ can be distinguished from one another by experiment, at least in principle. Photoemission differs from an Auger process in which an electron rather than a photon carries energy out of an interaction since photons and electrons are observably different. These are different transitions (sometimes denoted as different processes). On the other hand different pathways sometimes can lead to the same transition. The transition of an electron from the ground state to a state with a certain energy may in some cases occur either by direct excitation of the electron or by excitation to an intermediate bound state followed by an Auger decay leading to the same final state. Amplitudes (sometimes called mechanisms) for the various pathways leading to the same final state are added coherently. When one sums over observably different transitions, the sum is incoherent.

Transitions from interaction with an incident photon. There are a variety of ways in which an atom may interact with an incident photon depending on whether the process is elastic (no transition) or inelastic (a transition occurs) and on the number of photons in the final (and initial) states. The more important processes are listed at the end of chapter 9.

Photoannihilation, the Einstein photo effect, also called photoionization, photoexcitation or photo absorption, is the inverse of photoemission.

¶ The continuum to which a shape resonance decays is normally the same channel that supports the bound state. A Feshbach resonance decays into a completely different continuum whose ionization energy lies lower in energy.

Matrix elements for photoannihilation are similar to those for photoemission.

Transitions with incident charged particle without radiation. Single electron transitions are conventionally classified (Cf. section 2.1.2) as elastic (no transition), excitation (to a bound state of the target), transfer (to the bound state of the projectile) and ionization (excitation into the continuum). Using this convention there are six possible types of two electron transitions: double excitation, double transfer, double ionization, ionization-excitation, transfer-ionization and transfer excitation. All of these two electron transitions have been observed experimentally and calculated theoretically (Cf. section 7.4). For transitions of more than two electrons, some observations exist, but relatively few detailed calculations exist.

Transitions with incident charged particle with radiation. There have been some studies of single electron transitions with the simultaneous emission of a photon. For example, radiative electron transfer dominates over non-radiative transfer at high velocities [Briggs and Dettman, 1974] because transfer is a quasi-forbidden process at high velocity, i.e. capture of a free stationary electron is forbidden by conservation of energy and momentum. Not many studies have been made of few and many electron transitions with emission of one or more photons.

Transitions with multi-centered systems of charged particles. As the number of possible electronic states increases, many processes become possible and simply keeping track of various processes itself becomes a problem. This topic is treated approximately in sections 4.2 and 4.3 and also in chapter 8.

3.6 Appendix

1 *Separation of ψ_{el}*

The total Hamiltonian H of (3.1) may be expressed,

$$H = H_N(\nabla_R, R) + H_T(\nabla_r, r) + V(R, r) , \tag{3.36}$$

where H_N is the Hamiltonian for the projectile, H_T is the Hamiltonian of the isolated atomic target and V is the interaction connecting the two. Defining $\Psi(R,r,t) = \chi_N(R,t)\psi_{el}(r,R)$, then the equation $i\frac{\partial\Psi}{\partial t} = H\Psi$ becomes,

$$
\begin{aligned}
i\frac{\partial(\chi_N\psi_{el})}{\partial t} &= (H_N\chi_N)\psi_{el} - 1/M(\nabla_R\chi_N)\cdot(\nabla_R\psi_{el}) \qquad (3.37)\\
&\quad -\chi\nabla_R^2\psi_{el}/2M + \chi_N(H_T\psi_{el}) + \chi_N V\psi_{el} \,.
\end{aligned}
$$

The $\nabla_R^2\psi_{el}$ term is neglected because it is $O(1/M)$ smaller than the other terms.

The resulting equation is solved by choosing χ_N so that,

$$
i\partial\chi_N/\partial t = H_N\chi_N \,, \qquad (3.38)
$$

so that $\chi(R,t)$ is a wavefunction peaked about the classical trajectory $\vec{R} = \vec{R}(t)$. This trajectory is usually well approximated by the trajectory for elastic scattering. Next it is noted that $1/M(\nabla_R\chi_N)\cdot(\nabla_R\psi_{el}) = i\vec{v}\chi_N\cdot(\nabla_R\psi_{el})$ so that (3.38) becomes,

$$
0 = \chi_N(R,t)\ [H_T\psi_{el}(r,R) + V\psi_{el}(r,R) - i\vec{v}\cdot\nabla_R\psi_{el}(r,R)] \,. \qquad (3.39)
$$

If $R \neq R(t)$, one has $0 = 0$ because χ_N is small. For $R = R(t)$ one has $V(r,R(t)) = V(r,t)$ and $\psi_{el}(r,R(t) = \psi_{el}(r,t)$. Using $\vec{v}\cdot\nabla_R = \mathrm{v}\frac{\partial}{\partial Z} = \mathrm{v}\frac{\partial}{\partial t} = \frac{\partial}{\partial t}$ in (3.39), one has,

$$
i\frac{d\psi_{el}(r,t)}{dt} = (H_T + V)\,\psi_{el}(r,t) \equiv H_{el}\psi_{el}(r,t)\,. \qquad (3.40)
$$

This equation is used to determine the electronic wavefunction ψ_{el}.

There are a variety of derivations of the separation of the electronic from the nuclear terms [Delos and Thorson, 1972; Bransden and McDowell, 1992; Anton, 1994]. The most recent and most complete is due to Anton [1994], whose method and result is similar to that above which follows McGuire and Weaver [1986]. These results give a correction proportional to order (p/P), where $p(P)$ is a typical momentum for the electron (incident projectile). This differs from the Born Oppenheimer separation which gives a correction proportional to $(m/M)^{1/4}$, where $m(M)$ is the mass of the electron (nucleus). The physical idea behind the separation above is that the trajectory $\vec{R}(t)$ of a fast or heavy nucleus is not affected much by the target electrons. In turn small changes in $\vec{R}(t)$ in turn do not much affect the activity of the target electrons.

4
Independent electron approximation

The many body problem of N fully correlated particles is very difficult to solve exactly. Since many systems of practical interest are composed of many constituent particles, there have been various attempts to develop effective practical methods for evaluating properties of many body systems approximately . One such approach for many electron systems is the independent electron approximation. In this approximation the probability amplitude and the probability are calculated independently for each of the N electrons. Thus the N electron problem is effectively reduced to the simpler problem of solving the one electron problem N times. This simplifies the N electron problem considerably.

In the first section of this chapter the relatively simple independent electron approximation is developed in detail for an atomic target with a frozen electron cloud without exchange which interacts with a simple projectile with a point charge. In the second section this simple version of the independent electron approximation is generalized to multi-centered atomic targets. The full independent electron approximation, corresponding to the time dependent Dirac Fock approximation, is discussed in the final section.

4.1 Single atoms

Validity conditions. The independent electron approximation serves as a useful starting point for understanding many electron systems when multi-electron effects are weak. This independent electron approximation [Hansteen and Mosebekk, 1972; McGuire and Weaver, 1977; Hazi, 1981; Sidorovitch and Nikolaev, 1983] is exact under the following conditions.

(i) The projectile is a point charge.

(ii) The internuclear motion is separated from the electronic motion and treated as elastic classical scattering.

(iii) The $|\vec{r}_k - \vec{r}_j|^{-1}$ electron-electron interactions are approximated by single-electron (i.e., mean field) potentials.

Condition (i) is fully satisfied in collisions with moderately fast protons, bare ions or fast electrons. Condition (ii) is well satisfied for incident heavy ions and may be satisfied for high-velocity incident electrons whose de Broglie wavelength is small compared to atomic distances. Condition (iii) corresponds to ignoring effects of electron-electron interactions during the collision. This condition is considered in further detail in sections 3.2, 3.3, 6.1, 6.2, 7.1 and 8.1. Note that the independent electron approximation does not depend on the strength of the interaction V with the projectile, but depends only on the weakness of the electron-electron interaction.

1 Full atomic Hamiltonian

The Hamiltonian for scattering of an atom of nuclear charge Z_T with N electrons by a particle of charge Z and mass M is,

$$\begin{aligned} H &= -\frac{\nabla^2}{2M_r} + \frac{ZZ_T}{R(t)} - \sum_{j=1}^{N} \frac{Z}{|\vec{R}(t) - \vec{r}_j|} \\ &\quad + \sum_{j=1}^{N} \left[\frac{-\nabla_j^2}{2} - \frac{Z_T}{r_j} + \sum_{\substack{k \\ (k>j)}} \frac{1}{|\vec{r}_k - \vec{r}_j|} \right] \\ &= K + V + H_T \,, \end{aligned} \tag{4.1}$$

where M_r is the reduced mass of the projectile and $K = -\frac{\nabla^2}{2M_r}$ and

$$V(t) = \frac{ZZ_T}{R(t)} - \sum_{j=1}^{N} \frac{Z}{|\vec{R}(t) - \vec{r}_j|} = \frac{ZZ_T}{R(t)} - \sum_j V_j(t) \,, \tag{4.2}$$

where it is assumed that the trajectory of the projectile $\vec{R}(t)$ is well defined, e.g., $\vec{R}(t) = \vec{b} + \vec{v}t$, in accord with the results of section 3.2.1.

Correlated H_T. The $1/r_{kj} = |\vec{r}_k - \vec{r}_j|^{-1}$ electron-electron Coulomb correlation potentials now appear in (4.1) where H_T may be written as a sum of operators H_T^j which will later reduce to single particle operators when correlations are ignored. Specifically, from (4.1),

$$H_T = \sum_{j=1}^{Z_2} \left[-\frac{\nabla_j^2}{2} - \frac{Z_2}{r_j} + \sum_{\substack{k \\ (k>j)}} |\vec{r}_k - \vec{r}_j|^{-1} \right] \equiv \sum_{j=1}^{Z_2} H_T^j \,. \tag{4.3}$$

Because

$$\sum_{\substack{k \\ (k>j)}} |\vec{r}_k - \vec{r}_j|^{-1} \tag{4.4}$$

is not a single-electron operator, the operators H_T^j are not single-electron operators.

2 Uncorrelated Hamiltonian

The key step in the independent electron approximation is to remove the correlation between electrons. If the electron-electron interactions which give rise to correlation in (4.3), namely,

$$\sum_{\substack{k, \\ (k>j)}} |\vec{r}_k - \vec{r}_j|^{-1} \tag{4.5}$$

are approximated by an average (or mean field) potential*, i.e.,

$$\sum_{\substack{k \\ (k>j)}} |\vec{r}_k - \vec{r}_j|^{-1} \simeq v_j(r_j) \,, \tag{4.6}$$

then the H_T^j terms are indeed single-electron operators and correlation disappears.

Thus, the correlation potential v is defined by the sum of the differences between the $1/r_{kj} = |\vec{r}_k - \vec{r}_j|^{-1}$ Coulomb potentials between electrons and the corresponding approximate mean field potentials v_j, namely,

$$v \equiv \sum_{j=1}^{N} \left(\sum_{\substack{k, \\ (k>j)}} |\vec{r}_k - \vec{r}_j|^{-1} - v_j(r_j) \right) \,. \tag{4.7}$$

Sum of single particle potentials. Now if electron-electron correlations are ignored, then $[H_T^i, H_T^j] = 0$ since the H_T^j are single particle operators.

* In many cases use of a mean field potential can considerably reduce the effects of the residual correlation. An illustrative estimate is given at the end of section 6.2.5.

Recalling from (4.2) that $V(t)$ is a sum of single particle potentials $V_j(t)$ (plus a term depending only on R that is now omitted because the electronic eigenstates are orthogonal), then H_T becomes a sum of commuting single particle terms. Consequently, in the interaction picture (Cf. section 3.3) one has,

$$\begin{aligned} V_I(t) &= e^{H_T t} V e^{-iH_T t} \\ &= \sum_j \exp\left[i \sum_k H_T^k t\right] V_j \exp\left[i - \sum_l H_T^l t\right] \\ &= \sum_j e^{iH_T^j t} V_j e^{-iH_T^j t} = \sum_j V_I^j(t)\,. \end{aligned} \tag{4.8}$$

Thus, without electron-electron correlation, $V_I(t)$ is a sum of single particle operators $V_I^j(t)$.

Product of single particle evolution operators. As a consequence of ignoring correlation, the evolution operator reduces to a product of evolution operators, i.e.,

$$\begin{aligned} U(t, t_0) &= T\exp\left[-i \sum_j \int_{t_0}^{t} V_I(t_1) dt_1)\right] \\ &= \prod_j T\exp\left[-i \int_{t_0}^{t} V_I(t_1) dt_1\right] \\ &= U_0(t, t_0) = \prod_j U_j(t, t_0)\,. \end{aligned} \tag{4.9}$$

Now the electrons evolve independently during the collision.

Product of single particle probabilities. Finally, if electron exchange (discussed below) is ignored, then the probability amplitude is a product of independent-electron amplitudes, namely,

$$a^{fi} = < f|\psi_i> = < f|U|i> = \prod_j < f_j|U_j|i_j> = \prod_j a_j^{fi}\,, \tag{4.10}$$

where ψ_i represents the full electron wavefunction which reduces to $|i>$ as $t \to -\infty$. Furthermore, since H_T is a sum of single electron terms, $|i> = \prod_j |i_j>$ and $< f| = \prod_j < f_j|$. Consequently, the transition probability $|a^{fi}|^2$ is a product of independent-single electron probabilities in the independent distinguishable-electron (or time dependent Hartree) approximation, namely,

$$P = \prod_j P_j \,. \tag{4.11}$$

This simplifies the many body problem considerably.

This independent electron approximation corresponds to a simple generalization of the static Hartree product wavefunction to a dynamic probability amplitude using a single basis set, $\{\prod_j |s_j>\}$. If the effective screening among the electrons changes during the collision, then different basis states would be required at different times during the collision. In the development above it is assumed that the electron orbitals are frozen during the collision.

Electron exchange. Electron symmetries are in principle included in the definition of an uncorrelated wavefunction used here . However, other authors [e.g., Balescu, 1975; Becker, 1988] regard the effects of electron symmetries as corrections to the uncorrelated limit and refer to these corrections as 'Pauli correlation'. Now the effect of electron exchange may be included following Reading and Ford [1980]; Ford *et al.*, [1979]; Becker *et al.*, [1980] in a two electron system in which $|a,b> \rightarrow |\alpha,\beta>$. Here $|s>$ denotes the asymptotic one electron state and $|\psi(s)>$ denotes the complete state. In the independent distinguishable electron approximation (without electron exchange) above, the probability for a two electron transition quite simply is,

$$\begin{aligned} P(\alpha,\beta) &= |<\alpha,\beta|\psi(a),\psi(b)>|^2 \\ &= |<\alpha|\psi(a)>|^2|<\beta|\psi(b)>|^2 = P(\alpha)P(\beta)\,. \end{aligned} \tag{4.12}$$

The inclusive probability P_a for producing the vacancy 'a' in the ground state of A can be found using closure, namely,

$$P_a = \sum_{\alpha\neq a}\sum_{\beta} P(\alpha)P(\beta) = \sum_{\alpha\neq a} P(\alpha) = 1 - |<a|\psi(a)>|^2\,. \tag{4.13}$$

If more than one state on A is initially occupied, then one might attempt to impose exclusion by using,

$$P_a = 1 - \sum_{occupied\ \alpha} |<\alpha|\psi(a)>|^2\,. \tag{4.14}$$

In the independent electron approximation with electron exchange the probability for $|a,b> \rightarrow |\alpha,\beta>$ properly is,

$$P(\alpha,\beta) = | < \alpha,\beta | \frac{1}{\sqrt{2}} \left(\psi(a),\psi(b) - \psi(b),\psi(a)\right) > |^2 \,. \tag{4.15}$$

It is then straight forward (but a little tedious) to show [Ford and Reading, 1980; Becker, 1988] that the inclusive probability is,

$$\begin{aligned} P_a &= \sum_{\alpha\neq\beta}\sum_{\beta\neq a} P(\alpha,\beta) \qquad (4.16) \\ &= 1 - | < a|\psi(a) > |^2 - | < (a|\psi(b) > |^2 \\ &= 1 - \sum_{occupied\, k} | < k|\psi(a) > |^2 \,. \end{aligned}$$

This result is similar to the result obtained from the independent distinguishable electron approximation of (4.11) and is intuitively obvious as suggested by (4.14) above.

Now consider the inclusive probabilities for producing vacancies in both a and b. For distinguishable electrons, $P_{ab} = P_a P_b$. For indistinguishable electrons (again after some algebra) it may be shown [Ford, 1981],

$$P_{ab} = P_a P_b \; - \; | \sum_{occupied\, k} < a|\psi(k) >< \psi(k)|b > |^2 \; \leq \; P_a P_b \,. \tag{4.17}$$

Knudsen and Reading [1992] point out that (4.17) may be obtained from the determinant of the matrix $\delta_{a,b} - \sum_{occupied\, k} < a|\psi(k) >< \psi(k)|b >$, which may be generalized to three or more vacancies. The term subtracted in (4.17) accounts for electron exchange in occupied states, i.e., so called 'Pauli blocking' , is the same as that obtained by Lüdde and Dreizler [1985]. A field theoretical approach using creation-annihilation operators is given by Wells *et al.* [1992] based on the formulation of Reinhardt *et al.* [1981].

Reduction of coupled channels. Consider a system comprised of two electrons, 1 and 2, each of which can be in one of two states j and k. This is a four channel problem. Transitions between states is caused by an interaction V. For simplicity electron symmetry is neglected here (i.e., the electrons are distinguishable). Now let Ψ represent the full wavefunction and Φ represent the asymptotic wavefunction for the two correlated electrons. Then, the possible transition probability amplitudes are defined by,

$$\begin{aligned} a_{jj} &= (\Phi_{jj}, \Psi)\,, \qquad (4.18) \\ a_{jk} &= a_{kj} = (\Phi_{jk}, \Psi)\,, \end{aligned}$$

$$a_{kk} \quad = \quad (\Phi_{kk}, \Psi) \, .$$

The first index corresponds to electron 1 and the second index to electron 2.

In the independent electron approximation, $\Psi = \psi^1 \cdot \psi^2$ and $\Phi = \phi^1 e^{-iE_j^1 t} \phi^2 e^{-iE_k^2 t}$ (here setting $E_j^1 = E_j^2 = E_j$) so that,

$$\begin{aligned} a_{jj} &= (\phi_j^1 e^{-iE_j t} \phi_j^2 e^{-iE_j t}, \psi^1 \psi^2) = a_j^1 a_j^2 \, , \\ a_{jk} &= (\phi_j^1 e^{-iE_j t} \phi_k^2 e^{-iE_k t}, \psi^1 \psi^2) = a_j^1 a_k^2 \, , \\ a_{kk} &= (\phi_k^1 e^{-iE_k t} \phi_k^2 e^{-iE_k t}, \psi^1 \psi^2) = a_k^1 a_k^2 \, . \end{aligned} \tag{4.19}$$

Applying unitarity yields,

$$\begin{aligned} |a_{jj}|^2 + |a_{jk}|^2 &+ |a_{kj}|^2 + |a_{kk}|^2 \\ &= |a_j^1|^2 \left(|a_j^2|^2 + |a_k^2|^2 \right) + |a_k^1|^2 \left(|a_j^2|^2 + |a_k^2|^2 \right) \\ &= |a_j^1|^2 + |a_k^1|^2 = 1 \, , \end{aligned} \tag{4.20}$$

where $|a_j^\ell|^2 + |a_k^\ell|^2 = 1 \quad (\ell = 1, 2)$.

Now substituting into $H\Psi = i \ \dot{\Psi}$ the wavefunction $\Psi = \sum_f \ a_{if} \Phi_f$ with $H = H_T + V$ one obtains the usual coupled equations for these four channels, namely,

$$\begin{aligned} i \, \dot{a}_{jj} &= 2V_{jj} \, a_{jj} + 2V_{jk} \, e^{-i\omega t} a_{jk} \, , \\ i \, \dot{a}_{jk} &= V_{jk} \, e^{-i\omega t} a_{jj} + (V_{jj} + V_{kk}) a_{jk} + V_{jk} \, e^{i\omega t} a_{kk} \, , \\ i \, \dot{a}_{kk} &= 2V_{kj} \, e^{i\omega t} a_{kj} + 2V_{kk} a_{kk} \, , \end{aligned} \tag{4.21}$$

where $V = V^1 + V^2$, the $\{\Phi_f\}$ are orthogonal and $\omega = (E_j - E_k)$. Using (4.20) and $V_{jk}^\ell = < \phi_j^\ell | V^\ell | \phi_k^\ell >$ then one has, after a little algebra,

$$i \, \dot{a}_j^\ell = V_{jj}^\ell a_j^\ell + V_{jk}^\ell e^{-i\omega t} \, a_k^\ell \qquad (\ell = 1, 2) \, . \tag{4.22}$$

This is the usual coupled equation for the amplitude for each single electron.

Thus, in the independent electron approximation, the coupled channel equation for two electrons in two identical states may be solved in terms of two independent coupled channel equations, one for each electron [McGuire, 1982c]. This analysis has been extended to non-orthogonal basis sets for application to electron capture [Hall *et al.*, 1983]. It may be possible to generalize this result to more channels and more electrons.

3 Binomial distributions

In the independent electron approximation without exchange the probability P_N for transitions in a system of N electrons is a simple product of the transition probabilities P_j for each of the electrons, i.e.,

$$P_N = \prod_j^N P_j \,. \tag{4.23}$$

If each of the probabilities, P_j, is different, (4.23) cannot be simplified further. If some of the probabilities are the same, then one obtains the multinomial distribution of section 1.4.

In some cases it is sensible to use a model in which only two states are possible. One state is usually the initial state and a transition is said to occur if the electron is in the other state after the interaction. If $P_j = P$ is the probability that a transition occurs for the jth electron, then the probability that the electron does not make a transition is,

$$Q_j = 1 - P_j = 1 - P = Q \,. \tag{4.24}$$

If there are N electrons with the same probability P of making the transition, then the probability that n of these electrons make the transition and the remaining $N - n$ do not make the transition is,

$$P_N^n = \binom{N}{n} P^n (1 - P)^{N-n} \,. \tag{4.25}$$

This the binomial distribution discussed in sections 1.1-3. Here $\binom{N}{n}$ is the binomial coefficient that is equal to the number of different ways to arrange N objects taken n at a time.

Extension of this binomial distribution to include shake is given in section 7.1.1.

Depth and distribution of ions implanted in a solid. As an example, consider an ion of charge Z and energy E_{inc} incident on a surface of a solid composed of identical atoms with an interatomic spacing d. The problem is to find the average depth D of the deposition of the ions and the width ΔD of the distribution about the average value.

It is assumed that: i) the deposition is one dimensional because most of the ions scatter in the forward direction, ii) the probability for energy loss may be characterized by a known single velocity dependent probability per atomic electron P with a constant energy loss ΔE, and iii) each atom has N electrons. The distance D_j between the $j-1$ and the jth interaction depends on the probability that a collision has occurred. If the probability of energy loss (via ionization) is $< P_j >$, then, energy will be lost in every $1/ < P_j >$ collisions so that $D_j = d/ < P_j >$. If $< P_j > \geq 1$, as may be the case for multi-shell atoms, then it is required that the minimum value of D_j be d for a single atomic spacing. Thus,

$$\begin{aligned} D_j &= max(d, < d_j >) \\ d_j &= d/ < P_j >= d/NP_j \end{aligned} \tag{4.26}$$

where P is replaced by P_j since the velocity decreases with the number of collisions j. The energy of the ion at the jth step is,

$$E_j = E_{inc} - NP_j\Delta E \geq 0. \tag{4.27}$$

When $E_j \to 0$, then $j = N_{max}$ and the range D is found from

$$D = \sum_j^{N_{max}} D_j \tag{4.28}$$

and the distribution about that value is found from

$$\Delta D = \sum_j^{N_{max}} D_j \Delta n_j \tag{4.29}$$

where $\Delta n_j^2 =< n_j^2 > - < n_j >^2= NP_j(1 - P_j)$ is a property of binomial distributions (Cf. Appendices to this book).

Multiple ionization of atoms. As a second example, consider ionization of an electron in the K shell of a target atom. Here P is the probability that ionization occurs and $1 - P$ is the probability that ionization does not occur. Then the cross sections, σ^n, for ionizing n of the 2 electrons in the K shell are,

$$\begin{aligned} \sigma^2 &= \int P_2^2(\vec{b})d\vec{b}\,, \\ \sigma^1 &= 2\int P(\vec{b})[1 - P(\vec{b})]d\vec{b}\,, \\ \sigma^0 &= \int [1 - P(\vec{b})]^2 d\vec{b}\,. \end{aligned} \tag{4.30}$$

The reason for the factor of $2 = \binom{2}{1}$ in the middle equation above is that there are two ways in which one electron may be ionized (i.e. either of the 2 electrons may be ionized).

In many atoms there are electrons in different shells, K,L, ...S. In this case the cross section for ionizing n_K of the N_K, n_L of the N_L, ... and

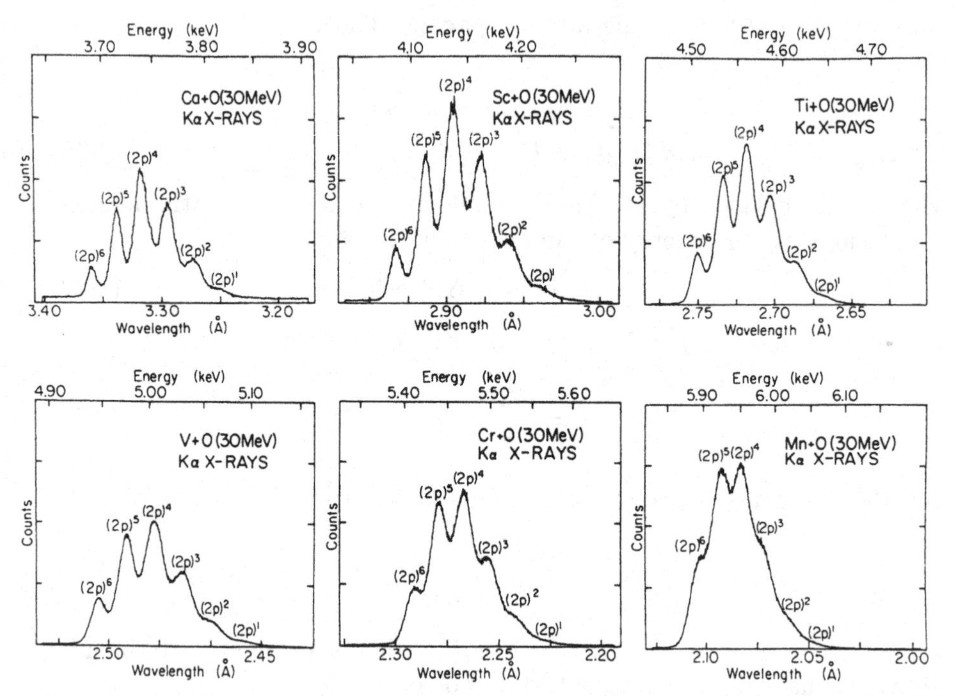

Fig. 4.1. Experimental yield versus target binding energy corresponding to 0 - 6 L shell vacancies with ionization of 1 K shell electron in an oxygen target [Kauffman *et al.*, 1973].

n_S of the N_S electrons is,

$$
\begin{aligned}
\sigma^{n_k\ n_L\ \dots n_S} &= \int \binom{N_K}{n_K} (P_K(b))^{n_K}(1-P_K(b))^{N_K-n_K} \qquad (4.31)\\
&\times \binom{N_L}{n_L}\Big(P_L(b))^{n_L}(1-P_L(b))^{N_L-n_L}\\
&\times \ \dots\dots \binom{N_S}{n_S} (P_M(b))^{n_S}(1-P_M(b))^{N_S-n_S}\, d\vec{b}\,,
\end{aligned}
$$

where P_K is the probability of ionizing an electron in the K shell, P_L the probability for an electron in the L shell, etc.

As an example, consider ionization of 1 of 2 electrons in the K shell, n of 8 electrons in the L shell and a sum over all final outcomes in the M shell. If $P_K << 1$ and $P_L(b) \simeq P_L(0)$ in the region in which $P_K(b)$ is

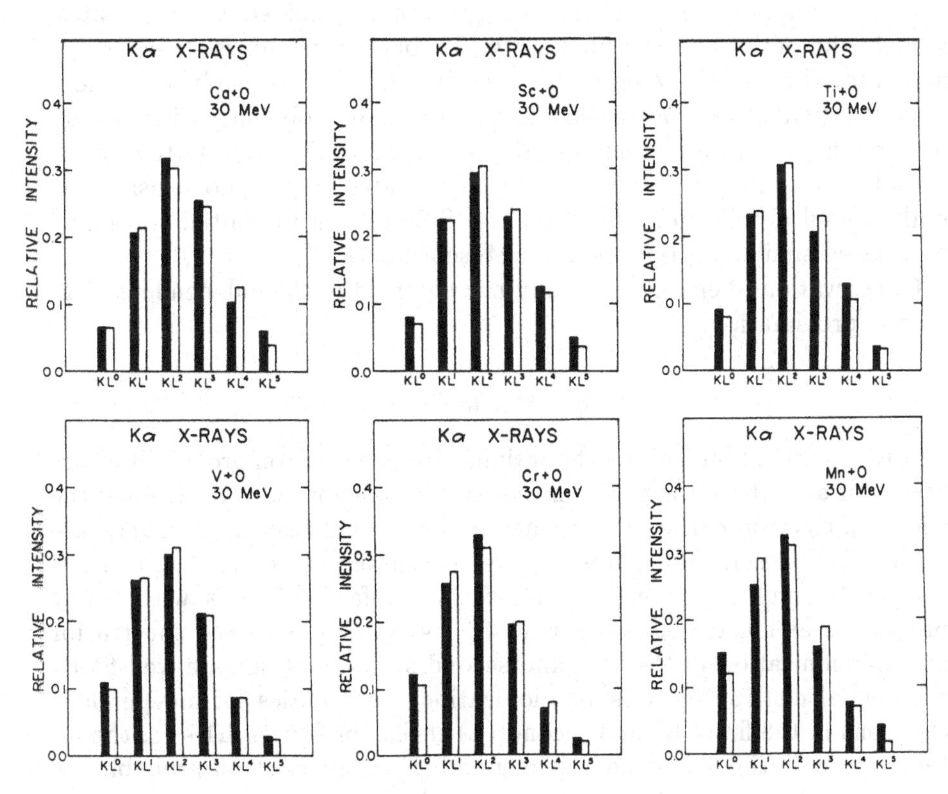

Fig. 4.2. Comparison of observed relative satellite intensities (solid bars) from figure 4.1 to a best least squares fit to a binomial distribution (open bars) of the L shell probability P_L for ionization of a K shell electron and 0 - 6 L shell electrons in oxygen [Kauffman *et al.*, 1973].

non-negligible, then

$$\begin{aligned}
\sigma_{1K,nL} &= \int \binom{2}{1} P_K(b)(1 - P_K(b)) \binom{8}{n} P_L(b)^n (1 - P_L(b))^{8-n} \\
&\times \sum_{n_M=0}^{N_M} \binom{N_M}{n_M} P_M(b)^{n_M} (1 - P_M(b))^{N_M - n_M} \, d\vec{b} \\
&\simeq \int 2P_K(b) \binom{8}{n} P_L(0)^n (1 - P_L(0))^{N-n} \, d\vec{b} \\
&= \sigma_K \binom{8}{n} P_L(0)^n (1 - P_L(0))^{N-n} , \qquad (4.32)
\end{aligned}$$

where σ_K is the cross section for ionizing 1 electron from the K shell. A sum rule for the binomial distribution given in section 1.3 has been used to do the sum over n_M.

In the binomial distribution the electron is confined to two states, namely the initial and the final state. In some cases this is a reasonable model for the atomic system. However, it is often useful to include more than two final states. Extension of the binomial probability distribution to three final states is described at the end of section 1.3. Extension to an arbitrary number of independent final states may be done using the multinomial distribution of section 1.4. Subshells in the initial state may be treated similarly [Hansteen and Mosebekk, 1972].

Conservation of energy is usually not enforced in the independent electron approximation.

4 *Comparison to experiment*

Simple statistical binomial distributions of single electron probabilities are useful for describing the probabilities for multiple ionization whenever the effects of electron correlation is small and effects of electron exchange are small. This description tends to give reasonable agreement with observations for multiple inner shell ionization in fast collisions with heavy projectiles as illustrated in figures. 4.1 - 4.2 where observed spectra for multiple ionization in the first and second shells of atoms are well fit by a binomial distribution of single ionization probabilities for independent electrons in the first (K) and second (L) shells. In figures 4.1 - 4.2 the interaction with the projectile is strong and the single electron probabilities are not much smaller than unity. This independent electron approximation does not depend on the strength of the perturbing interaction, but only on the weakness of electron correlation and neglect of electron exchange. It is surprising that the simple binomial distribution also gives a good description of multiple ionization in outer shells where electron correlation is not small as illustrated in figure 4.3. Binomial distributions may also be used to describe multiple electron capture from the inner shells of atomic targets in terms of the probability for capture of a single electron [Schlachter *et al.*, 1990].

The validity of the binomial distributions rests on neglect of electron correlation in both initial and final states. In many cases it is not clear that one may ignore correlation in the final state since the ejected electrons tend not to leave the collision region quickly, but often leave with a speed roughly half of their initial orbit speed. One simple way to deal with this effect is to increase the binding energy of the electrons in the target as the number of electrons leaving the target increases. In this model the binding energy increases as the remaining electrons adjust to the absence

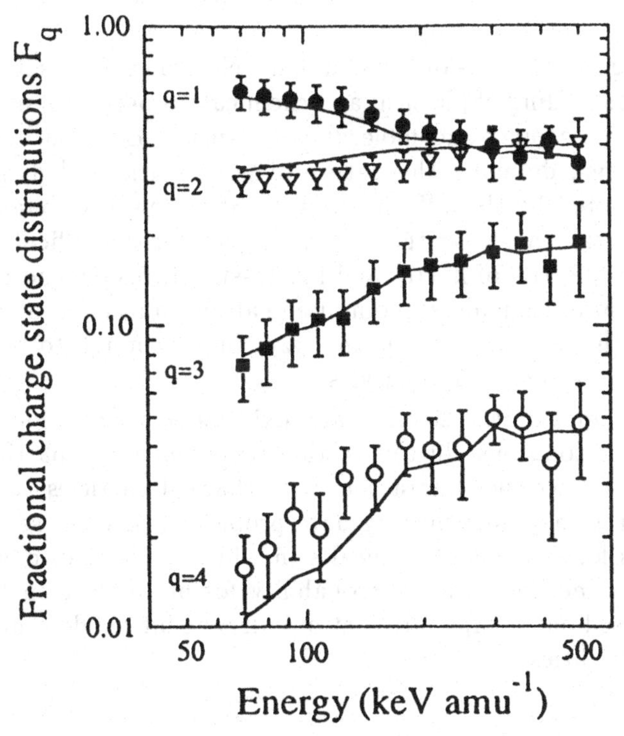

Fig. 4.3. Observed charge state fractions (data points) for $p + \mathrm{Fe} \rightarrow \mathrm{H} + \mathrm{Fe}^{q+}$, i.e., capture of one electron plus ionization of $q-1$ additional electrons together with calculated ionization probabilities P_n^N based on a binomial distribution [Shah *et al.*, [1995] where N is the total number of electrons available in a subshell and $n = q - 1$.

of the removed electrons. This simple improvement has been used by Olson [Cf. Berg *et al.*, 1988] and by Crothers [1991] to give better fits to experimental data. It has been called the 'independent event model', since each ionization 'event' is treated independently, but the electrons are not regarded as independent. The distributions used correspond to the multinomial distributions described in section 1.1.4.

A discussion of the more complete time dependent Dirac Fock (TDDF) and time dependent Hartree Fock (TDHF) methods is given in section 4.3. These more complete methods are used for slow ion-atom and atom-atom collisions.

4.2 Systems of atoms

Understanding interactions of few and many electron systems is central to detailed understanding of physical and chemical properties of microscopic and macroscopic atomic and molecular systems. Even on the scale of individual atoms, detailing the nature of both static and dynamic observables is limited by the difficulty of evaluating few and many electron effects. In general, the larger the system the greater the difficulty. While, in principle, properties of micro- and macro-structures depend on atomic properties, in practice understanding large atomic and molecular systems is limited by the lack of methods that are simple enough to be used for large systems of atoms and molecules.

The simple independent electron approximation is now widely used to describe atomic collisions. In this section an independent particle model is described for molecules interacting with charged particles, so that one may, under certain conditions, evaluate probabilities, cross sections and reaction rates for systems of molecules in which more than one electron is active. This method yields a probability for multi-electron transition that is expressed as a simple product of independent single center, single electron probabilities.

1 Full Hamiltonian for systems of atoms

Consider a molecule or cluster denoted by $C_1^{N_1} C_2^{N_2} ... C_I^{N_I} ... C_N^{N_N}$ where $C_I^{N_I}$ is one of N different subclusters. A subcluster is that part of a cluster which retains phase information about its constituent atoms or centers. In a given subcluster the same center C_I occurs N_I times. Thus, the total number of centers is $N_C = \sum_{I=1}^{N} N_I$. Each center has one or more electrons. It is the activity of one or more of these electrons which is the focus of attention here. In this model both the static and the dynamic properties of these electrons will be defined within each center independently. Each center, C_I^k, of the sub-cluster I has the same nuclear charge Z_I and nuclear mass, M_I, and its center of mass is located a distance $\vec{R}_I^k$ from the center of mass of the molecule, where the index k varies from 1 to N_I. Each individual center, C_I, carries n_I electrons. This molecule interacts with a projectile of charge Z_P and mass M, moving at a velocity $\vec{v}$, as illustrated in figure 4.4.

The total Hamiltonian of this system is [McGuire *et al.*, 1996b],

$$H = K + V + H_T \, . \tag{4.33}$$

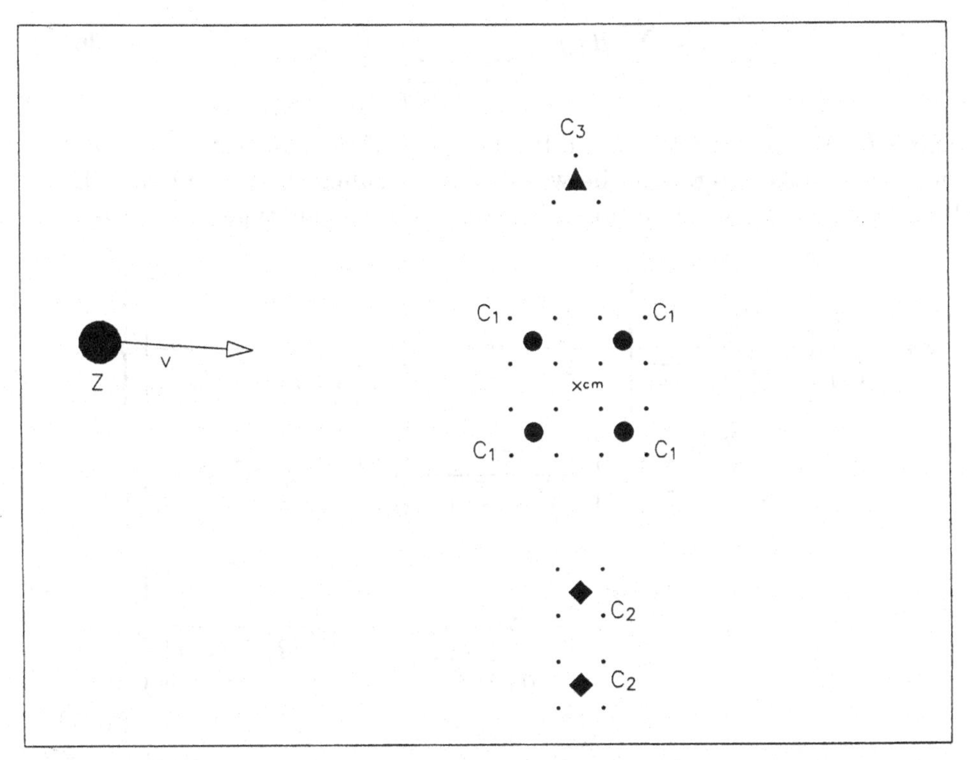

Fig. 4.4. A cluster of atomic centers in a collision with a projectile of charge Z and velocity v. In the target cluster there are four identical subcluster C_1 centers, two subcluster C_2 centers and one subcluster C_3 center. The center of mass is denoted by 'x_{cm}'.

Here

$$K = -\frac{\nabla^2}{2M} \tag{4.34}$$

is the kinetic energy of the projectile in the center of mass of the molecule, V the interaction of the projectile with the molecule given by,

$$V = Z_P \sum_{I=1}^{N} \sum_{k=1}^{N_I} \left\{ \frac{Z_I}{|\vec{R} - \vec{R}_I^k|} - \sum_{i=1}^{n_I} \frac{1}{|\vec{R} - \vec{R}_I^k - \vec{r}_I^{\,k,i}|} \right\} , \tag{4.35}$$

where $\vec{R}$ is the position of the projectile with respect to the center of mass of the molecule and $\vec{r}_I^{\,k,i}$ is the coordinate of the ith electron of the center

C_I^k with respect to its nucleus, as illustrated in figure 4.4.

Also, in (4.33) H_T is the Hamiltonian of the static molecule given by,

$$H_T = \sum_{I=1}^{N} H_{T,I} + \sum_{I=1}^{N} \sum_{\substack{K=1 \\ (K \neq I)}}^{N} W_{I,K} , \tag{4.36}$$

where $H_{T,I}$ represents the static Hamiltonian of the subcluster of type I and $W_{I,K}$ is the interaction between different subclusters I and K. The Hamiltonian $H_{T,I}$ in (4.36) for subclusters of the same type is given by,

$$\begin{aligned} H_{T,I} = \sum_{k=1}^{N_I} & \left\{ -\frac{\nabla^2_{\vec{R}_I^k}}{2M_I} + \sum_{i=1}^{n_I} \left[-\frac{\nabla^2_{\vec{r}_I^{k,i}}}{2} - \frac{Z_I}{r_I^{k,i}} + \frac{1}{2} \sum_{\substack{l=1 \\ (l \neq i)}}^{n_I} \frac{1}{|\vec{r}_I^{k,i} - \vec{r}_I^{k,l}|} \right] \right\} \\ & + \frac{1}{2} \sum_{k=1}^{N_I} \sum_{\substack{j=1 \\ (j \neq k)}}^{N_I} \left\{ \frac{Z_I^2}{|\vec{R}_I^k - \vec{R}_I^j|} + \sum_{i=1}^{n_I} \left[-\frac{Z_I}{|\vec{R}_I^k - \vec{R}_I^j - \vec{r}_I^{j,i}|} \right. \right. \\ & \left. \left. + \sum_{\substack{l=1 \\ (l \neq i)}}^{n_I} \frac{1}{|\vec{R}_I^k + \vec{r}_I^{k,i} - \vec{R}_I^j - \vec{r}_I^{j,l}|} \right] \right\} , \end{aligned} \tag{4.37}$$

and the interaction $W_{I,K}$ between subclusters of different types is given for $I \neq K$ by,

$$\begin{aligned} W_{I,K} = & \frac{1}{2} \sum_{k=1}^{N_I} \sum_{j=1}^{N_K} \left\{ \frac{Z_I Z_K}{|\vec{R}_I^k - \vec{R}_K^j|} - \sum_{i=1}^{n_K} \frac{Z_I}{|\vec{R}_I^k - \vec{R}_K^j - \vec{r}_K^{j,i}|} \right. \\ & \left. + \sum_{i=1}^{n_I} \left[-\frac{Z_K}{|\vec{R}_I^k + \vec{r}_I^{k,i} - \vec{R}_K^j|} + \sum_{l=1}^{n_K} \frac{1}{|\vec{R}_I^k + \vec{r}_I^{k,i} - \vec{R}_K^j - \vec{r}_K^{j,l}|} \right] \right\} . \end{aligned} \tag{4.38}$$

In order to develop a model with independent centers, one regards the first line in (4.37) above as the sum of the Hamiltonians of each individual center that belongs to the subcluster I. Line two contains the sum of the interactions between these atomic centers: i) nucleus-nucleus interaction between C_I^k and C_I^j, ii) the interaction term between the nucleus of C_I^k and the electrons of C_I^j, and iii) electron-electron interaction between electrons of C_I^k and those of C_I^j. The same kind of terms appear in (4.38) but now they correspond to the interactions between centers of different subclusters, i.e. C_I^k and C_K^j.

The Schrödinger equation to solve is given by,

$$(H - E)\Psi_{i,f}^{+,-} = 0 , \tag{4.39}$$

where E is the total energy of the system and $\Psi_i^+(\Psi_f^-)$ is the exact solution of (4.39) with correct outgoing (incoming) conditions corresponding to the entry (exit) channel.

Following section 3.6.1 one may separate the nuclear wavefunction of the point projectile from the electronic wavefunction and study the wavefunction of the electrons corresponding to,

$$\left(H_{el} - i\frac{\partial}{\partial t}\right)\psi_{i,f}^{+,-} = 0 , \tag{4.40}$$

where the Hamiltonian governing the evolution of the electrons in the molecule, H_{el}, is defined as

$$H_{el} = H_T + V - \sum_{I=1}^{N}\sum_{k=1}^{N_I} \frac{Z_P Z_I}{\mid \vec{R} - \vec{R}_I^k \mid} = H_T + V' , \tag{4.41}$$

and $\psi_{i,f}^{+,-}$ is the time dependent wavefunction with correct outgoing and incoming conditions that describes the electronic motion. The Born-Oppenheimer approximation is used and the $\vec{R}_I^k$ is regarded as fixed. Next it is assumed that the projectile motion may be treated classically [McGuire and Weaver, 1986] so that the projectile trajectory $\vec{R}(t)$ is well defined. The simplest (but not the only possible) trajectory is $\vec{R}(t) = \vec{b} + \vec{v}t$, where $\vec{b}$ is the impact parameter of the projectile relative to the center of mass of the molecule.

In (4.41) the potential V' is the sum of the interactions of the projectile with each of the target electrons given by,

$$V' = \sum_{I=1}^{N}\sum_{k=1}^{N_I}\sum_{i=1}^{n_I} V_I^{k,i} = \sum_{I=1}^{N}\sum_{k=1}^{N_I}\sum_{i=1}^{n_I} \frac{-Z_P}{\mid \vec{R}(t) - \vec{R}_I^k - \vec{r}_I^{k,i} \mid} . \tag{4.42}$$

If one defines

$$V_I^k = \sum_{i=1}^{n_I} V_I^{k,i} \tag{4.43}$$

and

$$V_I = \sum_{k=1}^{N_I} V_I^k \, , \tag{4.44}$$

the potential V' can be written as a sum of interactions $V_I(V_I^k)$ of the projectile with each subcluster of type I (center C_I^k), namely,

$$V' = \sum_{I=1}^{N} V_I = \sum_{I=1}^{N} \sum_{k=1}^{N_I} V_I^k \, . \tag{4.45}$$

It is now advantageous to work in the intermediate representation where one may take advantage of the fact that the eigenfunctions of H_T are known (or nearly known). In the intermediate representation the evolution operator $U(t, t_0)$ is governed by,

$$i\frac{dU(t,t_0)}{dt} = V'(t)U(t,t_0) \, , \tag{4.46}$$

where

$$V'(t) = e^{iH_T t} V' e^{-iH_T t} \, . \tag{4.47}$$

Here $V'(t)$ is not a sum of single electron (or single center) operators because H_T in (4.36) is not a sum of single electron (or single center) terms due to the correlation interactions between the electrons (or centers). Eq(4.40) may be formally solved using the time ordering operator T, namely,

$$U(t,t_0) = T \exp\left[-i\int_{t_0}^{t} V'(t)dt\right] \, . \tag{4.48}$$

The probability amplitude for transition electrons in the asymptotic initial state ϕ_i to the asymptotic final state ϕ_f of the molecule or cluster is found by projecting the full electronic wavefunction of (4.40) satisfying initial boundary conditions ψ_i^+ onto the asymptotic electronic wavefunction ϕ_f, namely,

$$A = \left\langle \phi_f \mid \psi_i^+ \right\rangle = \langle \phi_f \mid U(+\infty, -\infty) \mid \phi_i \rangle \, . \tag{4.49}$$

The probability $P(\vec{b})$ for a transition from ϕ_i to ϕ_f is given by the absolute square of A, and the corresponding cross section is found by a two dimensional integration over the impact parameter $\vec{b}$, namely,

$$\sigma = \int P(\vec{b})d\vec{b} = \int \mid A \mid^2 d\vec{b} \, . \tag{4.50}$$

This result holds for an arbitrary number of centers and an arbitrary number of electrons.

2 Independent subsystems

The independent subcluster approximation. In this subsection the independent subcluster approximation is developed where the wavefunction of the electron in each of the different subclusters evolves independently from the others.

Since $V' = \sum_I V_I$ from (4.45), the evolution operator from (4.48) might appear to factor into a product $\prod_I U_I$. However, $U(t, t_0)$ does not factor because H_T given by (4.36) contains interactions between the different subclusters. That is, the exact H_T generates dynamic correlation between subclusters. Specifically, the $H_{T,I}$ do not commute with the $W_{I,K}$ terms in (4.36). To obtain the independent subcluster approximation we neglect in H_T the interactions between the different subclusters, represented by $W_{I,K}$ in (4.36). When the $W_{I,K}$ are neglected, H_T reduces to a sum of commuting single cluster terms, namely,

$$H_T \cong \sum_{I=1}^{N} H_{T,I} \,. \tag{4.51}$$

With this approximation, one has, using (4.47),

$$V'(t) = e^{iH_T t} V' e^{-iH_T t} = \sum_{I=1}^{N} \left[e^{iH_{T,I}t} V_I e^{-iH_{T,I}t} \right] \equiv \sum_{I=1}^{N} V_I(t) \,, \tag{4.52}$$

where $V_I(t)$ now operates on a single subcluster.

Using (4.48) for the evolution operator, one now has,

$$\begin{aligned} U(t, t_0) &= T \exp\left[-i \sum_{I=1}^{N} \int_{t_0}^{t} V_I(t) dt \right] \\ &= \prod_{I=1}^{N} T \exp\left[-i \int_{t_0}^{t} V_I(t) dt \right] \\ &\equiv \prod_{I=1}^{N} U_I(t, t_0) \,. \end{aligned} \tag{4.53}$$

By neglecting the $W_{I,K}$ terms in the full Hamiltonian that interconnect the subclusters, the evolution operator, $U(t, t_0)$, has become a product of single subcluster evolution operators $U_I(t, t_0)$.

Now consider an event in which excitation occurs in the molecule or cluster in an interaction with the charged projectile. The asymptotic electronic wavefunction of the molecule or cluster, ϕ, is now a product of single subcluster wavefunctions, ϕ_I, i.e.,

$$\phi_{i,f} = \prod_{I=1}^{N} \phi_{Ii,f} \,. \tag{4.54}$$

Then, from (4.49) and (4.53) using the orthogonality of the ϕ_I and single cluster nature of the U_I operators, one has that,

$$A = \left\langle \prod_{I=1}^{N} \phi_{If} | \prod_{I=1}^{N} U_I(t,t_0) | \prod_{I=1}^{N} \phi_{Ii} \right\rangle \equiv \prod_{I=1}^{N} A_I \,. \tag{4.55}$$

Here A_I is the probability amplitude for a particular transition in the subcluster of type I. In the case that one can identify in which subcluster the transition occurs, (4.55) may also be obtained for ionization and electron capture processes.

The above result is stronger (i.e. has fewer approximations) than subsequent results. If at this point the A_I can be evaluated, then (4.50) can be used to find the final transition probabilities and cross sections for the molecule or cluster, and the further reduction to individual centers (as illustrated in figure 4.4) is not needed.

The independent center approximation. In this subsection the centers within each subcluster are decoupled. Also included is the optional possibility of expressing the probability amplitude, A_I, for each subcluster as a sum of products of the probability amplitudes, A_I^k, for electronic transitions on each constituent center, k. In this sum phases due to the translation between the centers are retained.

In order to obtain the independent center approximation we introduce an average potential, $\mathcal{V}(\vec{r}_I^{\,k,i})$, so that the Hamiltonian $H_{T,I}$ given by (4.37) results approximated by,

$$\begin{aligned} H_{T,I} &\cong \sum_{k=1}^{N_I} \left\{ -\frac{\nabla^2_{\vec{R}_I^k}}{2M_I} + \sum_{i=1}^{n_I} \left[-\frac{\nabla^2_{\vec{r}_I^{\,k,i}}}{2} + \mathcal{V}(\vec{r}_I^{\,k,i}) + \frac{1}{2} \sum_{\substack{l=1 \\ (l \neq i)}}^{n_I} \frac{1}{|\vec{r}_I^{\,k,i} - \vec{r}_I^{\,k,l}|} \right] \right\} \\ &\equiv \sum_{k=1}^{N_I} h_{I,k} \,, \end{aligned} \tag{4.56}$$

where the potential $\mathcal{V}(\vec{r}_I^{k,i})$ results from approximating the following terms,

$$\mathcal{V}(\vec{r}_I^{k,i}) \cong -\frac{Z_I}{r_I^{k,i}} + \frac{1}{2}\sum_{\substack{j=1\\(j\neq k)}}^{N_I}\left[-\frac{Z_I}{|\vec{R}_I^j - \vec{R}_I^k - \vec{r}_I^{k,i}|} + \sum_{\substack{l=1\\(l\neq i)}}^{n_I}\frac{1}{|\vec{R}_I^j + \vec{r}_I^{j,i} - \vec{R}_I^k - \vec{r}_I^{k,l}|}\right]. \tag{4.57}$$

Note that it is not necessary to include the nucleus-nucleus interaction between C_I^k and C_I^j in $\mathcal{V}(\vec{r}_I^{k,i})$ if the nuclei of the centers are regarded as frozen during the collision.

Then, the $h_{I,k}$ terms defined in (4.56) are indeed single center operators. Recalling that $V_I = \sum V_I^k$ from (4.44) one has (as in the previous section),

$$U_I(t,t_0) = \prod_{k=1}^{N_I} U_I^k(t,t_0), \tag{4.58}$$

where

$$U_I^k(t,t_0) = T\exp\left[-i\int_{t_0}^{t} V_I^k(t)dt\right]. \tag{4.59}$$

Now the centers evolve independently. At this point one could simply express the subcluster probability amplitude as a simple product of probability amplitudes for each of the constituent centers, i.e. $A_I = \prod_{k=1}^{N_I} A_{Ik}(\vec{R}_I^k)$.

Now consider optional symmetry effects due to identical, but otherwise uncorrelated centers. The basic effect of this symmetry is interference of scattering amplitudes analogous in some cases to classical scattering of waves by N_I identical centers. The wavefunction ϕ_I is written as,

$$\phi_I = S\prod_{k=1}^{N_I}\varphi_k(\vec{R}_I^k), \tag{4.60}$$

where $S = \frac{1}{\sqrt{N_I!}}\sum_Q \delta_Q Q$, is the general symmetry operator used by Goldberger and Watson [1964]. Q is the permutation operator; $\delta_Q = 1$ for the symmetric operator S and $\delta_Q = \pm 1$ if S is the antisymmetric operator. Also, in (4.60), $\varphi_k(\vec{R}_I^k)$ is a single center eigenfunction of the Hamiltonian $h_{I,k}$ of (4.56). For example, if $N_I = 2$ then $\phi_I = \frac{1}{\sqrt{2}}[\varphi_1(\vec{R}_I^1)\varphi_2(\vec{R}_I^2) \pm \varphi_1(\vec{R}_I^2)\varphi_2(\vec{R}_I^1)]$.

The probability amplitude A_I for a particular transition in the subcluster I results in,

$$\begin{aligned} A_I &= \left\langle \phi_{If} \mid \prod_{k=1}^{N_I} U_I^k(t,t_0) \mid \phi_{Ii} \right\rangle \\ &= \left\langle S_f \prod_{k'=1}^{N_I} \varphi_{k'f}(\vec{R}_I^{k'}) \mid \prod_{k=1}^{N_I} U_I^k(t,t_0) \mid S_i \prod_{k=1}^{N_I} \varphi_{ki}(\vec{R}_I^k) \right\rangle , \end{aligned} \quad (4.61)$$

where the index i and f refer to the initial and final wavefunctions respectively. Within the independent center approximation one has that only terms corresponding to the same $\vec{R}_I^k$ are non-zero. It is now assumed that the transition from state φ_{ki} to state φ_{kf} can be distinguished while one ignores in which center it has occurred. That is, off-diagonal exchange terms ($k \neq k'$) are neglected in the probability amplitude. Then, A_I results approximated by,

$$A_I \cong S \prod_{k=1}^{N_I} \left\langle \varphi_{kf}(\vec{R}_I^k) \mid U_I^k(t,t_0) \mid \varphi_{ki}(\vec{R}_I^k) \right\rangle \equiv S \prod_{k=1}^{N_I} A_{Ik}(\vec{R}_I^k) , \quad (4.62)$$

where S is the symmetric (antisymmetric) operator if the symmetry of S_i and S_f is the same (different).

Now consider the single center probability amplitude, $A_{Ik}(\vec{R}_I^k)$ corresponding to the transition $\varphi_{ki} \to \varphi_{kf}$. The amplitude A_I^k, evaluated at $\vec{R}_I^k$ and as a function of the impact parameter $\vec{b}$, is related to another amplitude evaluated at some other point, $\vec{R}_0$, in space by a phase due to translation in time [Shingal and Lin, 1989; Wang *et al.*, 1989], namely,

$$A_{Ik}(\vec{R}_I^k) = e^{i\delta_I^k} A_{Ik}(\vec{R}_0) , \quad (4.63)$$

where

$$\delta_I^k = q_z^k (R_{Iz}^k - R_{0z}) , \quad (4.64)$$

with q_z^k the z-component of $\vec{q}$ the momentum transferred to the projectile in the transition $\varphi_{ki} \to \varphi_{kf}$. The z axis is taken parallel to the velocity of the incoming projectile at large distances. Also, $(R_{Iz}^k - R_{0z})$ is the z-component of $(\vec{R}_I^k - \vec{R}_0)$. Here one chooses the $\vec{R}_0$ as the center of mass of the molecule and set $\vec{R}_0 = 0$. For a given center C_I^k the impact parameter in $A_{Ik}(0)$ is the impact parameter of the projectile relative to the nucleus of C_I^k, that is $\vec{b}_I^k = \vec{b} - \vec{R}_{I\perp}^k$ where $\vec{R}_{I\perp}^k = \vec{R}_I^k - \vec{R}_{Iz}^k$. For heavy projectiles with $M \gg m_e$ one has, $q_z^k \cong q_{min}^k$ where Q_{min}^k is the minimum momentum transferred to the projectile. If an excitation process takes place in center C_I^k, one has that $q_{min}^k = \Delta E^k/2\mathrm{v}$ where ΔE^k is the energy gain of the electrons in the transition $\varphi_{ki} \rightarrow \varphi_{kf}$ and in the electron capture case $q_{min}^k = \mathrm{v}/2 - \Delta E^k/\mathrm{v}$ as given by McDowell and Coleman [1970].

From (4.62) and (4.63) one obtains,

$$A_I = S \prod_{k=1}^{N_I} e^{i\delta_I^k} \; A_{Ik}(\vec{b}_I^k) = S \prod_{k=1}^{N_I} \exp[\, i \; q_{min}^k \; R_{Iz}^k \,] \, A_{Ik}(\vec{b}_I^k) \, . \tag{4.65}$$

Eq(4.65), the main result of this subsection, leads to the geometric structure factor, discussed below. Here A_{Ik} is the probability for a transition within the the kth center of a subcluster of type I. In the limit that the symmetry of the centers disappears (or becomes unimportant), one obtains the usual product of probability amplitude for independent centers, namely, $A_I \rightarrow \prod_{k=1}^{N_I} e^{i\delta_I^k} A_{Ik}(\vec{b}_I^k)$.

The independent electron approximation. In this subsection the correlation between electrons on each independent center is removed and the independent center independent electron approximation is obtained.

As is done in section 4.1.2 one introduces a mean field potential V_{ef} so that the single center Hamiltonian $h_{I,k}$ given by (4.56) is approximated by,

$$h_{I,k} \cong \sum_{i=1}^{n_I} \left[-\frac{\nabla^2_{\vec{r}_I^{k,i}}}{2} + V_{ef}(\vec{r}_I^{k,i}) \right] \equiv \sum_{i=1}^{n_I} h_{I,k}^i \, , \tag{4.66}$$

where $V_{ef}(\vec{r}_I^{k,i})$ gives a mean field approximation to the non-local electron-electron interactions, namely,

$$V_{ef}(\vec{r}_I^{k,i}) \cong \mathcal{V}(\vec{r}_I^{k,i}) + \frac{1}{2} \sum_{\substack{l=1 \\ (l \neq i)}}^{n_I} \frac{1}{\mid \vec{r}_I^{k,l} - \vec{r}_I^{k,i} \mid} \, . \tag{4.67}$$

In (4.66) the kinetic energy of the nucleus center has been neglected. This is valid in high velocity collisions for heavy projectiles where the collision is sufficiently fast so that the centers are effectively frozen in place during the collision.

From (4.43) and (4.48), and using the fact that $h_{I,k}^i$ are single electron Hamiltonian terms, the evolution operator given by (4.59) reads,

$$U_I^k(t, t_0) = \prod_{i=1}^{n_I} U_I^{k,i}(t, t_0) \, , \tag{4.68}$$

where

$$U_I^{k,i}(t,t_0) = T \exp\left[-i \int_{t_0}^{t} V_I^{k,i}(t)dt\right] . \tag{4.69}$$

Now the Hamiltonian of center C_I^k given by (4.66) is a sum of independent terms for each electron. Then, the electronic wavefunction for this center φ_k is a product of wavefunctions ϕ_k^i for each electron. As a consequence, the probability amplitude A_{Ik} of (4.62) is a product of single electron probability amplitudes a_{Ik}^i, namely,

$$A_{Ik} = \prod_{i'=1}^{n_I} \left\langle \phi_{kf}^{i'} | U_I^{k,i'}(t,t_0) | \phi_{ki}^{i'} \right\rangle \equiv \prod_{i=1}^{n_I} a_{Ik}^i . \tag{4.70}$$

In principle, the (anti)symmetry factor, $\mathcal{S}$, is not included in (4.70). The effects of the exchange symmetry of electrons in atomic collisions has been considered in section 4.1.1. At high collision velocities these exchange effects are often small and may then be neglected. It is often the case in large many electron systems that there are 'passive' electrons which are not relevant to the processes under study. If these 'passive' electrons are decoupled from the 'active' electrons under consideration, then the total probability for all possible final states of the decoupled passive electrons sums to unity and their presence may be neglected (Cf. section 3.3.5).

The geometric structure factor. The physical nature of the phase terms in (4.63) may be more easily understood by transforming from the impact parameter representation to the wave picture. The probability amplitude, $A(\vec{b})$, is generally related to the transition matrix, $T(\vec{q})$, by the relationship, $A(\vec{b}) = \mathrm{v}^{-1} \int e^{i\vec{q}_\perp \cdot \vec{b}}\, T(\vec{q})\, d\vec{q}_\perp$, where $\vec{q}_\perp$ is the component of the momentum transfer, $\vec{q}$, perpendicular to the asymptotic velocity of the incoming particle. Following this transformation, the amplitude A_I of (4.65) is given by,

$$T_I = T_I(0) \left(\mathcal{S} \sum_{k}^{N_I} e^{-i\vec{q}\cdot\vec{R}_I^k} \right) , \tag{4.71}$$

where $T_I(0)$ is the transition amplitude for a single center, C_I, located at $\vec{R}_0 = 0$. The cross section, $d\sigma$, differential in $\vec{q}$, is proportional to the square of T, so that,

$$d\sigma_{N_I} = d\sigma_I | \sqrt{N_I}\mathcal{S} \sum_{k=1}^{N_I} e^{-i\vec{q}\cdot\vec{R}_I^k} |^2 \equiv d\sigma_I G_{N_I} , \tag{4.72}$$

where G_{N_I} is the geometrical structure factor containing interference between the N_I identical centers and $d\sigma_I$ is the differential cross section for scattering from a single center, C_I. In those cases when $\mathcal{S}$ uses only plus operators (e.g. the spatial gerade wavefunctions of H_2), then $G_{N_I} = N_I + \sum_{k=1}^{N_I} \sum_{j>k} cos(\delta_I'^j - \delta_I'^k)$, where $\delta_I'^k = \vec{q} \cdot \vec{R}_I^k$. At this point (4.72) is equivalent to (4.50) using (4.63) - (4.65) because the integral of $|T(\vec{q}_\perp)|^2$ over $\vec{q}_\perp$ gives the same total cross section as the integral of $|A(\vec{b})|^2$ over $\vec{b}$. If the transverse momentum transfer, $\vec{q}_\perp$, is neglected (e.g. $q_\perp << q_z$), then $\delta_I'^k = \delta_I^k$ given by (4.64) and the approximations that follow. The factor G_{N_I} is the same as the well known geometrical structure factor [Kittel, 1956, p. 54] obtained for classical scattering of waves from N_I identical centers. Such a structure factor has also been used to analyze neutron scattering from heavy nuclei [Placzek *et al.*, 1951], where the cases of ideal lattices, crystals with thermal disorder and liquids are considered. A related analysis has been done for gluons [Duke and Owens, 1984]. It is easily shown (for $\mathcal{S}$ as plus) that as $\vec{R}_I^k \to 0$, $G_{N_I} \to N_I^2$; and as $\vec{R}_I^k \to \infty$, $G_{N_I} \to N_I$. The first limit is fully coherent and the latter is fully incoherent.

There are some additional points concerning these phases to be noted. First, addition of phases is consistent with the addition of energies. In the case of independent multiple transitions on a single center the probability amplitude is a product of independent amplitudes (e.g. (4.70)) so that the total phase is the sum of the individual phases. Each phase is linear in the transition energy as discussed after (4.64). The total transition energy is the sum of the individual transition energies. So the total phase for multiple transitions must be the sum of the individual transition phases in the independent particle approximation. Second, if the mass, M_I^k, of each of the centers is the same, then $\sum_k \delta_I^k = q_z \sum_k (R_{Iz}^k - R_{0z}) = 0$ since $\sum_k M_I^k (\vec{R}_I^k - \vec{R}_0) = 0$ because $\vec{R}_0$ is the center of mass. If some of the masses differ (e.g. due to different isotopes), then $\sum_k \delta_I^k$ sums to a overall non-zero constant phase. Third, phases occur because symmetrization gives a superposition of single electron atomic wavefunctions. Any superposition may lead to phase contributions. Finally, it is noted that in most cases where multiple transitions occur it is easier to approximately evaluate $A(\vec{B})$ than $T(\vec{q})$ using this method because $T(\vec{q})$ is not a simple product of simpler terms.

Non-localized projectiles. Interaction of a molecule or cluster with a heavy charged point particle was considered above. In many cases of physical and chemical interest, the projectile may carry electrons. In these cases rigorous application of these methods is more difficult because the classical trajectory of the incoming projectile, introduced above (4.42), is not well defined since the wavelength of the projectile electrons may be comparable to the size of the interaction region. Also, application of the independent electron approximation is difficult, especially for neutral, or nearly neutral projectiles, because often it is not sensible to neglect the interaction between the electrons on the target with the electrons on the projectile since this interaction leads to screening of the nuclear charge Z of the projectile by the n projectile electrons which may be significant. This problem is considered in sections 8.1 - 8.2. In first order perturbation theory in the wave picture one may simply replace the square of the projectile nuclear charge Z^2 by the square of an effective charge $Z^{eff\ 2}(\vec{q})$ which depends on the momentum transfer $\vec{q}$ of the projectile. For small $\vec{q}$ which corresponds to collisions at distances large compared to the distance of the projectile electrons from their nucleus, $Z^{eff\ 2} \rightarrow (Z-n)^2$, so that the projectile acts as a point particle of charge $(Z-n)$. For large $\vec{q}$ where the projectile electrons are well separated from the projectile nucleus, $Z^{eff\ 2} \rightarrow Z^2 + n$, which corresponds to independent scattering by the various charged particles on the projectile. Both limits are physically sensible. In particular, for one electron projectile ions in their ground state, $Z^{eff\ 2} = Z^2 + 1 - 2Z/(1 + \frac{q}{2Z}^2)^2$.

A relatively simple model containing the same physically sensible limits is easily obtained by making a classical transformation from the momentum transfer $\vec{q}$ to the impact parameter $\vec{b}$ and using $Z^{eff}(\vec{b})$ in place of the corresponding bare charge Z in the probability amplitude of (4.50) and the approximations that follow. It should be cautioned, however, that the correct scattering probability may not exceed unity, and also that a rigorous first order amplitude (Cf. section 8.2) is given by a convolution over virtual impact parameters $\vec{b}'$ of the probability amplitude for a point charge $A(\vec{b}')$ over $Z_P^{eff}(\vec{b}' - \vec{b})$.

Extension of these methods to projectiles that themselves have multiple centers may also be possible. In limiting cases where the projectile centers are well separated or very compact, it is valid to neglect interactions between centers on the target and interactions between centers on the projectile and to include interactions between centers of the target and the projectile. Symmetry between like centers on the projectile and target may also be included.

Summary.

A molecule or cluster with $N_C = \sum_I^N N_I$ centers, C_I is denoted by $C_1^{N_1} C_2^{N_2} ... C_I^{N_I} ... C_N^{N_N}$, where $C_I^{N_I}$ denotes a subcluster of N_I identical centers, C_I. The key approximations are as follows.

1. *The independent subcluster approximation*: The interaction between different subclusters is neglected so that $A \approx \prod_I^N A_I$.

2. *Coherent scattering from identical centers within subclusters*: Scattering of the wavefronts of the projectile from multiple identical independent centers is included within each subcluster so that $A_I = \mathcal{S} \prod_{k=1}^{N_I} e^{i\delta_I^k} A_{Ik}(\vec{b}_I^k)$, where $\mathcal{S}$ is the symmetry operator. The δ_I^k phase terms are related to the usual geometrical structure factor.

3. *The independent electron approximation*: The correlation between electrons within each center is neglected so that $A_{Ik} \approx \prod_{i=1}^{n_I} a_{Ik}^i$, and $\delta_I^k \approx \sum_i^{n_I} \delta_I^{ki}$.

The independent center independent electron approximation (ICEA), in the absence of the coherence terms in step 2 above, corresponds in the non-relativistic limit to a simple generalization of the static Hartree product wavefunction to a dynamic probability amplitude using a single time independent basis set, $\{\psi_{el} = \prod_j \psi_{(single\ orbital)_j}\}$ with frozen electron orbitals.

Application of this method to scattering of H_2 by fast ions and some comparison to observations of Coulomb explosions of molecule has been done using this method [Wang *et al.*, 1989; Wang, 1992].

4.3 Time dependent Dirac Fock

The uncorrelated limit in atomic scattering is most often defined to be the time dependent Dirac Fock (TDDF) approximation where the correlation potential v of (4.7) is neglected. In this limit the $1/r_{ij}$ Coulomb interactions between electrons are included within the the approximation of a single Slater determinant ansatz, which retains the antisymmetrization of the many electron wavefunction [Löwdin, 1959, 1995]. The TDDF is the extension to time dependent cases of the static relativistic Dirac Fock (DF) approximation commonly used as the uncorrelated limit for static wavefunctions. The non-relativistic limits are the time dependent Hartree Fock (TDHF) and Hartree Fock (HF) approximations.

TDHF and TDDF calculations in atomic scattering [Kuerpick *et al.*, 1994,1995; Schaut *et al.*, 1991; Kulander, 1987a,b; Gramlich *et al.*, 1986; Webster *et al.*, 1978; Froese Fischer, 1996, 1977] are solutions of the time-dependent Schrödinger or Dirac equation of the scattering system in the Hartree Fock or Dirac Fock limit. Thus the $1/r$ Coulomb interaction between the electrons is fully taken into account, but the wavefunction is chosen as a time dependent single Slater determinant only. This many particle wavefunction,

$$\Psi^{TDDF}(\vec{r}_1, \vec{r}_2, ..., \vec{r}_N, t) = \frac{1}{\sqrt{N!}} \begin{vmatrix} \psi_1(\vec{r}_1, t) & \dots & \psi_N(\vec{r}_N, t) \\ \psi_1(\vec{r}_2, t) & \dots & \psi_N(\vec{r}_N, t) \\ \cdot & & \cdot \\ \cdot & & \cdot \\ \psi_1(\vec{r}_N, t) & \dots & \psi_N(\vec{r}_N, t) \end{vmatrix}, \quad (4.73)$$

is build out of a set of time dependent, single particle wavefunctions ψ_i which are solutions of the single particle time-dependent TDHF or TDDF equations which (in case of TDDF) read,

$$\begin{aligned} h^{TDDF}\psi_i &= \left(K + V^N(\vec{r}_1, t) + V^C(\vec{r}_1, t)\right)\psi_i \\ &+ \int V^{exchange}(\vec{r}_1, \vec{r}_2, t)\psi_j(\vec{r}_2, t)d\vec{r}_2\psi_i(\vec{r}_1, t) \\ &= i\frac{\partial}{\partial t}\psi_i \qquad i = 1, 2, ..., N. \end{aligned} \quad (4.74)$$

Here the h^{TDDF} is the single particle relativistic Hamiltonian. K is the relativistic kinetic operator, V^N is the electron nuclear interaction, V^C is the direct part of the full Coulomb interaction between electrons, and $V^{exchange}$ is the exchange part of the full Coulomb interaction between electrons.

The advantage of this ansatz of a Slater determinant is the fact that the wavefunction is fully antisymmmetrized as it should be and the Coulomb interaction is fully taken into account. The only part which is missing is the part which would come in a complete calculation by using a sum of Slater determinants to describe the total many particle wavefunction. But this part is usually expected to be small for large atoms. Only in two electron systems like He is a large effect expected. The difference of the single Slater-determinant ansatz and the many Slater-determinant ansatz is what people in quantum chemistry usually call correlation [Löwdin, 1959, 1995; Desjardins *et al.*, 1995].

In practice actual solutions of these TDHF or even better TDDF equations are difficult to evaluate. To this time only the density functional community uses these equations (4.74), but they solve them in terms of the density functional approximation, where they use expressions for the total energy of the scattering systems in terms of the various density functionals [Perdew *et al.*, 1992; Dreizler and Gross, 1990].

Kuerpick *et al.* [1994,1995] have solved the single particle TDDF equations (4.74) and then used this information to obtain the many particle wavefunction (4.73) incorporating the useful method of inclusive probabilities of Lüdde and Dreizler [1985 (Cf. sections 3.3.5 and 4.1). A difficulty in these calculations is with the Dirac Fock equations, which for a quasi-molecular scattering system are hard to solve. Here the exchange part of the electron-electron Coulomb interaction is approximated by a local density functional which is taken to be proportional to $\rho^{1/3}$ where ρ is the electron density. Surprisingly, this approximation is thought to be even better than the Hartree or Dirac Fock because part of the residual correlation interaction due to a many Slater-determinant ansatz is already included. Continuum states in TDDF and TDHF are often not easy to evaluate and also are not easy to separate from bound states. These uncertainties in the uncorrelated limit generally restrict the precision with which correlation may be evaluated.

In actual calculations [Kuerpick *et al.* 1995] the molecular orbitals (which are the basis functions for the time dependent calculations at each internuclear distance) are calculated with a fixed number of electrons in the quasi-molecular orbitals. This approximation is reasonable for a system with many electrons and heavy nuclei (such as $Ar^{+12} + Ca$). However, the error is expected to increase in systems with smaller nuclear charges. In these smaller systems changes in the occupation of the molecular levels are expected to become more significant [Kuerpick *et al.* 1995].

5
Statistical methods

Quantum descriptions of many electron transitions rapidly tend to become impractical as the number of transitions increases because the of large number of coupled quantum states that one must consider. The difficulty of dealing with the dynamics of many quantum states is apparent from the fact it is not at present possible to describe two electron transitions in helium for impact by highly charged particles at intermediate and low velocities where the cross sections are relatively large. Statistical methods are sometimes effective for describing the many body problem in general and many electron transitions in particular. Also of utility are optical potential methods [Dreizler *et al.*, 1986], density functional techniques [Perdew *et al.* 1992; Dreizler and Gross, 1990; Kohland and Dreizler, 1986] and fluid dynamic methods [Horbatsch and Dreizler, 1986].

5.1 Statistical energy deposition model

In the statistical energy deposition (SED) model developed by Russek [1963], it is assumed that energy E_T obtained by an atomic (or molecular) target during a collision is statistically distributed to all of the N electrons in the system. One simply counts the number of ways that E_T can be distributed among the N equivalent electrons. If any electron has an energy greater than the ionization energy E_{ion} then it is ionized. P_N^n is the probability function for ionizing n of the N electrons. The following mathematical development is largely due to Tulkki [1987].

According to Tulkki [1987],

$$P_N^n(m) = \binom{N}{n} \sum_{j=0}^{m-rn} K_n(j)\, Q_{N-n}^r\,(M - rn - j)/K_n(m) \qquad (5.1)$$

gives the probability that when m units of energy are statistically distributed among N electrons, n and only n electrons will receive r units of energy each allowing their ionization. In (5.1)

$$K_\mu(\nu) = \binom{\mu+\nu-1}{\nu} \quad (\nu \geq 0)\,, \tag{5.2}$$

$$K_\mu(\nu) = 0 \quad (\nu < 0)\,,$$

$$Q_\mu^\lambda(\nu) = \sum_{i=0}^{[\frac{\nu}{\lambda}]} (-1)^i \binom{\mu}{i} K_\mu(\nu - i\lambda)\,, \tag{5.3}$$

where $[\frac{\nu}{\lambda}]$ is the integer part of $\frac{\nu}{\lambda}$.

The task is to determine

$$P_N^n(E_T/E_{ion}) = \lim_{\substack{m\to\infty \\ r\to\infty}} P_n^{(N)}(m)\,, \tag{5.4}$$

such that $x = m/r = E_T/E_{ion}$ stays finite. Tulkki observed that the summation over j can be carried out exactly. Substituting (5.2) and (5.3) in (5.1) and observing from (5.2) that $j \leq m - rn - ri$ gives,

$$P_N^n(m) = \frac{\binom{N}{n}}{\binom{m+N-1}{m}} \sum_i (-1)^i \binom{N-n}{i} L_n^{(N)}(m, r; i)\,, \tag{5.5}$$

where

$$L_n^{(N)}(m, r; i) = \sum_{j=0}^{m-rn-ri} \binom{j+n-1}{j} \times \binom{m-rn-j-ri+N-n-1}{m-rn-j-ri}\,. \tag{5.6}$$

Using the abbreviations,

$$\begin{aligned} a &= n-1\,, \\ b &= m-rn-ri+N-n-1+a\,, \\ c &= N-n-1+a\,. \end{aligned} \tag{5.7}$$

Eq(5.6) can be written as,

$$\sum_{j=0}^{b-c} \binom{j+a}{j} \binom{b-a-j}{c-a} = \binom{b+1}{c+1} . \tag{5.8}$$

This sum rule is given by Vilenkin [1971]. Thus (5.5) reduces to,

$$P_N^n(m) = \frac{\binom{N}{n} \sum_{i=0}^{[\frac{m-rn}{r}]} (-1)^i \binom{N-n}{i} \binom{m-rn-ri+N-1}{N-1}}{\binom{m+N-1}{N-1}} , \tag{5.9}$$

which is exact. Now let m and r go to infinity so that,

$$[(m-rn)/r] \to [x-n] \qquad (x-n-1 < i_{max} \leq x-n) \tag{5.10}$$
$$m - rn - ri + n - 1 = r(x - n - i + \tfrac{N-1}{r})$$

for all i except for $i = x - n$ which gives a negligible contribution if it occurs.

Using the formula,

$$\lim_{\substack{\mu \to \infty \\ \nu \text{ fixed}}} \binom{\mu+\nu}{\nu} = \frac{\mu^\nu}{\nu!} \tag{5.11}$$

in (5.9) yields the Russek formula

$$P_N^n(x) = \binom{N}{n} \sum_{i=0}^{k} (-1)^i \binom{N-n}{i} \left(1 - \tfrac{n+i}{x}\right)^{N-1} , \tag{5.12}$$

$$x = E_T/E_{ion} , \tag{5.13}$$

where k must be chosen such that

$$n + k \leq x < n + k + 1 . \tag{5.14}$$

Eq(5.13) implies that the energy has been shared statistically between the $N-n$ electrons which remain and the n electrons which are ejected. Thus it corresponds to a distribution where the ions are left in excited states. Since decay of excited states often occurs by autoionization, the SED distribution is not usually observed experimentally. Hence, justification of this simple method is incomplete [Aberg *et al.*, 1984].

Tolstikhin's derivation. Tolstikhin* has derived the SED result using

---

* Contributed to this book by O. Tolstikhin.

the microcanonical distribution of electrons in the intermediate multiply excited atomic state, which is formed in the collision and which further decays via ejecting of a number of electrons and/or photons. The presence of such an intermediate state in the collisional process is a key point of the energy deposition model. The assumption of the statistical equilibrium achieved in this state, thus the application of the microcanonical distribution to its description, is consistent with the original ideas of Russek [1963].

Following Tolstikhin, let ΔE be the total excitation energy deposited to the electronic subsystem of the target atom during the collision. Then, according to the microcanonical distribution the probability dW of formation of the state, where i-th electron has gained an excitation energy within the interval from ϵ_i to $\epsilon_i + d\epsilon_i$, is given by,

$$dW = A\delta\left[(\epsilon_1 + \epsilon_2 + \ldots + \epsilon_N) - \Delta E\right] d\epsilon_1 d\epsilon_2 \ldots d\epsilon_N \,, \tag{5.15}$$

where N is the number of atomic electrons, and A is a normalization constant defined by,

$$A \int_0^\infty \delta\left[(\epsilon_1 + \epsilon_2 + \ldots + \epsilon_N) - \Delta E\right] d\epsilon_1 d\epsilon_2 \ldots d\epsilon_N = 1 \,. \tag{5.16}$$

The range of integration in this normalization integral extends over all non-negative values of the excitation energies ϵ_i, including both excitations to discrete and continuum states. There is a simple way to calculate integrals of the type (5.16) which will be used further in derivation of (5.21), namely,

$$\begin{aligned} &\int_0^\infty d\epsilon_1 d\epsilon_2 \ldots d\epsilon_N \delta\left[(\epsilon_1 + \epsilon_2 + \ldots + \epsilon_N) - \Delta E\right] \qquad (5.17) \\ &= \frac{1}{2\pi} \int_{-\infty}^\infty d\tau \int_0^\infty d\epsilon_1 d\epsilon_2 \ldots d\epsilon_N \exp\left\{i\left[(\epsilon_1 + \epsilon_2 + \ldots + \epsilon_N) - \Delta E\right]\tau\right\} \\ &= \frac{1}{2\pi} \int_{-\infty}^\infty d\tau \left(\frac{i}{\tau + i\eta}\right)^N \exp\{-i\Delta E\tau\} = \Theta(\Delta E)\frac{\Delta E^{N-1}}{(N-1)!} \,. \end{aligned}$$

with $\eta \to 0$. Thus, for the normalization constant A in (5.16) one has,

$$A = \frac{(N-1)!}{\Delta E^{N-1}} \,. \tag{5.18}$$

Now, having the normalized distribution function (5.15), one can calculate probabilities of different processes as integrals over the corresponding regions of the phase space.

Table 5.1. *Comparison of IEA and SED probability distributions for n = 6.*

	Max P_6^n	
n	IEA	SED
1	40%	70%
2	33%	59%
3	31%	43.5%

For example, following Russek's development one may define the probability $P_N^n(\Delta E)$ of n-electron ionization as a probability to find the atom in a state, where any n and only n electrons have excitation energies more than I, while excitation energy of each of the remaining $(N-n)$ electrons is less than I. With this definition one has,

$$P_N^n(\Delta E) = A\binom{N}{n}\int_I^\infty d\epsilon_1 \ldots d\epsilon_n \quad (5.19)$$
$$\times \int_0^I d\epsilon_{n+1}\ldots d\epsilon_N \delta\left[(\epsilon_1+\epsilon_2+\ldots+\epsilon_N)-\Delta E\right] .$$

Calculating this integral one obtains,

$$P_N^n(\Delta E) = \quad (5.20)$$
$$\binom{N}{n}\sum_{k=0}^{N-n}\Theta\left(\Delta E-(n+k)I\right)(-1)^k\binom{N-n}{k}\left[1-\frac{n+k}{\Delta E/I}\right]^{N-1},$$

which coincides with the result of (5.12).

The binomial $P_N^n(b)$ and the SED $P_N^n(\Delta E_m)$ are compared in figures 5.1 and 5.2. The impact parameter b and energy transfer ΔE_m may in some cases be related classically [Cocke, 1979]. In any case the binomial and SED P_N^n are not the same. For example, the maximum values of P_N^n differ for $0 \neq n \neq N$ as seen in Table 5.1.

The SED distribution tend to be more successful at low collision velocities, while the binomial distributions are more often successful at high collision velocities (Cf. section 4.1.3).

In chemistry the basic concept of SED has been modified by using quantum transition probabilities (e.g. harmonic oscillator probabilities for vibrational transitions in molecules). This modifies the P_N^n probability functions. This procedure [Rosenstock *et al.*, 1952] goes by the name of quasi-equilibrium theory.

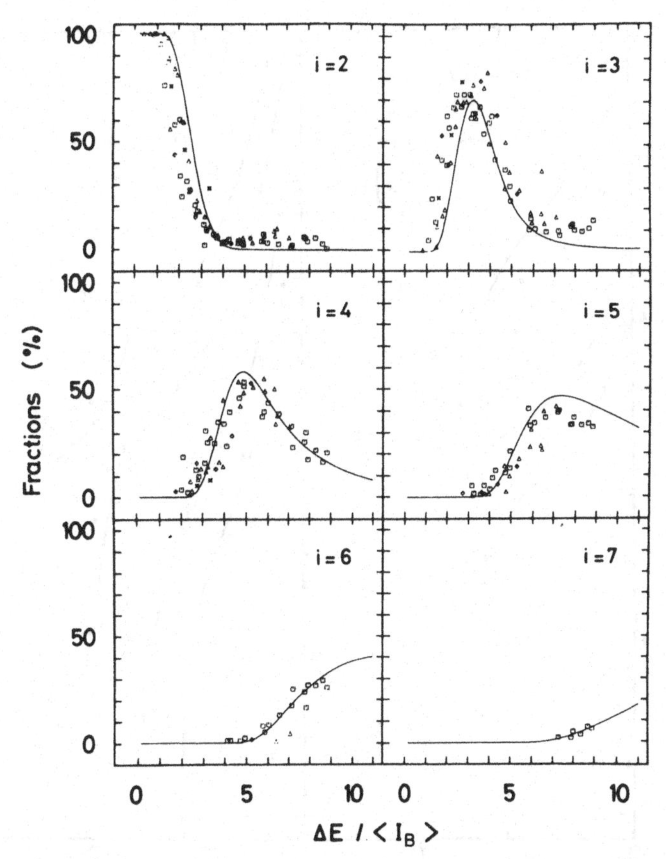

Fig. 5.1. The probability, P_N^i, for removing i of N electrons for $N = 8$ in the statistical energy deposition model [Müller *et al.*, 1985] as a function of the energy transfer, ΔE, into the system.

5.2 Fokker Planck method

The Fokker Planck method [Balescu, 1975] is a statistical method, widely used in nuclear physics, for analyzing distributions of events as a function of time. This method may also be applied to atomic collisions. In an ion-atom collision the atomic charge can change during a collision according to a probability distribution $W(N,t)$ which changes during the time of the collision $0 < t < T$. Initially the number of electrons $N = N_o$ is well defined, corresponding to $W(N, t \leq 0) = \delta(N - N_o)$. During the collision the width of the distribution, Γ, increases (Cf. figure 5.3). When the collision is over and the systems have readjusted, $W(N, t > T)$ no longer changes. This is the basic picture which directly corresponds to observed data in some cases.

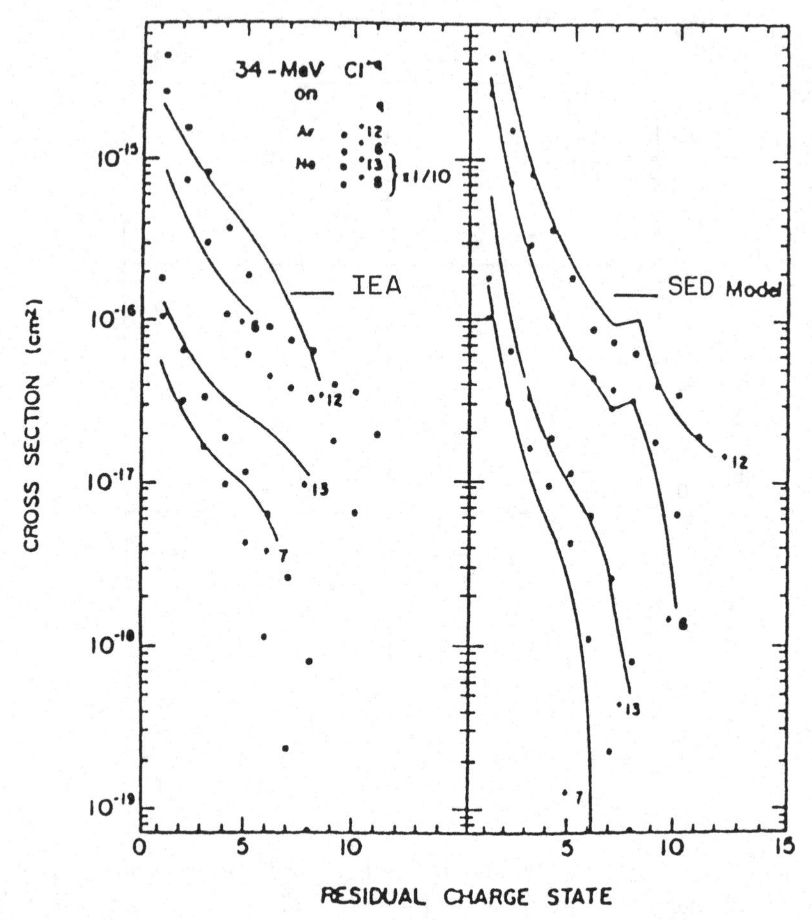

Fig. 5.2. Comparison of multiple ionization data to a binomial distribution (IEA) and to the statistical energy deposition model (SED) [Cocke, 1979]. The data is for 34 MeV Cl^{+q} on Ar and Ne targets for different values of the charge $+q$. Both IEA and SED models fit this data comparably well.

1 Fokker Planck equation

The equation governing the probability distribution $W(N,t)$ is the Fokker Planck equation,

$$\frac{dW(N,t)}{dt} = \frac{-d}{dN}\left[\mathrm{V}(N,t)\, W(N,t)\right] + \frac{d^2}{dN^2}[D(N,t)\, W(N,t)]\,. \quad (5.21)$$

This is a diffusion equation where V is a velocity coefficient and D is a diffusion coefficient. This equation is the same as the Schrödinger Equation $i\, d\Psi/dt = H(-\vec{\nabla}, \nabla^2)\Psi$ with time rotated by 90° in the complex plane.

Eq(5.21) corresponds to (92) of Balescu who gives a clear derivation and explanation, with $W = W_1$, $N = y$, $\mathrm{V} = A$, and $D = 1/2B$. The basic assumption is that the process is a Markov process: a statistical process where in each step information regarding previous steps is lost (i.e., the memory time is short). The first term on the right hand side of (5.21) corresponds to a frictional slowing corresponding, as is later illustrated, to the fact that the system is losing electrons. The second term corresponds to diffusion, or the time rate of change of the width Γ in figure 5.3. Note that if V and D go to zero, then $dW/dt = 0$ and W remains frozen in time.

Mean values of physical observables may be determined from W, e.g.,

$$\begin{aligned} < N^n > &= \int dN \, N^n \, W(N,t) \qquad \text{(moments of } N\text{)} \,, \qquad (5.22) \\ < E > &= \int dN \, E_N \, W(N,t) \qquad (E \text{ is energy, for example}) \,. \end{aligned}$$

The following mathematical properties are noted:

$$\begin{aligned} W(N,t) &\geq 0 \qquad \text{(non-negative probability distribution)} \,, \\ \int dN \, W(N,t) &= 1 \qquad \text{(conservation of total probability)} \,. \qquad (5.23) \end{aligned}$$

Finally, V and D are related to the first two moments (corresponding to a linear second order partial differential equation) via

$$\mathrm{V} = \frac{d}{dt} < N > \quad , \qquad (5.24)$$

$$D = \frac{d}{dt} \{< N^2 > - < N >^2\} \quad . \qquad (5.25)$$

2 Application to atomic collisions

The statistical distribution function W may be applied to atomic collisions by relating V and D to atomic collision quantities. There are various ways in which this may be done. For example one may introduce transition rates and Fermi-Dirac distributions for colliding atomic systems, as has been done in nuclear collisions. An alternate method, used here, is to consider one shell of an atom that is losing electrons (e.g., via direct Coulomb ionization) with a single electron probability, p, that is the same for all electrons within that atomic shell.

With this picture in mind, one may express V and D in terms of the single electron transition probabilities, $P(b,t)$ by using the definitions of V and D from (5.24) and (5.25). Noting that $n+N=N_o$ where n=number of vacancies, N=number of electrons and N_o=total number of states in the atomic shell it is easily shown that,

$$\frac{d}{dt} < N >= \frac{-d}{dt} < n > , \tag{5.26}$$

and

$$\frac{d}{dt}\{< N^2 > - < N >^2\} = \frac{d}{dt}\{< n^2 > - < n >^2\} . \tag{5.27}$$

If one uses, for example, a binomial distribution for the probability of producing n vacancies from N electrons $P_N^n = \binom{N}{n} P^n(1-P)^{N-n}$, it is straightforward to show that,

$$\begin{aligned} < n > &= NP , \\ < n^2 > - < n >^2 &= NP\,(1-P) , \end{aligned} \tag{5.28}$$

where $P = P(\vec{b},t)$ is a single electron probability for an atomic transition. Then,

$$\begin{aligned} \mathrm{V} &= -d/dt\ < n >= N\ dP/dt , \\ D &= +d/dt\ \{< n^2 - < n >^2 >\} = N\ dP/dt(1-2P) . \end{aligned} \tag{5.29}$$

Now V and D are known functions of N and t. They may be used in the Fokker Planck equation (5.21) to find $W(N,t)$. Note that after the collision, V and D go to zero, so that $W(N,t>T)$ does not change.

Next, let us consider some limitations of this method. The Fokker Planck equation assumes that a Markov process is valid, i.e., the system forgets its history before the collision. This implies that the collisions are well separated in space and time holds. From the atomic point of view, each shell is treated independently, each $\vec{b}$ is treated independently, and $P(t)$ may be computed accurately. Also, electrons only leave, and do not return. (This can be corrected by a more complicated model with particles moving back and forth, as has been done in nuclear physics.) Since the Einstein relation is not used, W does not go over to a Maxwellian distribution (Cf. Balescu, 1975. This may not always be a limitation). It is also required that $N_o >> 1$, so that $\int_0^{No} dN\ W(N,t) \simeq \int_0^\infty dN\ W(N,t)$ or $\int_{No}^\infty dN\ W(N,t) << 1$. Note

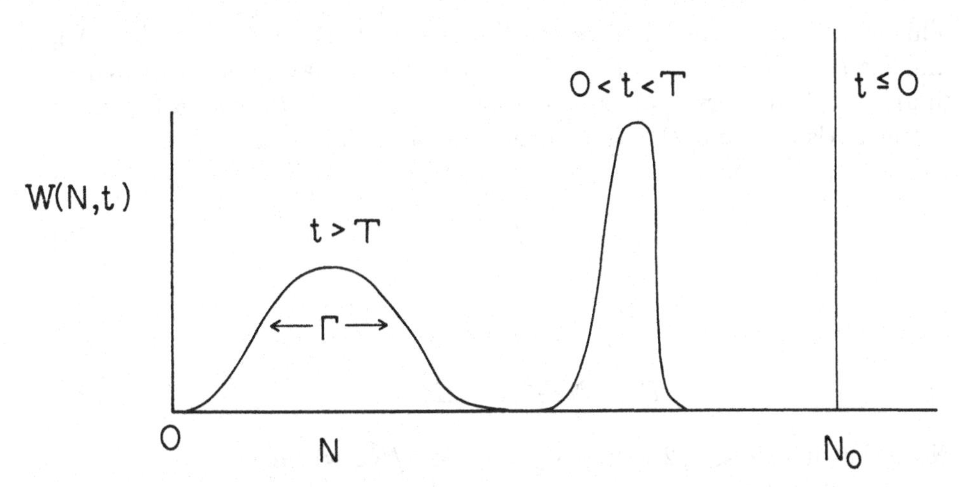

Fig. 5.3. The Fokker Planck distribution function $W(N,t)$ before, during and after a collision occurring during time T.

that $1 = \sum_n \binom{N}{n} p^n(1-p)^{N-n} = \int_0^\infty dN\ W(N,t)$ corresponds to conservation of total probability.

3 Simple model

If V and D are constant in (5.21), then an analytic solution is known to the Fokker Planck equation. Specifically, consider

$$\mathrm{V} = \begin{cases} -\mathrm{V}_o & 0 < t < T \\ 0 & \text{otherwise} \end{cases},$$

$$D = \begin{cases} D_o & 0 < t < T \\ 0 & \text{otherwise} \end{cases}. \tag{5.30}$$

The Fokker Planck equation then becomes,

$$\frac{dW}{dt} = -\mathrm{V}_o \frac{dW}{dN} + D_o \frac{d^2W}{dN^2}. \tag{5.31}$$

The solution is for $0 < t < T$,

$$W(N_1,t) = (4\pi\, D_o t)^{-1/2} \exp\left[\frac{-(N_o - N + V_o t)^2}{4D_o t}\right] . \tag{5.32}$$

This is a Gaussian distribution in N. As $t \to 0$, $W(N,t) \to \delta(N_o - N)$, and for $t > T$, $W(n, t > T) = W(N,T)$. The behavior is illustrated in figure 5.3. This corresponds to taking $P(t)$ and dP/dt as step functions is time. Also V_o and D_o are independent of N. Thus take,

$$-V_o = N_o \frac{P_o}{T} , \tag{5.33}$$

$$D_o = N_o \frac{P_o}{T}(1 - 2P_o) . \tag{5.34}$$

Note that if $P_o << 1/2$ then $-V_o = D_o \equiv \alpha/No$ whence,

$$W(N, t > T) = (4\pi\alpha)^{-1/2} \exp\left[-\frac{(N_o - N + \alpha)^2}{4\alpha}\right] \quad (N_o > N, \alpha). \tag{5.35}$$

This distribution depends only on one parameter $\alpha = N_o P_o$ which is equal to $< N_o - N >=< n >$, the average number of vacancies in the spectrum. This model for $P_o << \frac{1}{2}$ makes a specific prediction between the position of the peak of the Gaussian distribution and the width of the Gaussian distribution, namely that (since $-V_o = D_o$),

$$< N_o - N >=< N^2 > - < N >^2 = \alpha =< 16 ln2)\Gamma^2 , \tag{5.36}$$

consistent with a binomial distribution when $p << 1/2$. A more sophisticated model may be able to explain deviations from a binomial distribution.

Note that this model gives some insight into the meaning of V and D in the Fokker Planck equation. The 'diffusion coefficient' D is simply the time rate of change of the half width (with a factor of $[16ln2]^{1/2}$), and the 'frictional velocity' V tells us how fast the peak of the distribution moves from $0 < t < T$.

Since observed data are often reduced using best fits to Gaussian distribution, and since Fokker Planck is based on Gaussian distribution, the Fokker Planck method is well suited for analysis of experimental data, although explicit understanding in terms of microscopic atomic properties is now lacking.

Information theory [Aberg *et al.*, 1984; Blomberg *et al.*, 1986; Aberg, 1987; Aberg *et al.*, 1987] is a statistical method, related to the Fokker Planck method in its origins, for extracting the maximum information from experimental data with noise by using the smallest number of parameters. Useful interrelations between information theory, multinomial distribution, binomial distributions and other statistical methods are documented by Eadie *et al.* [1971].

5.3 Maximum entropy method

The maximum entropy method (MEM)was developed in the 1950 s by Jaynes [1957,1979]. MEM [Hägg, 1993; Hägg and Goscinski, 1993] is based on information theory [Aberg *et al.*, 1987; Press *et al.*, 1986]. It is a method to determine the probability $P(x)$ for an event to occur as a function of an arbitrary variable (or set of variables) x. $P(x)$ is not determined from the Schrödinger Equation, Newton's Laws or other such differential equations. $P(x)$ is determined by maximizing the entropy, $S = -\sum_x P(x)\, lnP(x)$, subject to constraints chosen by the person using the method. The idea of MEM is to provide a best fit to data without using any 'fundamental' equation or concept† other than maximizing entropy. MEM has been variously used in image processing, radioastronomy, crystallography, NMR spectroscopy and solid state physics. The distribution of charge states following ion-atom collisions is considered below following Hägg and Goscinski [1993,1996].

In this example one seeks a theoretical form for the probability $P(q)$ of obtaining an atom in the charge state q following the collision of the atom with a charged projectile. The form of $P(q)$ is found by maximizing the entropy S given by the standard expression [Landsberg, 1990],

$$S = -\sum_q P(q)\ln(P(q))\,, \tag{5.37}$$

subject to certain constraints.

The first constraint is conventionally the constraint of normalization,

$$\sum_q P(q) = 1\,. \tag{5.38}$$

The remaining constraints are found by determining M moments, m_k, from experimental data, namely,

† Perhaps there is no fundamental equation or concept in nature.

$$m_k = \sum_q q^k P^{exp}(q) \qquad k = 1, ..., M. \tag{5.39}$$

Here P^{exp} is taken from experiment. The corresponding constraints for the theoretical $P(q)$ are,

$$\mu_k = \sum_q q^k P(q) \qquad k = 1, ..., M. \tag{5.40}$$

The constraints that determine the theoretical form of $P(q)$ are imposed by setting,

$$m_k = \mu_k \qquad k = 0, 1, ..., M \;, \tag{5.41}$$

where $k = 0$ denotes the constraint of (5.38).

If the entropy S in (5.37) is maximized with these constraints, then one obtains the expression [Jaynes, 1979],

$$P(q) = \exp\left(\sum_{k=0}^{M} \lambda_k q^k\right) \;, \tag{5.42}$$

where λ_k are Lagrange parameters chosen to satisfy the constraints imposed. This yields a functional form for P(q).

This method is useful when M is significantly smaller than the number of data points observed. Hägg and Goscinski [1993] have used this method to reproduceably describe data for charge state distributions in beam-foil experiments.

5.4 Intermediate systems

The methods described in this chapter are statistical methods which endeavor to describe observable quantities using methods based on simple counting (i.e., statistics). The methods used in other parts of this book are based in solving a differential equation, such as the Schrödinger Equation[‡], to solve the problem exactly (or approximately so). Both methods should in principle ultimately give the same correct result. In practice it has often been difficult to detail the interrelation between these two approaches.

[‡] With increasing numbers of interconnected atomic systems description of properties involving large numbers of quantum states becomes tedious. It may be that quantum mechanics is too detailed to give a useful description of many large atomic, molecular and condensed matter systems.

The statistical parameters are not always clearly related to the atomic parameters. This gap between the statistical and atomic methods presents a significant and useful problem.

At nano-dimensions there is an interesting and potentially useful opportunity to understand the transition from quantum to statistical properties. In some cases this opens to study problems that are exact and linear quantum mechanically and exact but non-linear and chaotic classically[§] [McDaniel and Mansky, 1994]. One method that combines advantages of statistical mechanics for averages of large quantum systems is random matrix theory which predicts fluctuations in observables described as Ericson fluctuations [Ericson, 1963]. Heller [1995] has given examples of such processes in so called 'mesoscopic systems' which form a bridge between quantum and statistical classical systems. Observations of chaotic behavior in quantum corrals have been observed and analyzed by Marcus [1993].

§ Physicists typically try to understand macroscopic phenomena in terms of microscopic particles (which in some cases have not been directly observed, e.g., electrons and protons). In these cases it is natural to ask: what is the classical limit of quantum mechanics? Lindhardt [1988] has suggested a more direct question: how may one describe the process in which an experimentalist creates conditions to observe quantum coherence in the midst of often noisy statistical classical macroscopic phenomena? Weaver [1990] points out that quantum mechanics provides an apparently necessary bridge from a classically prepared macroscopic system to classical detection. That an experimentalist is able to coax (often reversible) coherent quantum processes from a sea of fluctuating (usually irreversible) macroscopic events seems worthy of tribute.

6

Correlated multi-electron electron transition probabilities

6.1 Correlation

There is a significant difference between complex and merely large. This difference is related to the the notion of correlation which defines the rules of interdependency in large systems. The relevant question is: how may one make complicated things from simple ones? Biological systems are complex because the atomic and molecular subsystems are correlated. From the point of view of atomic physics correlation in condensed matter, chemistry and biology is determined at least in part by electron correlation in chemical bonds and the complex interdependent structures of electronic densities. Understanding correlation in this broad sense is a major challenge common to most of science and much of technology. This is sometimes referred to as the many body problem. In a general sense correlation is a conceptual bridge from properties of individuals to properties of groups or families.

The concept of correlation arises in many different contexts. 'Individual' may mean an individual electron, an individual molecule or in principle an individual person, musical note or ingredient in a recipe. In this book individual refers to electron for the most part. In this case the interaction between individuals is well known, namely $1/r_{12}$. However, that does not mean that electron correlation is well understood in general. Although much has been done to investigate correlation in various areas of physics, chemistry, statistics, biology and materials science, in many cases little is well understood except in the limit of weak correlation. In this weak limit one may begin to address the question of what correlation is and how it works. A more fruitful question for either more complex systems or exact calculations of strongly correlated simple systems may be 'what are the patterns (or rules) of correlation?'. Perhaps both the challenge and the reward in understanding correlation lies in finding better ways to ask the question.

Correlation may be defined and studied in both space and time. Examples of time correlation include time ordering effects [section 7.1.1] and memory. Most studies have been confined to two particle or pair correlation since experimental studies of higher order correlation are often difficult. The emphasis in this text is on spatial correlation.

Correlation may be defined mathematically as the deviation from a product of single particle terms, e.g., $\psi(r_1, r_2, ...r_n) - \psi_1(r_1)\psi_2(r_2)...$-....$\psi_n(r_N)$. So correlated means non-separable or interdependent. Physically, correlation between N electrons occurs because the electrons influence one another, i.e. they interact with one another. This interaction is called the correlation interaction v and is the difference between the Coulomb interaction between the electrons and the mean field of the electrons. Thus, correlation in a wavefunction is the difference between the given wavefunction and a Hartree Fock wavefunction (a product wavefunction found using a mean field for the electrons symmetrized to include the identical nature of the electrons). The electron symmetry is neglected in the simpler Hartree approximation. In practice one defines correlation by defining the uncorrelated limit. For time dependent systems considered in this book the uncorrelated limit is defined by the time dependent Dirac Fock (TDDF) approximation, which is the relativistic generalization of the Hartree Fock approximation. Any quantity that is uncorrelated in one representation (e.g. b space) is also uncorrelated in the conjugate space of the Fourier transform (e.g. q space).

Mathematical correlation has been studied in a broad variety of systems. There are, for example, statistical studies of correlation in the spatial patterns of H_2O molecules in water [Stanley, 1996], sequencing of codons in DNA [Peng *et al.*, 1995a], heartbeat intervals in healthy and diseased hearts [Peng, *et al.*, 1995b] feeding patterns of birds [Viswanathan *et al.*, 1996], corporate growth [Stanley *et al.*, 1996], and even a possible description of free will [Stanley, 1996]. In these studies correlation has been used to bring 'order out of randomness'. The surprising strength of these long range correlations and their effect on scale invariance has been ascribed to a multiplicity of pathways between individuals. Without the multiplicity of pathways the correlation would decrease exponentially with distance rather than as a more slowly decaying finite power law. Memory, a key element of living consciousness , is also non-localized . Memory is not contained in just one single nerve cell, but the information is is stored in a number of cells which interact with one another [van Hemmen, 1996]. Thus correlation is used to understand complex systems. Like correlation complexity is defined by its opposite. 'Complex' means 'not simple'. The connection of these more general examples of statistical correlation to electron correlation has not yet been carefully studied.

The term 'electron correlation' seems to have been first used in physics in 1933 by Wigner and Seitz*†. The modern definition of correlation in static atomic systems was established by Löwdin [1959, 1995]. Many of the methods now used to determine correlation in electronic wavefunctions were introduced by Kutzelnigg, Del Re and Berthier [1968], who point out some earlier uses of correlation in mathematical statistics. Significant contributions have been made by Davidson [1976] and by Froese Fischer [1996, 1977]. Correlation in single electron transitions has been reviewed by Webster *et al.* [1978] and by Crowe [1987]. Electron correlation in molecules and solids is discussed in the book of Fulde [1995]. Ziesche [1995] has suggested a connection between correlation, which may be used to describe the degree of mixing in a dynamic system, and entropy. Since understanding properties of macroscopic materials in terms of microscopic atomic properties is interesting, it is sensible to note stochastic uses of correlation [Balescu, 1975; Huang, 1987] since statistics are ultimately useful in dealing with large numbers of atoms. Correlation is often introduced in stochastic and statistical mechanics [Balescu, 1975] as a generalization of the notion of statistical standard deviation of a distribution,

$$\sigma^2 =< (x- < x >)^2 > \ . \tag{6.1}$$

If σ^2 is not zero then the distribution is not localized at a single value $< x >$, but rather the distribution is spread out. This leads to the idea that correlated functions are also spread out, i.e., not confined to a single x_i (e.g. denoting a single particle), but connect x_i and x_j. Thus, if $\mathcal{P}$ is some physical property, then $\mathcal{P}- < \mathcal{P} >$ is a measure of the correlation of the property $\mathcal{P}$.‡

BBGKY hierarchy. Consider the one particle distribution function given by,

---

* Wigner [Rhoades, 1986, p.35] commented that, 'Physics does not even try to give us complete information about the events around us – it gives information about the correlation between those events.'

† Somewhat earlier Mozart elegantly noted,
'Nor so I hear in my imagination the parts successively,
but I hear them, as it were, all at once.'

‡ In a general sense theoretical physics has moved from strict causality (perfect correlation) in the nineteenth century to quantum mechanics (variable correlation) in the twentieth century by expanding the idea and use of correlation. The notion of chaos, which usually implies weak correlation, is a further expansion of the concept of correlation. Thus, correlation ranges from fully causal δ-like distributions to totally random distributions. It has been suggested that the notion of a single exact result is an ideal limit of perfect correlation between variables, and further that this notion of unique truth may be unnecessarily restrictive [McGuire, J.H., 1995; Fish, 1982; Derrida, 1988, 1976].

$$f_1(r_i) = \rho(r_i) = \psi^* \psi \,. \tag{6.2}$$

If $f_2(r_1, r_2) = f_1(r_1) f_1(r_2)$, the two particle distribution function f_2 is uncorrelated. In general, however, f_2 is correlated and may be written as,

$$f_2(r_1, r_2) = f_1(r_1) f_1(r_2) + g_2(r_1, r_2) \,, \tag{6.3}$$

where g_2 is called the two particle correlation function. For three particles one may generally write,

$$\begin{aligned} f_3(r_1, r_2, r_3) &= f_1(r_1) f_1(r_2) f_1(r_2) \\ &+ f_1(r_1) g_2(r_2, r_3) + f_1(r_2) g_2(r_1, r_3) + f_1(r_3) g_2(r_1, r_2) \\ &+ g_3(r_1, r_2, r_3) \,, \end{aligned} \tag{6.4}$$

and so forth for $f_4, f_5 ... f_N, ...$ This is called a cluster expansion of the distribution function f_N in terms of the jth order correlation functions g_j. The $N+1$ particle distribution function can be generated from the N particle distribution function via the two particle correlation potential $v(r_i - r_j)$. This is referred to as the BBGKY (Bogoliubov-Born-Green-Krikwood-Yvon) hierarchy [Balescu, p86].

Generalized correlation coefficient. If $\mathcal{P}$ is a one particle operator, then for a two electron system characterized by a distribution function f_2 the generalized correlation coefficient τ is defined as follows,

$$\tau_f = \frac{\int \int g_2(x_1, x_2) \mathcal{P}(x_1) \mathcal{P}(x_2) dx_1 dx_2}{(\int f_1(x) \mathcal{P}^2(x) dx) - (\int f_1(x) \mathcal{P}(x) dx)^2} \,. \tag{6.5}$$

Here, (i) $\tau = 0$ if the electrons are uncorrelated, and (ii) $|\tau| < 1$. Most importantly, τ is independent of choice of basis functions. This generalized correlation coefficient has been applied by Christensen-Dalsgaard [1988] to atomic structure, but application has not yet been made to dynamic correlation in atoms and molecules.

Degree of correlation K [Grobe *et al.*, 1994]. A canonical representation of an exact wavefunction $\Psi(r_1, r_2)$ for a two electron system is to expand Ψ in terms of single particle Slater determinants $\psi(r_1, r_2)$ [Cowan, 1981], namely,

$$\Psi(r_1, r_1) = \sum_i c_i \; \psi_i(r_1, r_2) \,. \tag{6.6}$$

Here Ψ may be regarded as an entangled state (i.e., sum) in the set of pure states $\{\psi\}$. The Slater determinant above describes orthonormal single particle orbitals for fermions and a product of identical orbitals for bosons. The normalization condition on Ψ requires that $\sum_i |c_i|^2 = 1$. The average probability P_i^{av} is given by $\sum_i |c_i|^4$. Note that $P_i^{av} \leq 1$ and that $P_i^{av} = 1$ corresponds to a wavefunction represented by a single orbital, which is uncorrelated. The inverse of this is an average 'number' of effectively non-zero mixing probabilities, which is a way to measure correlation (or extent of entanglement). Thus, the degree of correlation is defined as,

$$K = (P_i^{av})^{-1} = \left[\sum_i |c_i|^4\right]^{-1} . \tag{6.7}$$

Correlation mixes otherwise independent particle orbitals. K is a measure of the degree of mixing which corresponds to information loss from a 'pure' uncorrelated state. Consequently, the logarithm of K is sometimes called the Stückelberg entropy.

Grobe *et al.* [1994] have used this degree of correlation K to analyze both electron hydrogen scattering and photo detachment of electrons from atoms.

Ziesche [1995] has developed a quantum kinematic measure of correlation strength related to entropy and information theory as a distributive property of an N body system. Gersdorf *et al.* [1996] have used this method to describe the correlation entropy of the H_2 molecule.

Correlation is important in chemistry and biology. For example neither life nor death nor most of the processes in between may be described in terms of the transitions of a single isolated electron. Correlation is necessary.

6.2 Scattering correlation

Here effects of electron-electron Coulomb interactions during an atomic interaction are considered. These multiple electron effects are missing in the independent electron approximation and, for the most part, they may be regarded as electron correlation. To some extent correlation in the dynamics of interacting atoms is analogous to correlation in atomic structure. In both cases correlation arises from the $1/r_{ij}$ electron-electron interaction and both are defined in terms of an uncorrelated limit (time dependent Dirac Fock and (time independent) Dirac Fock, respectively). In both cases one attempts to determine the wavefunction. However, in the static case the emphasis is on evaluating energy levels, while in the dynamic case transition probabilities and cross sections are often sought. Since the probability amplitude is usually expressed in terms of a scattering operator acting between asymptotic states $|i>$ and $|f>$ it is convenient to consider separately multi-electron effects on the scattering operator and on the asymptotic wavefunctions. This division is somewhat arbitrary and does not explicitly address relaxation of excited final states (e.g. Auger and photon decay). For simplicity only a two-electron system is considered here. Generalization to a system of N electrons is straightforward.

1 Asymptotic states

The unperturbed Hamiltonian for a two electron system from (4.3) may be rewritten as,

$$\begin{aligned} H_T &= \left[\frac{-\nabla_1^2}{2} - \frac{Z_2}{r_1}\right] + \left[\frac{-\nabla_2^2}{2} - \frac{Z_2}{r_2}\right] + r_{12}^{-1} \\ &= \left[\frac{-\nabla_1^2}{2} - \frac{Z_2}{r_1} + v_1(r_1)\right] + \left[\frac{-\nabla_2^2}{2} - \frac{Z_2}{r_2} + v_2(r_2)\right] \\ &+ \left[r_{12}^{-1} - v_1(r_1) - v_2(r_2)\right] . \end{aligned} \tag{6.8}$$

Here the mean field potentials $v_1(r_1)$ and $v_2(r_2)$ has been introduced (Cf. (4.6)). A common example of such a potential is,

$$v_1(r_1) = s/r_1 , \tag{6.9}$$

where s is a screening factor (or screening parameter) (for example $s = 5/16$ in variational calculations of the ground state). The correlation potential v is defined by,

$$v(r) = \left[r_{12}^{-1} - v_1(\vec{r}_1) - v_2(\vec{r}_2)\right] . \tag{6.10}$$

Thus,

$$H_T \equiv H_T^1 + H_T^2 + v , \tag{6.11}$$

where H_T^1 and H_T^2 are one electron operators.

The asymptotic states, $|i>$ and $|f>$ are eigenstates of H_T. These asymptotic states are not known exactly for few and many electron systems. A number of methods are available to evaluate these asymptotic, static wavefunctions approximately, especially for the ground state and also for singly and doubly excited states of two electron systems. A relatively large amount of work has been done on the evaluation of atomic and molecular wavefunctions for few and many electron systems [Davidson, 1976; Froese Fischer, 1996, 1977; Chakravorty *et al.*, 1993; Berry, 1995, 1986]. One method commonly used to evaluate multi-electron static wavefunctions is the configuration interaction (CI) method, where added to the independent electron wavefunction are additional single electron terms called configuration interaction terms [Szabo and Ostlund, 1982, chapter 4]. Other methods strive to find 'approximate constants of motion' and 'pretty good quantum numbers' [Braier and Berry, 1994; Berry and Krause, 1986; Herrick *et al.*, 1980]. The method of hyperspherical coordinates has been developed for doubly excited states by Macek [1968] and by Lin [1986]. A different hyperspherical elliptic method has been developed by Tolstikhin *et al.* [1995] which gives analytic results in the limit of various infinite mass ratios for three charged particles (with two particles of opposite charge from the third). Extensive discussion of evaluation of bound state wavefunctions and of static electron correlation may be found elsewhere.

Continuum wavefunctions are not well known, especially for two strongly correlated continuum electrons (although some useful analytic expressions have been developed [Brauner *et al.*, 1989, section 2.5]). Configuration interaction methods using pseudostates in the continuum are also used (Cf. section 7.2) and other developments are emerging [Shakeshaft, 1995].

Usually $|i>$ and $|f>$ are approximate. Sometimes $|i>$ and $|f>$ are regarded as eigenfunctions of different approximate H_T^i and H_T^f because in some cases the mean field potential is physically different before and after an atomic interaction. In these cases $|i>$and $|f>$ are non-orthogonal. This non-orthogonality can result in non-vanishing matrix elements for certain (e.g. multiple) transitions. Since there is a physical cause for these non-zero matrix elements, they are retained (unlike spurious non-orthogonality terms which are often discarded). A simple example is the shake matrix element arising from a change of screening during the collision (e.g. s changes in (6.9) during the collision). If the correlation potential v is now ignored, the resulting probability amplitude for shake transitions of $N-1$ electrons following an interaction of one electron with the projectile is,

$$a = \langle f_1 | U_1 | i_1 \rangle \prod_{j=2}^{N} \langle f_j | i_j \rangle \, . \tag{6.12}$$

Now the ratio of double to single ionization cross sections is independent of the charge and the velocity of the projectile [Mittleman, 1966]. Amplitudes of this type may dominate multiple electron transitions at high collision velocities in some cases. Since these shake amplitudes are expressed as a product of single electron terms (because v=0), shake is usually considered not to be electron correlation (Cf. section 7.1.1), although shake is a many electron effect.

2 Scattering operator

The interaction potential is expressed in the interaction representation (Cf. section 3.2.2) as,

$$\begin{aligned} V_I(t) &= e^{-iH_o t} V(t) e^{iH_T t} \qquad (6.13) \\ &= e^{-i(H_T^1 t + H_T^2 t + vt)} \left[V_1(t) + V_2(t)\right] e^{i(H_T^1 t + H_T^2 t + vt)} \\ &= V_I^1 + V_I^2 = \sum_{j=1}^{N} V_I^j \, , \end{aligned}$$

because $[H_T^j, v] \neq 0$, it follows that V_I^j is a many electron operator. This is not an artifact of the intermediate representation. Rather it is one way to express the influence of the correlation interaction, v, occurring during the collision. Formulation in other representations (e.g. the Heisenberg representation) is usually more difficult (Cf. section 3.2.2).

It is sometimes useful to write the many electron scattering potential $V_I^j(t)$ as

$$V_I^j(t) = V_I^{j0}(t) + V_I^{jC}(t) \, , \qquad (6.14)$$

where V_I^{j0} is the limit of V_I^j as the correlation potentials, $|\vec{r}_i - \vec{r}_j|^{-1}$ go to zero, corresponding to the independent-electron approximation (Cf. section 4.1.2). The many body operator for scattering correlation V_I^{jC} is the difference of the full operator of (6.14) and the independent-electron operator with $v = 0$ in (6.14).

This scattering correlation potential may be used to express the amplitude for scattering correlation to first order , namely,

$$U - U_0 = -i \int_{t_0}^{t} V_I^C(t) dt \, , \qquad (6.15)$$

where the independent evolution operator, U_0, is found using only the $V_i^{j0}(t)$ term in (6.14). This amplitude for scattering correlation has been further considered by McGuire [1987] and also by Martin and Salin [1995,1996].

3 Analytic expressions

At moderately high velocity v and when correlation v is not too strong, the independent electron approximation usually provides a reasonable approximation since the first order term in Z/v is often relatively small (Cf. section 7.4). In the independent electron approximation the probability amplitude is a product of single electron amplitudes [Cf. (4.10) in section 4.1.2]. At large impact parameters b the single transition amplitudes for excitation or ionization are proportional to dipole matrix elements, $D^{if} =< f|\vec{r}|i >$ and the probability amplitude for excitation or ionization of two electrons is [McGuire, 1987],

$$a_{IEA} = a_1 a_2 = -\pi^2 (Z_P^2/\mathrm{v}^2) D_1^{if} D_2^{if} \; e^{-(q_0^1+q_0^2)b}/b^2 \; , \tag{6.16}$$

where $q_o = q_{min}$ is the minimum momentum transfer of the projectile. This term varies as $(Z/\mathrm{v})^2$.

At very high velocities the first order term in Z/v dominates. A qualitative estimate of this term is sometimes given by the simple shake contribution to the full first order term given in the next chapter in (7.24). With $|i >= |i_1 > |i_2 >$ and $< f| =< f_1'| < f_2'|$, and ignoring indistinguishability and correlation in the evolution operator U, one quickly has $a \simeq =< f_1'|U_1|i >< f_2'|i_2 >= a_1 < f_2'|i_2 >$. Using the large impact parameter approximation for a_1, one has,

$$a_{Shake} = a_1 < f_2'|i_2 >= -i\pi(Z/\mathrm{v}) D_1^{if} \; e^{-q_0^1 b}/b < f_2'|i_2 > \; . \tag{6.17}$$

Some simple estimates of $< f_2'|i_2 >$ are given in section 6.3 below.

If the collision is not too fast, then there may be sufficient time during the scattering event for the correlation interactions between electrons to influence the transitions. An approximate expression for the lowest order scattering correlation term may also be expressed in closed form at large impact parameters [McGuire, 1987], namely,

$$a_{sc} = -2i \frac{Z}{\mathrm{v}^3} Q^{if} [K_0(q_0 B) - (q_0 B) K_1(q_0 B)] \; , \tag{6.18}$$

where Q^{fi} is a quadropole matrix element and K is a Bessel function of the second kind. Since this term varies as Z/v^3, it tends to be small at large v and when electron correlation is weak.

These simple analytic expressions (which may be generalized to N electron transitions) give a simple qualitative picture of mechanisms for multiple excitation and ionization by charged particles. At very high velocities the small terms first order in Z/v dominate. At moderately high velocities when multi-electron effects are small, the independent electron

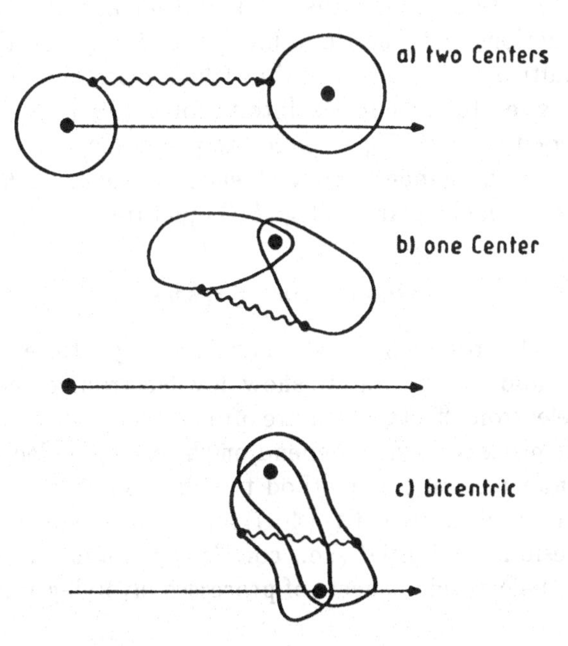

Fig. 6.1. Pictures demonstrating correlation at different centers in atomic interactions.

approximation dominates. As the collision velocity is further decreased and correlation becomes stronger, correlation occurring during the scattering event becomes significant. At low v the molecular orbital (MO) picture [Madison and Merzbacher, 1975] and possibly the statistical energy deposition model [Cf. section 5.1] come into play. Near threshold scattering correlation may overlap with Wannier threshold effects arising from the long range nature of the electron-electron interaction.

4 Multi-center correlation

Correlation effects in atomic interactions may occur in various ways. Three of these are illustrated in figure 6.1. Electrons on two distinct atomic centers may interact with one another as depicted in figure 6.1.a, corresponding to the atomic orbital (AO) picture in which each electron is identified with a single atomic nucleus. This is sometimes referred to as 'two center correlation' and is described in more detail in sections 8.1

- 8.4. A second case is 'one (or single) center correlation' in which the correlation is confined to the electrons about a single atomic nucleus, usually the target nucleus. This may be described in terms of 'initial state correlation' and 'final state correlation' as described by many body perturbation theory (MBPT) (Cf. section 7.3) and includes the effects of configuration interaction (CI) terms in the initial and final state many electron wavefunctions (Cf. sections 6.2.1 and 7.2.2). A third case is 'bicentric (or multicentric) correlation' in which electrons are associated with more than one atomic nucleus [Stolterfoht, 1988]. This picture is useful in low velocity interactions where the projectile and target form a 'quasi-molecule' which changes slowly at each internuclear R during the collision. This is the molecular orbital (MO) picture.

5 Dielectronic processes

The picture of dielectronic processes introduced by Stolterfoht in 1991 provides a useful and specific way to show how electron correlation works in terms of two electron effects which are often directly observable. Here a variety of atomic processes which depend on the electron-electron interaction are interconnected. In this method the electron-electron interaction is partitioned into a mean field (Cf. (4.6)) representing effective one-body aspects and a residual potential (i.e., correlation potential v) representing two-body or dielectronic aspects of processes involving more than one electron.

To illustrate the role of dielectronic processes in both the structure and dynamics of atoms Stolterfoht [1991] has used diagrams shown in figures 6.2 and 6.3. The role of configuration interaction (CI) in atomic structure is illustrated in figure 6.2. In the independent electron limit (e.g. Dirac Fock or Hartree Fock) single electron configurations are used. However, these single-configuration states are not eigenstates of the full atomic Hamiltonian. Consequently, interactions between the different single configurations (or channels) occur. Bound state CI occurs within the closed channels. Free-bound state CI occurs between the closed channels and single ionization channels. Coupling between single and double ionization channels is called transchannel CI while the coupling within double ionization channel is referred to as DI interchannel CI.

The effects of configuration interaction (CI) are associated with dielectronic processes which in turn are associated with various scattering mechanisms in figure 6.3. The difference is that in the case of a dielectronic process an active electron interacts with another active electron as a result of a collision with an incident heavy particle (e.g. autoexcitation), whereas in the case of the scattering mechanisms an electron in an external beam interacts with a target electron (e.g. electron impact

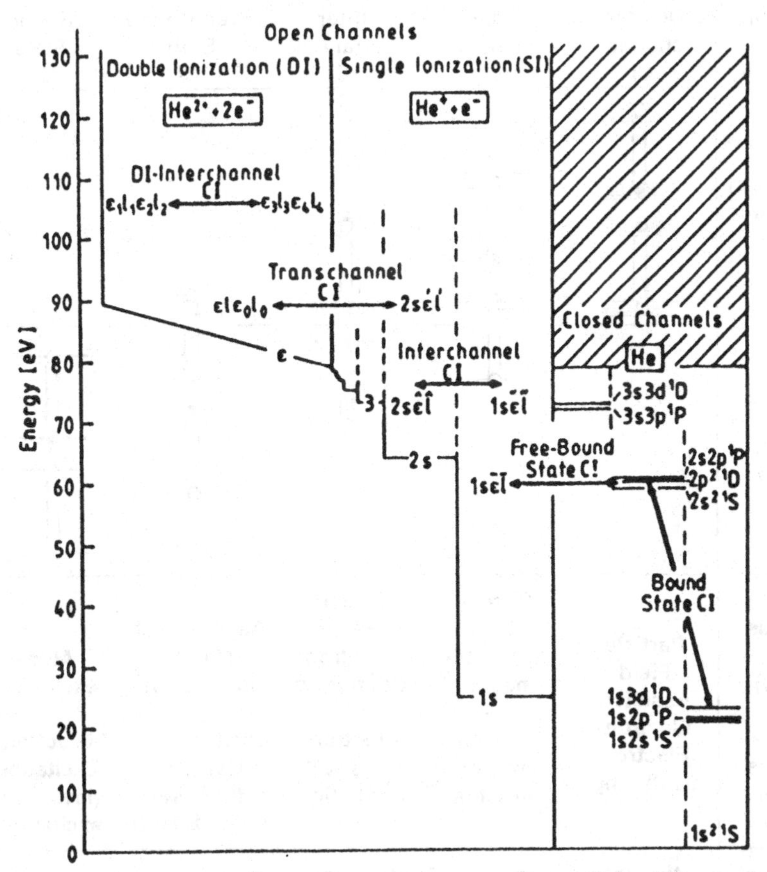

Fig. 6.2. Diagram of single configuration states illustrating different types of configuration interaction (CI) in helium.

ionization). In figure 6.3 the interchannel interaction is associated with super-elastic electron scattering and, inversely, with electron impact ionization. Similarly, the free-bound state CI is associated with the Auger effect or auto-ionization and, inversely, with the inverse Auger effect or dielectronic capture. To provide visualization of the central dielectronic nature of these processes, scattering events related to specific effects of the CI terms are given at the bottom of figure 6.3.

In his dielectronic analysis Stolterfoht gives rule of thumb estimates of effects of various contributions in atomic scattering. Inner shell nuclear effects are estimated as $Z^2 \times 10$ eV, while mean field effects are in the order of 1 - 20 eV and effects of the residual dielectronic (or correlation) tend to be about 0.1 eV. Hence the residual dielectronic effects are often relatively small.

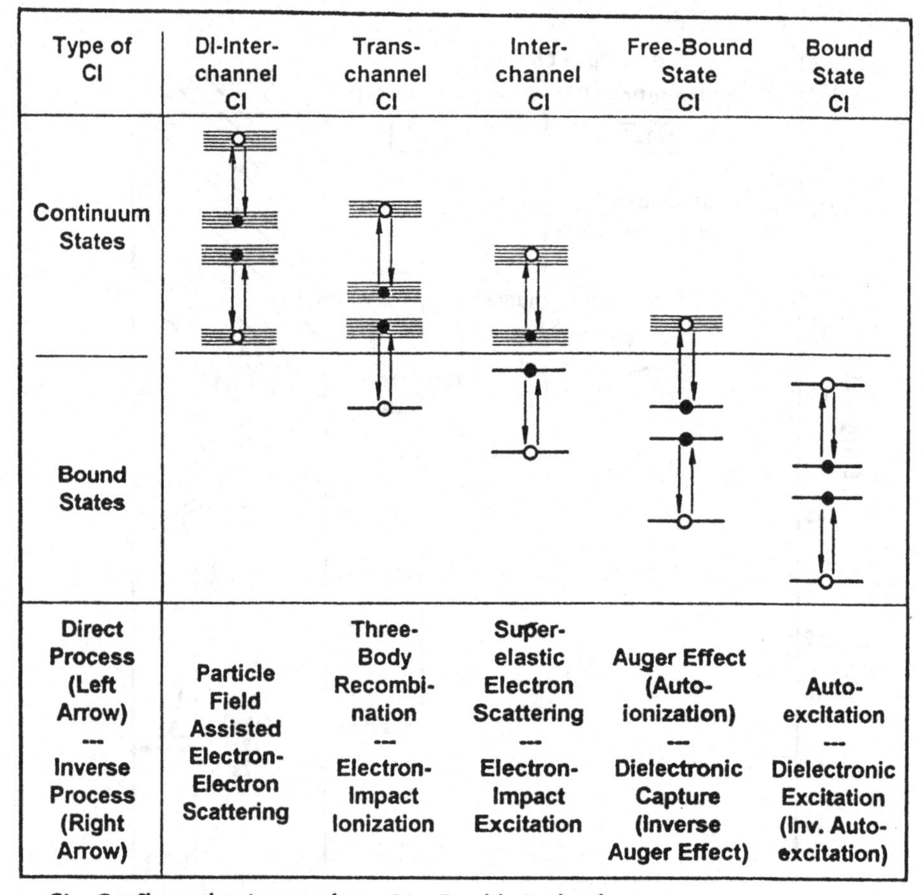

Fig. 6.3. Diagram relating various dielectronic (two electron) processes to various configuration interaction (CI) effects (listed at the top) and to various scattering mechanisms (listed at the bottom). The CI effects are the same as illustrated in figure 6.2. Direct and time reversed processes are indicated by left and right arrows, respectively. These arrows indicate transitions between single-electron configurations.

6.3 Shake probabilities

The idea of shakeoff was introduced to describe ionization of an atom whose nucleus undergoes beta decay. In beta decay the charge of the nucleus changes quickly producing a new set of eigenfunctions for the atomic states. Consequently, the old states are no longer eigenfunctions of the new atom, but rather a linear combination of the new eigenstates with a finite probability of being in a state of ionization. Thus an electron in an initial eigenstate may make a transition to the continuum following the sudden change of charge of the nucleus. This is called shakeoff. For a transition to an excited bound state 'shakeup' is used and for de-excitation 'shakedown' is used.

The term 'shake' is sometimes used differently by different authors. Consider, for example, shakeoff where the final state is a continuum state $|\vec{k}>$. The simple shakeoff amplitude is often taken to be $<\vec{k}|i'>$ which is non-zero because the electron screening differs for the initial and final sets of eigen states. However, in many body perturbation theory, this amplitude is expanded in powers of the correlation potential v and the lowest order shake term is $<\vec{k}|v|i>$. Sometimes the conventional product of independent electron orbitals is replaced by many electron (correlated) orbitals. Other authors think of shake as occurring only in a sudden limit where the outgoing electron leaves quickly so that $|\vec{k}>$ is represented by a plane wave. Usually the meaning of the term 'shake' is clear from context.

Although shake processes are due to the electron-electron interaction, most authors consider shake terms to be uncorrelated because the shake terms are expressed by an independent factor multiplying the remaining amplitude and are therefore uncorrelated (Cf. section 6.1).

1 *Simple shake factors*

Before considering generalized shake probabilities which employ correlated initial state asymptotic wavefunctions, it is convenient to begin with the simple shake probability amplitude defined [Aberg, 1967, 1976; Carlson, 1967] by $a_{shake} =< f'|i>$. In this simple shake picture the amplitude $< f'|i>$ is non-zero because H_i and H_f are different due to differences in electron screening. Thus simple shake corresponds to rearrangement in the final state due to a change in the charge screening of the target nucleus as a result of the excitation of an electron by the projectile. Note that the simple shake probability is technically uncorrelated because $a = a_1 \cdot a_{shake}$ where a_1 is the probability amplitude for scattering the first electron. Furthermore, the simple shake probability is usually independent of the projectile, i.e. shake represents a post collision decay.[§]

For simple shake processes, the transition amplitude for an arbitrary process in a two electron target may be written as,

---

§ If one assumes that the projectile is completely decoupled from the shake process, then it is difficult to account for the energy, ΔE, needed for the electronic 'shake' transition. However, if one regards shake as having a lifetime, Δt, such that $\Delta E \Delta t \simeq 1$, then the energy, ΔE, required may be provided within the limits of the Uncertainty Principle by the projectile at a distance $v\Delta t$ from the atomic target.

$$\begin{aligned} a &= < f'|U|i> = < f_2' f_1' | e^{i\int (V_2+V_1)dt} | i_2 i_1 > \\ &= < f_2'|U_2|i_2 >< f_1'|U_1|i_1 > = < f_2'|i_2 >< f_1'|U_1|i_1 > . \end{aligned} \tag{6.19}$$

This product form in a sense corresponds to an independent electron approximation. The role of the interaction V_1 is to change the screening seen by the second electron. Hence there is some physical interdependency. However, this change in screening is usually taken to be independent of V. In this sense the shake amplitude, $< f_2'|i_2 >$, is independent of what happens to the electron which interacts with the projectile, $< f_1'|U_1|i_1 >$.

For transition of a single electron, it is often the case that $< i_2'|i_2 > \simeq 1$ so that the passive electron may in effect be ignored. For the transition of two electrons, since the shake probability,

$$P_S = |< f_2'|i_2 >|^2 \tag{6.20}$$

is independent of the impact parameter, $\vec{b}$, one has,

$$\begin{aligned} \sigma^{++} &= \int d\vec{b} |< f_2'|i_2 >< f_1'|U_1|i_1 >|^2 \\ &= |< f_2'|i_2 >|^2 \int d\vec{b} |< f_1'|U_1|i_1 >|^2 = P_S \sigma^+ . \end{aligned} \tag{6.21}$$

This is easily generalized to the transition of many electrons.

Counting of final states for double ionization can be confusing. For distinguishable electrons, a single transition usually corresponds to a transition of either electron 1 and electron 2. In this case, $\sigma^+ = 2\int d\vec{b}| < f_1'|U_1|i_1 > |^2 = P_S\sigma^+$. It is conventional in this case to include a factor of 2 for double ionization as well corresponding to the different final states. However, such counting appears to conflict with the counting used in binomial distributions (also used for independent processes) which has no factor of 2 for the double transition. The resolution is that there are three probabilities: the shake probability P_S, the interaction probability P_V, and the probability for everything else $Q = 1 - P_S - P_V$. Now, $(1)^2 = (P_S + P_V + Q)^2 = P_S^2 + P_V^2 + Q^2 + 2P_SP_V + 2P_SQ + 2P_VQ$. This binomial expression has the factor of 2 on both single and the P_SP_V double transition. However, the P_S^2 term is unphysical, although it is usually small. In most shake applications $Q \simeq 1$.

In the simple shake picture the probability that the shake electron does not remain in its ground state is given using closure, namely,

$$\sum_{f' \neq i} |< f'|i >| = 1 - |< i'|i >|^2 . \tag{6.22}$$

Here $|i>$ and $|i'>$ differ due to a change in the screening of the target nucleus by electrons. For a target of bare nuclear charge Z and screening parameter s using ground state wavefunctions, one obtains with $Z' = Z - s$.

$$\begin{aligned} P_{S_{tot}} &= 1 - | <_i |i'> |^2 \\ &= |4(ZZ')^{3/2} \int_0^\infty dr\, r^2 e^{-(Z+Z')r}|^2 \\ &= 1 - \frac{[Z(Z')]^3}{[\frac{1}{2}(Z+Z')]^6} \\ &\cong 1 - \frac{1-3(s/Z)+3(s/Z)^2}{1-3s/Z+\frac{15}{2}(s/Z)^2} = \frac{3}{4}\frac{s^2}{Z^2}\,. \end{aligned} \tag{6.23}$$

For direct excitation and ionization the simple shake probability has the same dependence on Z as the IEA probability for direct excitation or ionization by the projectile, i.e., Z^{-2}.

The total shake probability is the probability that the electron goes to a state other than the ground state of the target. For helium with $Z = 2$ and $s = 5/16$, corresponding to a hydrogenic wavefunction with a variational charge Z' chosen to minimize the binding energy, $P_{S_{tot}} = 2.1\%$ while the shakeoff probability, $P_{SO} = P_{S_{tot}} - \sum_{bound\ states} P_{bound\ states} = 0.71\%$. It is straightforward to compute probabilities for shake into specific states. The simple shakeup probability [Jacobs and Burke, 1972] for excitation to the $2s$ state from the $1s$ state is given by,

$$\begin{aligned} P_{S_{up}} &= | < 2s|1s' > |^2 \\ &= |(ZZ')^{3/2} \int_0^\infty r^2\, dr 2e^{-Z'r}\frac{\sqrt{2}}{2}(1 - \tfrac{1}{2}Zr)e^{-Zr/2} \\ &= 2(ZZ')^3 \left[\frac{2}{(Z' + \frac{1}{2}Z)^3} - \frac{Z}{2}\frac{6}{(Z' + \frac{1}{2}Z)^4}\right]^2 \\ &\cong \frac{2^{11}}{3^8}\frac{s^2}{Z^2} = 0.312\frac{s^2}{Z^2}\,. \end{aligned} \tag{6.24}$$

Note the $2s$ excitation is about 0.76% for helium in this simple model. Most of the shake electrons usually are ionized electrons corresponding to shakeoff. The simple shakeoff probability in helium has been computed by McGuire and Heil [1987]. In general Aberg [1990] suggests that as a rule of thumb shakeup dominates over shakeoff for outer shell (valence) electrons whereas for inner shell electrons shakeoff dominates.

In addition to shakeup and shakeoff it may be possible to have shakeover where the electron collapses onto the projectile after the collision. The simple shakeover probability has been recently evaluated by McGuire *et al.* [1988], namely.

Table 6.1. *Average radii and screening constants for ground state and excited states of helium using a Hartree Fock Slater program.*

State		$< r >$	s
$(1s)^2$	$1s$	0.940	0.405
$(1s)(2s)$	$1s$	0.764	0.038
	$2s$	4.35	0.621
$(1s)(2p)$	$1s$	0.764	0.038
	$2p$	4.64	0.922
$(1s)(3s)$	$1s$	0.755	0.013
	$3s$	10.58	0.724
$(1s)(3p)$	$1s$	0.756	0.016
	$3p$	11.08	0.872

$$P_{S_{over}} = \frac{144}{\mathrm{v}^8} Z_p^5 Z^5 s^2 \left[\frac{1}{Z_p^2} - \frac{1}{Z^2} \right]^2 . \tag{6.25}$$

This simple shakeover probability is dependent on the properties of the projectile ($Z, \vec{b}$ and v), unlike shakeoff or shakeup. As of 1995, there was no experimental evidence for shakeover. However, using (6.25) for transfer ionization, it is easily shown that the ratio of total cross sections for transfer-ionization to single electron transfer are predicted to increase with collision energy if shakeover dominates over direct capture.

The simple shake probabilities are expressed in terms of the change in screening s. In a variational calculation it is easily shown that $s = \frac{5}{16}$ to first order in a $1/Z$ expansion by varying s in a hydrogenic wavefunction so as to minimize the binding energy of the state. For K-shell electrons $s = 0.3$ is commonly used as a fit of hydrogenic wavefunctions to more accurate numerical wavefunctions. Another estimate of the screening constant based on Hartree Fock Slater (HFS) wavefunctions has been discussed by Mukoyama [1986] where the screening constant is expressed in terms of the expectation value of the radius, i.e., $s =< r_H > / < r >$. Here r_H is the expectation value of r for a hydrogen atom, namely $\frac{1}{2}[3n^2 - l(l+1)]$ and $< r >$ is the mean value for a HFS wavefunction. Values of $< r >$ and the corresponding screening constant, s, for helium are listed in Table 6.1. Simple calculations sometimes useful for qualitative understanding have been done for other systems by Aberg [1967] and Carlson *et al.* [1968].

Table 6.2. *Experimental high energy ratios of cross sections for double to single electron transitions in helium. If not otherwise noted, the ratio is for the process in addition to ionization divided by single ionization.*

Process	Ratio	Theory
Photoionization (total σ)	1.7×10^{-2} ($\pm$ 10%)	sections 9.3 and 9.5
Compton scattering (total σ)	8×10^{-3} ($\pm$ 20%)	sections 9.4 and 9.5
Ionization by $p^{\pm}$, $e^{\pm}$, bare ions (total σ)	2.6×10^{-3}($\pm$10%)	section 7.4
Ionization by e^- (differential σ, small q)	same as photons	section 10.2
Ionization by p^+ (differential σ binary encounter,large q)	2×10^{-2} ($\pm$ 30%)	sections 10.2 and 10.3
Ionization by neutral atoms (total σ)	1×10^{-2}(?)	Wang *et al.* [1990]
Photoexcitation to $n = 2$ (total σ)	5%(?)	section 10.2
Excitation ($n = 2$) by charged particles	1×10^{-3}?	Nagy *et al.* [1995]; Franz and Altick [1995]
Capture	Velocity dependent?	section 6.3.1
Transfer-ionization/transfer	2-4 $\times 10^{-2}$(?)	section 6.3.1

2 Generalized shake factor

Simple shake, using simple product wavefunctions without exchange, depends entirely on the electron-electron screening in the initial and final states. When more accurate wavefunctions are used, the results are often sensitive to the correlation in these wavefunctions. The accuracy of the results is often significantly improved. Using exact wavefunctions [Aberg, 1976] introduced the generalized shakeoff probability defined by,

$$P_{Generalized\ Shake} = |a|^2 \tag{6.26}$$

$$= |\int e^{i(\vec{q}-\vec{k})\cdot\vec{r_1}}\phi_i^*(r_2, r_3, ...r_N)\Phi(r_1, r_2, r_3, ...r_N)|^2 .$$

Here $\Phi(r_1, r_2, ...r_N) = |i>$ is the exact wavefunction for the initial state and Φ_f^* is approximated by $e^{-i\vec{k}\cdot\vec{r_1}}\phi_i^* r_2, ...r_N$ for a fast outgoing electron of momentum $\vec{k}$. If the wavefunctions are not normalized to unity, then a normalization factor must be included. For excitation to a bound state $e^{-i\vec{k}\cdot\vec{r_1}}$ is replaced by ϕ_f. All multipoles are included in the $e^{i\vec{q}\cdot\vec{r_1}}$ operator from the vector potential $\vec{A}$. The generalized shakeoff probability depends on $\vec{q}-\vec{k}$, the momentum transferred to the residual target ion, where $\vec{q}$ is the momentum transfer of the projectile and $\vec{k}$ is the momentum of the outgoing electron(s). It is the variable $\vec{q}-\vec{k}$ that may be used to relate the generalized shake probability for different physical processes including, for example, ionization by charged particles, photoannihilation and ionization by Compton scattering (Cf. section 9.4.1 and chapter 10).

3 High energy ratios in helium

In the limit of high collision velocities (or energies) the cross section for a double electron transition is often proportional to the cross section for a single electron transition. This may occur if the transition probability for correlation is stronger than the probability for the brief interaction with the projectile. In the high velocity limit it is useful to study the ratio of double to single electron transition cross section since the single collisions effects tend to cancel in the ratio.

Observed values of some of the various ratios in He are given in Table 6.2. Observations in targets other than He are less complete [Barnett, 1990]. The ratios for He vary, in some cases by an order of magnitude. Theory suggests that some of these ratios are interrelated (Cf. sections 10.1 - 10.4). Of central importance is the value of the high energy photoionization ratio.

6.4 The Liouvillian

Change of little systems, such as individual atomic systems, is governed by the Hamiltonian H. Change of distributions of many particles is governed by the Liouvillian $\mathcal{L}$.

1 The Hamiltonian as a generator of dynamics for little systems

The Hamiltonian H specifies how a system of N particles interacts. Let the positions and momenta of these N particles be denoted by $q = \{q_1, q_2, ..., q_N\}$ and $p = \{p_1, p_2, ..., p_N\}$. Any property b of the system will be described in terms of q, p and the time t, i.e. $b = b(\{q_j\}, \{p_j\}, t) = b(q, p, t)$.

Classical dynamics: The time dependence of p and q is determined from the Hamiltonian $H = H(q, p)$ via Hamilton's equations, namely,

$$\begin{aligned} \dot{q} &= \partial H / \partial p \, , \\ \dot{p} &= -\partial H / \partial q \, . \end{aligned} \tag{6.27}$$

The time dependence of b is now given by,

$$\begin{aligned} \dot{b} = \frac{db}{dt} &= \sum_j \left(\frac{\partial b}{\partial q_j} \dot{q}_j + \frac{\partial b}{\partial p_j} \dot{p}_j \right) = \sum_j \left(\frac{\partial b}{\partial q_j} \frac{\partial H}{\partial p_j} - \frac{\partial b}{\partial p_j} \frac{\partial H}{\partial q_j} \right) \\ &\equiv [b, H]_p \, , \end{aligned} \tag{6.28}$$

where $[\ ,\ ]_p$ is called the Poission bracket.

One thus defines,

$$[\ , H]_p \equiv [H]_p \, . \tag{6.29}$$

The operator of (6.29) may be used to generate the time dependence of b. Consider, for any $b(t)$, the Taylor expansion of $b(t)$ about $b(t = 0) \equiv b$, namely,

$$b(t) = b + \dot{b}t + \frac{1}{2} \ddot{b}\, t^2 + ... + \frac{1}{n!} \overset{n}{b}\, t^n + ... \, . \tag{6.30}$$

From (6.28),

$$\dot{b} = [b, H]_p = [H]_p \, , \tag{6.31}$$

and applying the second derivative,

$$\ddot{b} = [\dot{b}, H]_p = [[b, H]_p]_p = [H]_p^2\, b \, , \tag{6.32}$$

so that

$$\overset{n}{b} = [H]_p^n\, b \, . \tag{6.33}$$

Then (6.30) becomes,

$$b(t) = \sum_{n=0}^{\infty} \frac{1}{n!} \overset{n}{b}\, t^n = \sum_{n=0}^{\infty} \frac{1}{n!} t^n [H]_p^n\, b \tag{6.34}$$

$$\equiv \; e^{t[H]_p}\, b \; = e^{t\frac{d}{dt}}\, b \; \equiv U(t)\, b \, .$$

Eq(6.35) defines $[H]_p$ as the generator of time dependence for any property $b(q,p)$.

Quantum dynamics: In quantum mechanics the property b is defined in terms of an expectation value (similar the average value $< b >$ defined later) of a quantum operator $\hat{b}$ acting on a quantum state $|s>$, namely $b =< s|b|s >$. The time dependence of $\hat{b}$ is generated in quantum mechanics according to,

$$\dot{\hat{b}} = \frac{d}{dt}\hat{b} = \frac{1}{i\hbar}[\hat{b}, H]\,, \tag{6.35}$$

in analogy with the classical result of (6.28), where $[\ ,\ ]_p$ is simply replaced by $\frac{1}{i\hbar}[\ ,\ ]$. Following the usual quantum procedure [Goldberger and Watson, 1964, p.50],

$$\begin{aligned} \hat{b}(t) &= e^{iHt/\hbar}\,\hat{b}e^{-iHt/\hbar} = U(0,t)\hat{b}U(t,0) \\ &\equiv e^{it/\hbar[H]}\,\hat{b}\,, \end{aligned} \tag{6.36}$$

where the last line looks like the classical result of (6.35).

The Hamiltonian is usually used for systems of small numbers of particles. Solving the commutator equation $\frac{d}{dt}\hat{b} = \frac{1}{i\hbar}[\hat{b}, H] = \frac{1}{i\hbar}[H]b$ generates $b(t) =< s|e^{iHt/\hbar}\hat{b}e^{-iHt/\hbar}|s >$ from $b = b(t = 0)$. In this way H may be used to generate the dynamics of small systems of particles.

2 The Liouvillian as a generator of dynamics for big systems

Classical dynamics: A macroscopic property $B(x,t)$ may be generated from the corresponding microscopic $b(q,p)$ by averaging over the probability density $F(q,p;t)$ for finding the large system at the point (q,p) in phase space, namely,

$$B(x,t) = \int dq\; dp\; b(q,p)F(q,p;t) \equiv < b >\,, \tag{6.37}$$

where $\sum_j$ is implied.

Since $F(q,p;t)$ is itself a property b of the microscopic system, one has from (6.35),

$$F(q,p;t) = e^{-[H]_p t}\; F(q,p) \equiv e^{\mathcal{L}t} F(q,p)\,, \tag{6.38}$$

where in accord with (6.28),

$$\mathcal{L}F = [H,F]_p = \sum_j \left(\frac{\partial H}{\partial q_j}\frac{\partial F}{\partial p_j} - \frac{\partial H}{\partial p_j}\frac{\partial F}{\partial q_j} \right) = \frac{\partial F}{\partial t} , \tag{6.39}$$

where $[\ ,H]_p$ has been replaced by $[H,\ ]_p$ giving a minus sign in (6.38). Eq(6.39) is called the Liouville equation and is as central to statistical mechanics as the Hamiltonian is to quantum mechanics [Balescu, 1975, p.41]. The Liouvillian $\mathcal{L}$ may be used to generate correlation from the uncorrelated limit in a many body system. For example $\mathcal{L}$ may be used to develop the BBGKY hierarchy.

Quantum dynamics: The quantum probability density f^W satisfies a similar Liouville equation [Balescu, 1975], namely,

$$\frac{\partial f^W}{\partial t} = \mathcal{L}f^W , \tag{6.40}$$

where f^W is the Wigner function (including the symmetry of identical particles).

Thus, the Liouvillian $\mathcal{L}$ generates the time dependence of the probability density f^W or F which enables one to express time dependent macroscopic variables $B(t)$ to time independent microscopic variables b. Thus $\mathcal{L}$ generates correlated dynamics for macroscopic systems from microscopic systems.

An example: Consider a quantum system governed by a many body Hamiltonian $H = H_T + v$, where v is the correlation potential. This system is perturbed by an interaction potential $V(t)$. In the interaction representation the electronic wavefunction ψ is given by,

$$\psi(t) = U(t,t_0)\psi(t_0) , \tag{6.41}$$

with

$$\begin{aligned} U &= Te^{-i\int_{t_0}^{t} V_I(t')dt'} , \\ V_I &= e^{-i(H_T+v)t}V(t)e^{i(H_T+v)t} . \end{aligned} \tag{6.42}$$

Now if $v \to 0$, the correlation disappears and $H \to H_T = \sum_j H_T^j$ which is a sum of single particle terms. In this uncorrelated limit the interaction potential reduces to a sum of single particle operators, namely,

$$V_i \to V_I^0 = e^{-iH_0 t} V(t) e^{iH_T t} = \sum_j V_I^{0\ j} , \tag{6.43}$$

and

$$\begin{aligned} U_I \to U_I^0 &= Te^{i\int V_I^0(t)dt} = Te^{-i\sum_j \int V_I^{0\ j} dt} \\ &= \prod_j Te^{-i\int V_I^{0\ j} dt} = \prod_j U_I^{0\ j} , \end{aligned} \tag{6.44}$$

so that the evolution operator is uncorrelated. Thus,

$$\psi(t) = U_I^0(t,t_0)\psi(t_0) = \prod_j U_I^{0\ j}(t,t_0)\psi(t_0) = \prod_j \psi_j(t) , \tag{6.45}$$

i.e., each particle evolves independently (Cf. chapter 4).

Dynamic (or scattering) correlation is defined as the difference between U_I (with the correlation potential v) and U_I^0 (without v).

Consider the full Liouvillian, $\mathcal{L}$, and the uncorrelated Liouvillian, $\mathcal{L}_0$ with $\mathcal{L}_c = \mathcal{L} - \mathcal{L}_0$. That is,

$$\begin{aligned} V_I &= e^{-i(H_T+v)t} V(t) e^{i(H_T+v)t} = e^{-i\mathcal{L}t} V , \\ V_I^0 &= e^{-iH_T t} V(t) e^{iH_T t} = e^{-i\mathcal{L}_0 t} V . \end{aligned} \tag{6.46}$$

Now,

$$V_I = e^{-i\mathcal{L}t} V = e^{-i\mathcal{L}t} e^{-i\mathcal{L}_0 t} V_I^0 . \tag{6.47}$$

The operator,

$$\mathcal{O} = e^{-i\mathcal{L}t} e^{-i\mathcal{L}_0 t} \tag{6.48}$$

connects the correlated potential operator V_I to the uncorrelated V_I^0, and satisfies the initial condition $\mathcal{O}(t=0) = 1$. The equation that this operator satisfies may be found by differentiating (6.48), namely,

$$\frac{\partial \mathcal{O}}{\partial t} = -ie^{-i\mathcal{L}t} [\mathcal{L} - \mathcal{L}_0] e^{i\mathcal{L}t} = -ie^{i\mathcal{L}t} \mathcal{L}_c e^{-i\mathcal{L}t} , \tag{6.49}$$

or equivalently,

$$\mathcal{O}(t) = 1 + \int_{t_0}^{t} e^{i\mathcal{L}t'}\mathcal{L}_c e^{-i\mathcal{L}t'}dt' \tag{6.50}$$
$$= 1 + \int_{t_0}^{t} \mathcal{O}(t')e^{i\mathcal{L}_0 t'}\mathcal{L}_c e^{-i\mathcal{L}_0 t'}dt' \; .$$

Dynamic correlation may be obtained by solving (6.49) or (6.51) for $\mathcal{O}(t)$ which generates the correlated V_I from the uncorrelated V_I^0, which in turn generates the correlated evolution operator $U_I(t, t_0)$ that gives the fully correlated electronic wavefunction $\psi(t)$ for all t.

7
Perturbation expansions

In this chapter an analysis for various order Born amplitudes is considered. Emphasis is given to the second order probability amplitude which is interpreted in terms of its real and imaginary parts. The main concepts are: time ordering, correlation in space, correlation in time, and dispersion relations between real and imaginary parts of the second order quantum amplitude. The second order term in Coulomb scattering of charged particles usually dictates properties of two electron processes such as Auger transitions, direct double excitation, transfer-ionization, etc. This second order term has some relation to second order terms for photoionization, Raman scattering and Compton scattering discussed in chapter 9. Expansions in the interaction potential V alone and in both V and the correlation potential v are considered.

7.1 Formulation

The probability $P(\vec{b})$ for a transition from an initial state $|i>$ to a final state $|f>$ may be expressed (Cf. section 2.1) in terms of the probability amplitude $a(\vec{b})$,

$$P(\vec{b}) = |a(\vec{b})|^2 , \tag{7.1}$$

where $\vec{b}$ is the impact parameter of the scattering event.

The probability amplitude $a(\vec{b})$ may be expressed (see (2.22)) in terms of the usual Born (or perturbation) expansion in powers of the interaction potential $V(t)$, namely,

$$a(\vec{b}) \;\; = \;\; lim_{t\to\infty} < \psi_f|\psi_i(t) > = < f|\; U(+\infty, -\infty)\; |i >$$

$$
\begin{aligned}
&= \ < f|T \exp\left(-i\int_{-\infty}^{+\infty} V(t)dt\right)|i> \\
&= \ < f|T \sum_{n=1}^{\infty} \frac{1}{n!}\left(-i\int_{-\infty}^{+\infty} V(t)dt\right)^n |i> \\
&= \ < f|\sum_{n=1}^{\infty} \frac{(-i)^n}{n!} \int_{-\infty}^{+\infty} dt_n \int_{-\infty}^{+\infty} dt_{n-1} .... \\
&\times \ \int_{-\infty}^{+\infty} dt_1 T[V(t_n)V(t_{n-1})...V(t_1)]|i> \\
&= \ < f|\sum_{n=1}^{\infty} (-i)^n \int_{-\infty}^{+\infty} V(t_n)dt_n \int_{-\infty}^{+t_n} V(t_{n-1})dt_{n-1} .... \\
&\times \ \int_{-infty}^{+t_2} V(t_1)dt_1|i> \ ,
\end{aligned}
\tag{7.2}
$$

where it is assumed that $< f|i >= 0$. T is the time ordering operator discussed in detail in 7.1.2. The time dependence of $V(t)$ is assumed to be explicitly known. For example, if the projectile is a particle with charge Z, and an internuclear trajectory, $\vec{R}(t)$, then $V(t) = Z/|\vec{R}(t) - \vec{r}|$, where $\vec{R}$ is the position of the target electron relative to its nucleus.

1 *Binomial distributions with shake*

The uncorrelated limit without exchange of the exact many electron transition probability leads to a binomial distribution of two state single electron probabilities in the limit that all of the single electrons probabilities are identical. This limit may include shake terms. It is done here to second order in the interaction potential, V. To obtain uncorrelated transition probabilities including shake in a system of N electrons one may use the following approximations (Cf. section 4.1):

1. Ignore correlation in both the initial and final states. Then, $|i >= \Pi_j|i_j >$ and $|f >= \Pi_j|f_j >$.
2. Include shake. Then, $< f_j|i_j > \neq \delta_{f_j i_j}$.
3. Ignore exchange. The wavefunctions are not asymmetrized. The electrons may be distinguishable.
4. Ignore scattering correlation in the evolution operator U. Then, $U = Te^{i\int V dt} = Te^{i\int \sum_j V_j dt} \simeq \Pi_j Te^{i\int V_j dt} = \Pi U_j$.

Consequently,

$$a \simeq \Pi_j < f_j|U_j|f_j >= \Pi a_j \tag{7.3}$$

This gives an independent electron approximation without exchange because a is a product of single electron probability amplitudes.

Second Born approximation. Consider the above amplitudes in the case of $N = 2$ (i.e., helium-like systems). Consider terms through second Born in the expansion of the scattering operator $U(\infty, -\infty)$ in powers of V (i.e., powers of Z/v, where v is the velocity of the projectile).

For single ionization one has $|f_2> = |i_2^{(f)}> \neq |i_2>$ so that,

$$\begin{aligned} a^+ &\simeq a_1\, a_2 = < f_1|U_1|i_1 >< i_2^{(f)}|U_2|i_2 > \simeq < f_1|U_1|i_1 >< i_2^{(f)}|i_2 > \\ &\simeq < f_1|1 + i\int V_1 dt|i_1 >< i^{(f)}|i_1 > = a_{1\,(1B)} < i_2^{(f)}|i_2 > \end{aligned} \tag{7.4}$$

where the term $< f_1|i_1 >< i_2^{(f)}|i_2 >$ is dropped because there can be no transition if nothing happens to the target electrons. This unphysical term aries from the shake requirement that $< f_j|i_j > \neq \delta_{f_j i_j}$. In a better development this would occur naturally. Here $a_{(1B)}$ indicates the first Born approximation which holds at moderately high velocity for single ionization. Often $< i_2^{(f)}|i_2 > \simeq 1$. Now,

$$\mathrm{P}^+ = |a^+|^2 \simeq |a_{(1B)}|^2 \sim Z^2/\mathrm{v}^2 \tag{7.5}$$

because (using $z = \mathrm{v}t$)

$$\begin{aligned} a_{(1B)} &= i\int < f_1|V_1|i_1 > dt = i\int < f_1|\frac{Z}{|\vec{R} - \vec{r}|}|i_1 > \frac{dz}{\mathrm{v}} \\ &= \frac{Z}{\mathrm{v}}\int < f_1|\frac{dz}{|\vec{R} - \vec{r}|}|i_1 > \sim \frac{Z}{\mathrm{v}} \end{aligned} \tag{7.6}$$

and $< i_2^{(f)}|i_2 >$ is independent of Z and v.

For double ionization it is convenient to assume that again electron 1 and electron 2 are distinguishable. Here electron 1 is the primary electron and electron 2 is either caused by shake or by a second interaction with the projectile. Then,

$$\begin{aligned} a^{++} &\simeq a_1\, a_2 = < f_1 f_2|U_1 U_2|i_1 i_2 > \\ &= < f_1 f_2|(Te^{i\int V_1 dt})\,(Te^{i\int V_2 dt})|i_1 i_2 > \\ &\simeq < f_1 f_2|(1 + i\int V_1 dt - 1/2(\int V_1 dt)^2) \\ &\times\ (1 + i\int V_2 dt - 1/2(\int V_2 dt)^2)|i_1 i_2 > \end{aligned}$$

$$
\begin{aligned}
&\simeq < f_1 f_2|1 + i\int V_1 dt + i\int V_2 dt - (\int V_1 dt)(\int V_2 dt)|i_1 i_2 > \\
&= < f_1 f_2|1 + i\int V_1 dt - (\int V_1 dt)(\int V_2 dt)|i_1 i_2 > \\
&= i < f_1|\int V_1 dt|i_1 >< f_2|1 + i\int V_2 dt|i_2 > \\
&= a_{1\ (1B)}(< f_2|i_2 > + a_{2\ (1B)}) \equiv a_{1\ (1B)}(a_{2\ Shake} + a_{2\ (1B)})
\end{aligned} \tag{7.7}
$$

where T is the time ordering operator (which enforces quantum causality). The shake probability amplitude is $a_{2\ Shake} =< f_2|i_2 >$. The V_2 term was dropped because of the assumption that electron 1 is the primary electron. Keeping the V_1^2 and V_2^2 terms does not change the analysis significantly. The $< f_1 f_2|i_1 i_2 >$ term is again dropped because it is unphysical. Note that a^{++} retains its product form indicative of an independent electron approximation. Now,

$$P^{++} \simeq |a_{1\ (1B)}|^2 |a_{2\ Shake} + a_{2\ (1B)}|^2 = P_1 P_2 = \Pi_j P_j \tag{7.8}$$

Note that P^{++} contains a Z^3/v^3 term that changes for $Z \to -Z$ (i.e., differs for p^+ and e^- projectiles). At high velocity the effect of the mass difference between p^+ and e^- is quite small. This gives different single transition probabilities P_j for p^+ and e^- due to the difference in their charges Z.

Binomial distributions occur when the single electron transition probabilities are the same and each electron has just two possible states (e.g. a filled ground state or a vacant ground state). Then $P_j = |a_j|^2$ is the probability that a transition occurs and $(1 - P_j)$ is the probability a transition does not occur. In a system of N electrons if P_j is independent of j, except for $j = 1$ which excludes shake, then the usual binomial distribution follows for the $N - 1$ electrons (Cf. sections 1.1.2 and 4.1.3). For $N > 2$ this means that binomial distributions with different values of P for p^+ and e^- may be used to analyze date in cases where the assumptions used are valid [Sant'Anna *et al.*, 1997]. The p^+ - e^- difference goes to zero as Z/v at high velocity. Note that the probability for the primary electron drops out in ratios of probabilities and cross sections for different degrees of ionization, e.g., σ^{n+1}/σ^n.

Limiting cases. Two limiting cases for multiple ionization are considered.

1. Strong interaction with the projectile. Then each of the U_j operators is different from unity, and $P = \Pi_j\ P_j$ where each of the P_j is the probability of ionizing the jth target electron in a direct interaction

with the projectile. This is the usual independent electron approximation without exchange and without any electron-electron effects (e.g. without shake terms). This usually holds when the interaction with the projectile is stronger than the (residual) correlation interaction (i.e., the difference between the $1/r_{ij}$ and the mean field $v(r_j) \simeq 1/r_{ij}$ terms. A perturbation expansion in V in not required.

2. Weak interaction with the projectile. In this case one obtains the first Born amplitude in V. That is, $\Pi_j U_j = \Pi_j \; Te^{i\int V_j dt} \simeq \Pi_j(1 + i\int V_j dt)_j \simeq 1 + \sum_j \int V_j dt$ where the kth electron is ionized by the projectile and the rest undergo shakeoff when $|f_j> \neq |i_j^{(f)}>$. In this case multiple ionization is dominated by shake terms. That is, $a \simeq \int <f_k|V_k|i_k> dt \;\; \Pi_{j\neq k} <f_j|i_j>$. This usually holds in the limit of high velocity if initial and final state correlation are neglible.

2 *Time ordering*

Time ordering in quantum amplitudes is governed by T the Dyson time ordering in (7.2) operator (as defined by (2.22) or Goldberger and Watson, 1964, p.48). T means that $V(t_{j-1})$ acts before $V(t_j)$, $(j \leq n)$. Setting $T = 1$ means that all time sequences are included (see figure 7.1) so that time ordering is not significant. In general time ordering can be significant in all higher order amplitudes. The effects of time ordering may be significant in any two (or many) electron transition, Auger transitions, Raman and second order Compton scattering, and second order Thomas processes such as those in mass transfer reactions. It is useful to use $T = 1 + (T - 1)$, where $(T - 1)$ represents the effects of time ordering, i.e. quantum causality. These $T - 1$ time ordering effects will be isolated mathematically and analyzed in this section. This analysis follows Nagy *et al.* [1996] for impact by charged particles and is also similar to that used by Heller *et al.* [1982] for Raman scattering by photons and by and Fano [1961] for analysis of resonances.

Meaning of T and the step function. Time ordering arises from causality: cause must occur before effect. The time ordering operator T enforces causality in the interaction process. Specifically, T keeps $t_j > t_{j-1}$ in the above integration, so that T may be represented in terms of the Heavyside step function $\Theta(t_j - t_{j-1})$,

$$\begin{aligned} \Theta(t_j - t_{j-1}) &= 1 \qquad t_j \geq t_{j-1} \,, \\ &= 0 \qquad t_j < t_{j-1} \,. \end{aligned} \tag{7.9}$$

Specifically,

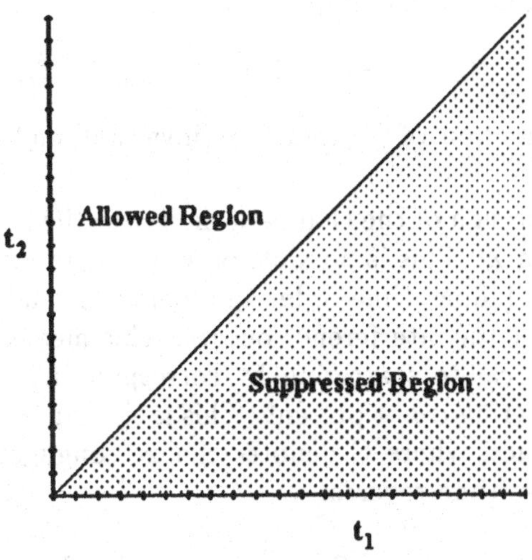

Fig. 7.1. Allowed and suppressed regions for the time ordering operator T in second order.

$$\frac{T}{n!} = \prod_j \Theta(t_j - t_{j-1}) , \tag{7.10}$$

which corresponds to the definition of T in (7.2). This insures that $V(t_{j-1})$ acts before $V(t_j)$ in (7.2).

Theorem: *Transition probabilities and cross sections do not depend on the sign of the interaction potential (i.e. $V \to -V$) if the effects of time ordering are ignored for transitions of well defined parity.*

Proof: If the effects of time ordering are ignored, then $T = T_{av}$ in (7.4) and,

$$\begin{aligned} U(\infty, -\infty) &= T_{av} \sum_{n=0}^{\infty} \frac{(-i)^n}{n!} \int_{-\infty}^{\infty} V_I(t_n) dt_n \cdots \int_{-\infty}^{+\infty} V_I(t_1) dt_1 \\ &= U_{even} + iU_{odd} , \end{aligned} \tag{7.11}$$

where U_{even} and U_{odd} are even and odd in V (or Z). Each factor of $\int_{-\infty}^{+\infty} V_I(t)dt$ yields a matrix element of $\int_{-\infty}^{\infty} e^{i(E_I - E_{I'})t} < I|V(t)|I' > dt$ which is purely real or purely imaginary depending on the parity of the states (real for $l + m - l' - m'$ even and imaginary, if this sum is odd). Because in the product of the integrals each intermediate state occurs twice, the phase of the product depends only on the parity of the initial and the final states. Thus, for a given transition U_{even} and U_{odd} are purely real (or purely imaginary). Consequently,

$$|a|^2 = |< f|U(\infty, -\infty)|i > |^2 = U_{even}^2 + U_{odd}^2 \tag{7.12}$$

which is an even function of V, and thus invariant under $Z \to -Z$ for charged projectiles.

Dependence on the sign of the charge of the projectile is obtained only if one has odd terms in Z. This is possible only by the interference between U_{even} and U_{odd}. Including time-ordering each term may have real and imaginary parts, thus the interference occurs. This means that scattering by particles is identical to scattering by corresponding anti-particles, if and only if, time ordering is not present. This is similar to CPT invariance in high energy physics, where C denotes charge congugation, P denotes parity and T denotes time reversal.

Counting time regions. The effect of the time ordering operator T is shown in the last two lines of (7.2) above – namely to exclude $t_j < t_{j-1}$. Now, if one sets $T = 1$ in (7.2) so that time ordering is ignored, then multiple counting occurs. In the case of $n = 2$ contributions from $t_2 > t_1$ are not excluded when $T = 1$, which leads to double counting unless the factor of $1/2!$ is included with T.

In general the factor of $n!$ compensates for the multiple counting.

Second order terms. If one ignores terms in the Born perturbation expansion in V for $n > 2$ in (7.2), then one has a second Born approximation given by,

$$\begin{aligned} a(\vec{b}) &= a^1 + a^2 + a^3 + ... + a^n + ... \approx a^1 + a^2 \\ &= -i \int_{-\infty}^{+\infty} < f|\, T\, V(t_1)|i > dt_1 \\ &+ \frac{(-i)^2}{2!} \int_{-\infty}^{+\infty} dt_2 \int_{-\infty}^{+\infty} dt_1 < f|\, T\, V(t_2)\, V(t_1)\, |i > . \end{aligned} \tag{7.13}$$

There is no time ordering in the first term since only one order is possible. Thus $T = 1$ in the first term above. (Note the factor of $-i$ carried by each power of V so that even powers of V and odd powers of V tend to be 90 degrees out of phase and tend not to interfere in the observable quantity $|a|^2$.)

One may now insert a complete set of intermediate eigenstates $|I >< I|$ which belong to the same set of eigenstates as $< f|$ and $< i|$ and which have energies, ϵ_I. Next the explicit time dependence is separated from all these states using $|s >= e^{-i\epsilon t}|\tilde{s} >$, which yields,

$$\begin{aligned}
a(\vec{b}) &\approx -i\int_{-\infty}^{+\infty} e^{i(\epsilon_f-\epsilon_i)t_1} < \tilde{f}|\ V(t_1)|\tilde{i} > dt_1 \\
&+ \frac{(-i)^2}{2!}\int_{-\infty}^{+\infty} dt_2 \int_{-\infty}^{+\infty} dt_1\ T\ e^{-i\epsilon_f t_2} \\
&\times \sum_{\tilde{I}} < \tilde{f}|V(t_2)|\tilde{I} > \ e^{-i\epsilon_I(t_2-t_1)} \ < \tilde{I}|V(t_1)|\tilde{i} > \ e^{-i\epsilon_i t_1} \\
&= -i\int_{-\infty}^{+\infty} e^{i(\epsilon_f-\epsilon_i)t_1} < \tilde{f}|\ V(t_1)|\tilde{i} > dt_1 \\
&+ (-i)^2\int_{-\infty}^{+\infty} dt_2 \int_{-\infty}^{+\infty} dt_1\ \Theta(t_2-t_1)\ e^{-\epsilon_f t_2} \\
&\times \sum_{\tilde{I}} < \tilde{f}|V(t_2)|\tilde{I} > \ e^{-i\epsilon_I(t_2-t_1)} \ < \tilde{I}|V(t_1)|\tilde{i} > \ e^{-i\epsilon_i t_1} \\
&= a^1 + a^2 \qquad (7.14)
\end{aligned}$$

where for the second Born term $\frac{T}{2!} = \Theta(t_2 - t_1)$ and $T_{av} = 1$ corresponding to the average value of $2\Theta(t_2 - t_1)$.

Time ordering identity. The following identity [Ziman, 1970, pp. 81 and 109] leads to insight into the nature of the second order amplitude,

$$\begin{aligned}
\frac{T}{2}e^{-i\epsilon(t_2-t_1)} &= \Theta(t_2-t_1)e^{-i\epsilon(t_2-t_1)} \\
&= \frac{i}{2\pi}\int_{-\infty}^{+\infty} e^{-i\omega(t_2-t_1)}\frac{1}{\omega-\epsilon+i\eta}d\omega \\
&= \frac{i}{2\pi}\int_{-\infty}^{+\infty} e^{-i\omega(t_2-t_1)}\left\{-i\pi\delta(\omega-\epsilon) \ + \ \mathcal{P}\frac{1}{\omega-\epsilon}\right\}d\omega \\
&= \frac{1}{2}[\,T_{av} + i(T - T_{av})\,]\,e^{-i\epsilon(t_2-t_1)}\,, \qquad (7.15)
\end{aligned}$$

where $\mathcal{P}$ denotes the principle part of the integral (i.e. excluding the point at $\omega = \epsilon$). Here one has separated the time ordering term $(T-1) = (T - T_{av})$ corresponding to the $\mathcal{P}$ term, from the non-time ordering term $T_{av} = 1$ corresponding to the $-i\pi\delta(\omega - \epsilon)$ term*. There is an obvious interpretation of the above equation. The T_{av} term includes only the contribution at $\omega = \epsilon$ corresponding to energy conservation in the intermediate states. The time ordering term corresponds to contributions from energy non-conserving intermediate states, i.e. virtual states.

* This development runs parallel to that for interactions with photons given by Heitler [1954] who introduces in section 8 $\zeta(x) = \mathcal{P}/x - i\pi\delta(x)$ corresponding to line 3 of (7.15) above.

Note that the T_{av} and the $T - T_{av}$ terms in (7.15) differ by 90 degrees in phase in the second order amplitude a^2, i.e.,

$$a^2 = |a^2_{T_{av}}| + i|a^2_{T-T_{av}}| . \tag{7.16}$$

Thus, interference properties depend on time ordering.

Time correlation. The average value of $T = 2\Theta$ in (7.15) is $T_{av} = 1$, since the average value of $\Theta = 1/2$. Consequently $(T - 1)$ represents fluctuations about the average value of T. This is related to the notion of correlation as $\sqrt{\text{Property} - < \text{Property} >_{av}}$ given in section 1.4.3 and to the mathematical definition of correlation in section 6.1 as a difference from the uncorrelated value Property$-$ $<$ Property $>_{uncorrelated}$. Thus, $(T - 1)$ is taken as a measure of correlation in time.

Since $T - 1 = 2(\Theta - < \Theta >_{av})$, the $T - 1$ operator is antisymmetric about $t_1 = t_2$. Specifically $(T - 1)(\tau) = sign(\tau) = \pm 1$, where $\tau = t_2 - t_1$. So in (7.13) $T - 1$ takes the difference between the matrix elements $< f|V(t_2)V(t_1)|i >$ and $< f|V(t_1)V(t_2)|i >$ about $t_1 = t_2$.

Classical correspondence. The mathematical requirement for time ordering is that the propagator (or Green function) is constrained to be zero for $t_2 < t_1$. This causality condition is the same classically and quantum mechanically. However, for classical particles $|\psi|^2(t) = \int U(t,t')|\psi|^2(t')dt'$, while for waves (e.g. quantum waves) $\psi(t) = \int U(t,t')\psi(t')dt'$ manages to preserve phase information. In both cases $U(t_2, t_1) = 0$ if $t_2 < t_1$. The reason that quantum observables may not be localized in time is due to the wave nature of wavepackets. Wavepackets, either classically or quantally, are governed by the band width theorem which relates the width of a packet in time Δt to its width in frequency $\Delta\nu$ (namely, $\Delta\nu\Delta t \geq 1$). It is this property that disturbs causality quantum mechanically†.

For one electron transitions in targets interacting with a projectile of charge Z, effects of Z^3 terms such as polarization of the target electron cloud (including those in classical particle calculations such as CTMC) arise from the interference between the first order amplitude and the $T - T_{av}$ contribution of the second order amplitude. These Z^3 effects may be associated with time ordering as a consequence of the theorem after (7.4).

Dispersion relation. It may be shown that the real and imaginary

† This point is further discussed in the introduction to chapter 10 of Goldberger and Watson [1964] who refer to Wigner's struggles with this topic.

parts of (7.15) may be connected by a dispersion relation. The Fourier transform of (7.15), $f(\epsilon)$, is analytic in the upper half of the complex ϵ plane (more specifically, it is a simple pole in the lower half complex ϵ plane). Consequently, closing the integral in the upper half plane the elementary theorem of Cauchy gives,

$$f(\epsilon) = \frac{1}{2\pi i} \int_{-\infty}^{+\infty} \frac{f(\omega)}{\omega - \epsilon} d\omega \,. \tag{7.17}$$

Then taking the real and imaginary parts, one has,

$$\begin{aligned} Re\{f(\epsilon)\} &= -\frac{1}{2\pi} \int_{-\infty}^{+\infty} \frac{Im\{f(\omega)\}}{\omega - \epsilon} d\omega \,, \\ Im\{f(\epsilon)\} &= \frac{1}{2\pi} \int_{-\infty}^{+\infty} \frac{Re\{f(\omega)\}}{\omega - \epsilon} d\omega \,. \end{aligned} \tag{7.18}$$

It is straightforward to show that this holds for the real and imaginary parts of the second order contribution to the probability amplitude $a(\vec{b})$ as well. This means that the energy-conserving (classical, non-time ordered) part of the second order amplitude is related to an integral over the energy-non-conserving (non-classical, time ordered) part, and vice versa.

A classical example of the application of this dispersion relation in the vicinity of a resonance is given by Jackson [1975, p. 286] for anomalous dispersion in a dielectric medium. A second example may be found in the case of the second order Thomas scattering peak in electron transfer [McGuire *et al.*, 1993] where the $T - T_{av}$ term contributes equally as much as the T_{av} term of (7.15), and where there is a classically underdamped shift in the impact parameter of the collision.

Higher order terms. A similar result may be generally obtained from the nth order Born term of (7.2) in a straightforward way, using,

$$\begin{aligned} \frac{T}{n!} \prod_j e^{-i\epsilon_j(t_j - t_{j-1})} &= \prod_j \Theta(t_j - t_{j-1}) e^{-i\epsilon_j(t_j - t_{j-1})} \\ &= \prod_j \frac{i}{2\pi} \int_{-\infty}^{+\infty} e^{-i\omega(t_j - t_{j-1})} \frac{1}{\omega_j - \epsilon_j + i\eta} d\omega \\ &= \prod_j \frac{i}{2\pi} \int_{-\infty}^{+\infty} e^{-i\omega_j(t_j - t_{j-1})} \\ &\times \left\{ -i\pi\delta(\omega_j - \epsilon_j) + \mathcal{P} \frac{1}{\omega_j - \epsilon_j} \right\} d\omega_j \end{aligned}$$

$$= \frac{1}{n!}[\,1 + (T-1)\,]\prod_j e^{-i\epsilon_j(t_j - t_{j-1})} \,. \qquad (7.19)$$

Note the phase factor of i between the 1 and the $(T-1)$ operators.

3 *Spatial correlation*

'Correlation' caused by the $1/r_{12}$ electron-electron correlation potential corresponds to 'spatial correlation' as distinct from 'time correlation' discussed above.

First order term in V. In a two electron transition if the projectile interacts once via the interaction V with an electron, then a two electron transition cannot occur unless the electrons interact with each other. That is, without electron correlation a two electron transition does not occur in first order in V since the matrix elements $< f_2|i_2 >< f_1|V(r_1)|i_1 >$ and $< f_2|V(r_2)|i_2 >< f_1|i_1 >$ are zero due to the orthogonality of the independent electrons, i.e. $< f_j|i_j >= 0 \;\; (j = 1, 2)$. If the electron correlation interaction is retained, then two electron transitions may occur when one electron interacts with the other electron when the eigenstate of the system is perturbed by one interaction with V. Thus, a non-zero first order term in V for a two electron transition occurs only when the $1/r_{12}$ electron-electron interaction plays a role (e.g. as in section 7.1.1).

Second order term in V. Two electron transitions may occur without electron correlation in the second order amplitude a^2 in (7.2). However, without correlation $a^2 = a_1^1 \cdot a_2^1$ corresponding to a simple product of two first order amplitudes as given by the independent electron approximation (Cf. (4.10) in section 4.1). As is clear from (7.13) there is no time ordering in the first order amplitudes, a_j^1. Several authors [McGuire and Straton, 1990b; Stolterfoht, 1991; Nagy *et al.*, 1996] have explicitly confirmed that in second order in V there is no time ordering without electron correlation. In this case the $a^2 = a_1^1 \cdot a_2^1$ term in (7.13) is real and reduces to the independent electron approximation for two electrons in accord with (4.10).

If the first order term in V is non-zero, it is imaginary if the corresponding matrix element of (7.13) is real. This is often the case. Under these circumstances the first order term a^1 and the second order term $a_1^1 \cdot a_2^1$ are 90 degrees out of phase and there is no interference between the first and second order terms. In order to have a difference in two electron transition probabilities and cross sections under $V \to -V$ (e.g. $Z \to -Z$), it is then necessary to have spatial electron correlation in the first order term and time correlation (which requires electron correlation) in the second order term .

In the second order term electron correlation is necessary for time ordering. However, in higher order terms in V this is generally not the casecorrelation!and time ordering in higher order perturbation theory.

From the discussion above, it is clear that differences in scattering of particles and anti-particles occur only when first order terms in V due to spatial electron correlation and the time ordering term in the second order amplitude are both non-zero. For two-electron processes, this means that both spatial correlation and time correlation are both required to have differences in scattering by particles and anti-particles. Observations of such Z^3 effects are discussed in section 7.4.1.

Strong fields and multi-photon processes. For weak electromagnetic fields one does not usually consider higher order processes. For highly charged projectiles the Coulomb distortion of the projectile field may be significant [Godunov *et al.*, 1996]. Also in intense laser fields multi-photon processes may occur. That is, multi-photon processes can be non-negligible just as the higher Born terms in V can be non-negligible in some cases. In these cases it may be possible to use multi-photon scattering to identify and to probe dynamic electron correlation using an analysis similar to that developed for potential scattering in the sections above. Again, it may be helpful and unifying to keep in mind that the V^n term in Coulomb scattering corresponds to a virtual n photon process.

7.2 Method of expansion in the scattering potential V

Fully correlated probabilities and cross sections for two electron transitions may be evaluated through second order in the interaction with the projectile using a technique developed by Straton, McGuire and Chen [1992]. The atomic wavefunction is approximated by a sum of products of one electron wavefunctions, with the coefficients chosen by diagonalizing the fully correlated many electron Hamiltonian. Thus, spatial correlation is included in both the asymptotic and scattering regions by using these configuration interaction (CI) wavefunctions for initial, intermediate, and final states [Szabo and Ostlund, 1982]. The many electron scattering amplitude is expressed as a product of transition amplitudes for one electron atoms. Straton [1991] has found a general analytical form for these amplitudes [see section 2.3.2]. Both the energy-conserving and energy-non-conserving parts of the second order amplitude are found, so that time ordering effects are included.

The approach below represents final bound and continuum states by means of pseudostate expansions. The pseudostates have the same mathematical form as hydrogenic single electron states, but coefficients including the effective nuclear charge are changed so that the energies of these states do not correspond to real atoms. If these pseudo energies are positive, then the pseudostates represent a band of continuum energy states with positive energies near the pseudo energy. Because these expansions have the same analytical character as hydrogenic orbitals, an exponential multiplied by a polynomial in the radius, the general analytic form found for hydrogenic transitions may be applied to transitions to such states. This technique again results in a one-dimensional integral for the energy-non-conserving second order term and analytical expressions for the energy-conserving second order and first order terms in the many electron transition amplitude.

1 Scattering amplitude

The probability amplitude for transition from an initial state $|i\rangle$ to a final state $\langle f|$ may be written as [McGuire, 1987],

$$a_{if} = \langle f|\ T\ exp[\ -i \int V(t)dt\ ]\ |i\rangle\ , \tag{7.20}$$

where V is the interaction between the projectile and the electrons in the atom. Then,

$$V(t) = \sum_j\ e^{iH_T t}\ V_j(t)\ e^{-iH_T t}\ , \tag{7.21}$$

in which,

$$V_j(t) = -\frac{Z}{|\vec{R}(t) - \nu\vec{r}_j|} + \frac{Z}{R}\ , \tag{7.22}$$

$\nu = M_t/(M_t + m_e)$, and T is the time ordering operator (Cf. section 7.1.2). The Born expansion of the amplitude is then given by,

$$a_{fi} = a_{fi}^{(0)} + a_{fi}^{(1)} + a_{fi}^{(2)} \cdots\ , \tag{7.23}$$

where $a_{fi}^{(0)} = \langle f|i\rangle$, which will be zero for inelastic scattering under consideration here. The first order amplitude is,

$$a_{fi}^{(1)} = -i\int_{-\infty}^{\infty} dt \langle f|e^{iH_T t}\ [V_1(t) + V_2(t)]\ e^{-iH_T t}|i\rangle\ . \tag{7.24}$$

For initial and final states given as CI wavefunctions, one has,

$$|i\rangle = \sum\nolimits_{\alpha_1\alpha_2} C_{\alpha_1\alpha_2} |\alpha_1\rangle|\alpha_2\rangle \,, \tag{7.25}$$

and

$$\langle f| = \sum\nolimits_{\beta_1\beta_2} \bar{C}_{\beta_1\beta_2} \langle\beta_1|\langle\beta_2| \,, \tag{7.26}$$

where the $C_{\alpha_1\alpha_2}$ and $\bar{C}_{\beta_1\beta_2}$ are the CI coefficients that give the normalized strength of the mixed single state configurations, $|\alpha_1\rangle|\alpha_2\rangle$ and $\langle\beta_1|\langle\beta_2|$. Then one obtains a weighted sum of one-electron scattering amplitudes,

$$A^{\lambda'}_{n'\ell'-m',nlm}{}^{\lambda}(\omega, b, \nu, -1) = \int_{-\infty}^{\infty} dt\; e^{i\omega t} \langle n'\ell'm'|V(t)|n\ell m\rangle \,, \tag{7.27}$$

in which $\lambda = Z_t/a_0$ and b is the impact parameter, and overlap integrals $\langle n'\ell'm'|n\ell m\rangle$ as the fully correlated two-electron amplitude,

$$\begin{aligned} a^{(1)}_{fi} &= \sum\nolimits_{\beta_1,\beta_2} \bar{C}_{\beta_1,\beta_2} \\ &\times \sum\nolimits_{\alpha_1,\alpha_2} C_{\alpha_1,\alpha_2}\; (-i) \left[A^{\lambda'\lambda}_{\beta_1,\alpha_1}(\omega)\langle\beta_2|\alpha_2\rangle + A^{\lambda'\lambda}_{\beta_2,\alpha_2}(\omega)\langle\beta_1|\alpha_1\rangle\right] . \end{aligned} \tag{7.28}$$

The second order amplitude may be put in the form of pair-products of one-electron transition amplitudes (7.27) if the time-ordered intermediate-state propagator is replaced by its integral representation [Ziman, 1970, pp. 81 and 109; Fetter and Walecka, 1971, p. 79],

$$\frac{T}{2}\; e^{-iE_n(t-t')} = \int_{-\infty}^{\infty} d\Omega\; e^{-i\Omega(t-t')} \left[\frac{1}{2}\delta(\Omega - E_n) + \frac{i}{2\pi} P\, \frac{1}{\Omega - E_n}\right] . \tag{7.29}$$

The second order amplitude is,

$$\begin{aligned} a^{(2)}_{fi} &= \sum\!\!\!\!\!\!\int_n \sum\nolimits_{\beta_1,\beta_2} \bar{C}_{\beta_1,\beta_2} \sum\nolimits_{\alpha_1,\alpha_2} C_{\alpha_1,\alpha_2} \\ &\times \left\{ \frac{(-i)^2}{2} \left[\sum_{\nu_1}\sum_{\nu_2'} \hat{C}_{\nu_1,\beta_2} \hat{C}_{\alpha_1,\nu_2'}\; A^{\lambda'\lambda}_{\beta_1,\nu_1}(E_f - E_n)\; A^{\lambda'\lambda}_{\nu_2',\alpha_2}(E_n - E_i) \right.\right. \\ &+ \left. \sum_{\nu_2}\sum_{\nu_1'} \hat{C}_{\beta_1,\nu_2} \hat{C}_{\nu_1',\alpha_2}\; A^{\lambda'\lambda}_{\beta_2,\nu_2,}(E_f - E_n) A^{\lambda'\lambda}_{\nu_1',\alpha_1}(E_n - E_i) \right] \\ &- \frac{i}{2\pi}\, P \int_{-\infty}^{\infty} \frac{d\Omega}{\Omega - E_n} \end{aligned}$$

$$\times \left[\sum_{\nu_1}\sum_{\nu_2'} \hat{C}_{\nu_1,\beta_2}\hat{C}_{\alpha_1,\nu_2'}\, A^{\lambda'\lambda}_{\beta_1,\nu_1}(E_f-\Omega)\, A_{\nu_2',\alpha_2}(\Omega-E_i) \right.$$

$$\left. + \sum_{\nu_2}\sum_{\nu_1'} \hat{C}_{\beta_1,\nu_2}\hat{C}_{\nu_1',\alpha_2}\, A^{\lambda'\lambda}_{\beta_2,\nu_2}(E_f-\Omega) A^{\lambda'\lambda}_{\nu_1',\alpha_1}(\Omega-E_i) \right] \Bigg\}, \quad (7.30)$$

where $\sum\!\!\!\!\int_n$ extends over bound and continuum intermediate states.

2 Pseudostates

Pseudostates, as discussed at the beginning of section 7.2, are one convenient method for approximating a multitude of high lying and continuum states by use of a single approximate wavefunction. Au and Drachman [1988] have shown excellent convergence of results for van der Waals coefficients and frequency-dependent polarizabilities with as few as 5 to 10 pseudostate terms representing both bound and continuum states. Drachman *et al.* [1990] have likewise shown rapid convergence for two-photon transitions in hydrogen using a finite set of pseudostates to represent the infinite sum over bound and continuum states for the second order terms. Here pseudostates are used not only for the sum in (7.30) but also for the final states, which are products of bound and continuum states.

As with the diagonalization of the many electron Hamiltonian to obtain the CI coefficients (7.25), calculation of the tables of pseudostate expansion coefficients may be sequestered from the calculation of the full scattering amplitude to alleviate computation time. To represent the excited bound state exactly, and to make sure the continuum states are orthogonal to the excited bound states, one must have the appropriate exponent and number of terms in the polynomial expansion. Bhatia [1992] has noted that to better represent the long-range nature of Coulomb waves, one may use complex pseudostates.

3 One electron amplitudes in pseudostates

Straton [1991 and section 2.2.2] has given the general state-to-state one electron transition amplitude in terms of derivatives of an analytic function,

$$D = \left(-i\frac{\partial}{\partial\omega}\right)^p A^j\, K_j(bA)\,, \quad (7.31)$$

where

$$A = \sqrt{\frac{\gamma^2}{\nu^2} + \frac{\omega^2}{v^2}}\,. \tag{7.32}$$

in which $\gamma = \frac{\lambda'}{n'} + \frac{\lambda}{n}$ and v is the projectile velocity. It may be shown [Gradshteyn and Ryzhik, 1980, (8.486.14) and (0.432.1)] that,

$$D = \sum_{r=0}^{p} \frac{p!(-b)^r(2\omega)^{p-2r}}{(p-2r)!2^r v^{2r}} A^{j-r}\, K_{j-r}(bA)\,. \tag{7.33}$$

The general form for one electron transitions is given by (2.37) in section 2.2.2.

Wherever one wishes to use pseudostates, one simply replaces either sum over the hydrogenic polynomial coefficients and normalization factor,

$$N_{n\ell}\,\lambda^{3/2} \sum_{s=0}^{n-\ell-1} \frac{(-1)^s(\lambda/n)^{s+\ell}}{(n-\ell-1-s)!(2\ell+1+s)!s!} \times 2^{s+\ell} \tag{7.34}$$

by the sum over the pseudostate polynomial coefficients and the normalization factor. Note the factor of $2^{s+\ell}$ that appears in the hydrogenic wavefunction that is removed in the Gaussian transformation (2.51).

Calculations of total cross sections for excitation-ionization in helium have been completed by Raecker *et al.* [1994], by Nagy *et al.* [1995] and by Franz and Altick [1995]. There is some difference with the observations of Fülling *et al.* [1992]. In particular the factor of two differences observed between protons and electrons at about 10 atomic units of velocity (about 2.5 MeV/amu) are not reproduced by these calculation.

7.3 Many body perturbation theory

Many body perturbation theory (MBPT) is a comparatively well defined technique developed by Bruekner [1955, 1959] and Goldstone [1957] for evaluating various properties of many body systems used in most areas of physics and some of chemistry. In atomic physics expansions are done in both the interaction V with the projectile and the electron-electron correlation interaction v. This feature tends to restrict the use of MBPT to systems in which both interactions are weak. Individual MBPT amplitudes may be expressed in terms of diagrams, as discussed below. Because of these features MBPT provides a common, well defined, conceptual language for describing observable properties of many body systems including the interactions of many electron atomic targets with various projectiles.

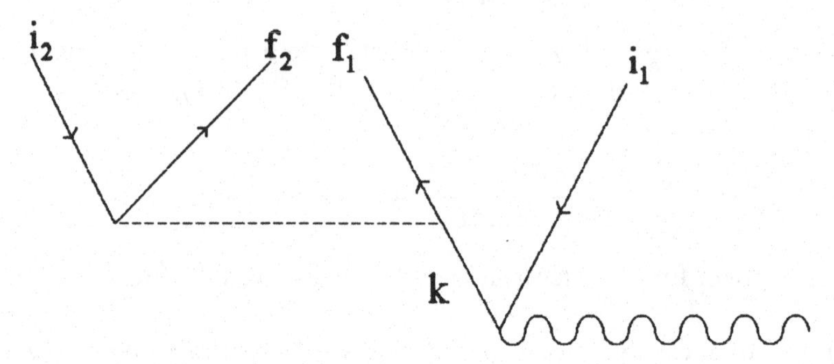

Fig. 7.2. Typical second order MBPT diagram for a transition from the ground state of a target (not shown) to a final state with two electrons excited.

MBPT is used in many areas of physics and chemistry. Here an elementary description of MBPT diagrams and matrix elements is given. A more complete and more general description of MBPT techniques is given by Fetter and Walecka [1971].

1 Diagrams and matrix elements

The scattering amplitude (or corresponding T matrix) in MBPT is found by summing contributions from MBPT diagrams. In a MBPT diagram such as figure 7.2 time increases in the upward direction. The ground state of the system is not shown in the description used here. Excited states are represented by particle-hole pairs where the hole represents the absence of a particle in the ground state and the hole propagates backward in time. Interaction of the atomic target with the projectile is represented by a wavy horizontal line. Interaction of a target electron with another target electron is represented by a horizontal dashed line. In figure 7.2 a second order MBPT diagram is shown in which the projectile interacts with the atom producing a particle-hole pair. The particle-hole pair propagates. Then there is a second particle-hole pair produced by the interaction of the excited electron with a second target electron. The final state consists of two excited electrons and two holes in the ground state. Since there are two interactions, this is a second order MBPT diagram which is first order in the interaction with the projectile and also first order in the electron-electron interaction.

There is a well defined procedure for writing the matrix element that corresponds to each MBPT diagram. This procedure is illustrated below

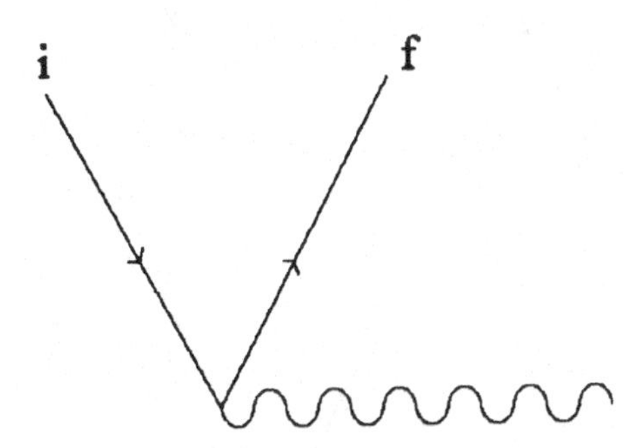

Fig. 7.3. The MBPT diagram for a transition from $|i>$ to $|f>$ via the interaction V.

with examples of first and second order MBPT amplitudes.

Lowest order term. The lowest order MBPT diagram to describe the amplitude for scattering from a initial state $|i>$ to a final state $|f>$ is shown in figure 7.3. In this simple case the system goes from $|i>$ to $|f>$ via the interaction V, and the corresponding MBPT T matrix element (proportional to the scattering amplitude f) is,

$$T = < f|V|i> \quad . \tag{7.35}$$

The evaluation of such first order matrix elements and cross sections (with appropriate normalization factors) is discussed in chapter 2 for incident charged particles and in chapter 9 for incident photons.

Second order terms. For the second order term shown in figure 7.2, one has excitation by the projectile of an intermediate state $|k>$, which propagates via a Green function $G^+(k)$, and then excitation of a second electron via the electron-electron interaction v. This corresponds to,

$$T = \sum_k < f_1 f_2|v|ki > G^+(\epsilon) < k|V|i> \quad . \tag{7.36}$$

In this case the sum over k runs over all intermediate states including continuum states. In general the Green function in the above representation is the inverse of the difference of energies before and after a time determined by a horizontal line between subsequent interactions. For example, in figure 7.2 $G^+(\epsilon) = (\epsilon_i + \epsilon_{P_i} - \epsilon_{P_f} - \epsilon_k + i\eta)^{-1}$ in the case that

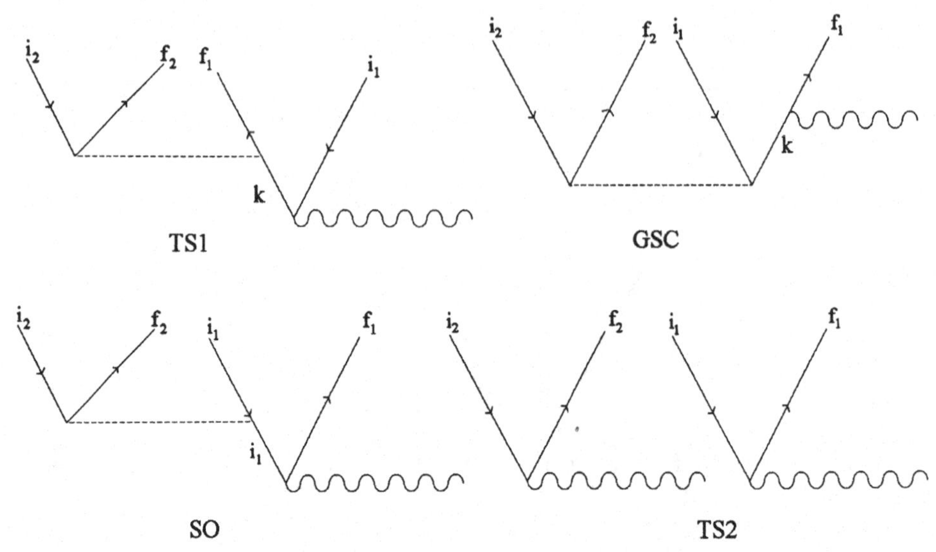

Fig. 7.4. Second order MBPT diagrams.

the projectile has incident and final energies ϵ_{P_i} and ϵ_{P_f}, respectively.

Usually there are three second order MBPT terms first order in the interaction potential V and also first order in the correlation potential v. There is one diagram second order in V without correlation. These diagrams are shown in figure 7.4. The first of these diagrams is a disconnected diagram where each electron is independently excited by the projectile. Since this is a two step process in which the projectile interacts twice, it has been called two-step 2 or TS2. It leads to the lowest order contribution to the independent electron approximation where each electron interacts with the projectile independently (i.e. without the influence of the electron-electron interaction). The T matrix element is represented by,

$$T_{TS2} = < f_2|V|i_2 > G^+(\epsilon_{i_1} - \epsilon_{f_1} + \epsilon_{i_2} - \epsilon_{f_2}) < f_1|V|i_1 > \ . \tag{7.37}$$

This leads to a probability amplitude that is a product of independent first order single electron amplitudes.

The remaining second order amplitudes are first order in V and also first order in the electron-electron interaction, v. The TS1 amplitude is the second order term discussed above. It is called a two-step 1 amplitude because it corresponds to a two step process in which the first interaction with the projectile is followed by an electron-electron interaction and is thus first order in V. The T matrix is given by,

$$T_{TS1} = \sum_k < f_1 f_2|v|ki > G^+(\epsilon_i + \epsilon_{P_i} - \epsilon_{P_f} - \epsilon_k) < k|V|i > \ . \quad (7.38)$$

The term labeled GSC is ground state correlation. In this case the interaction between the electrons in the initial state precedes the interaction with the projectile. The T matrix element is represented by,

$$T_{GSC} = \sum_k < f_1|V|k > G^+(\epsilon_{i_1} + \epsilon_{i_2} - \epsilon_{f_2} - \epsilon_k) < kf_2|v|i_1 i_2 > \ . (7.39)$$

The SO term is the shakeoff term. During the collision the effective charge seen by the second electron changes when the first electron is excited. In the final state this corresponds to the absence of an electron-electron interaction. This corresponds to a series of particle-hole interactions in the final state. The first order T matrix element shown in figure 7.4 is represented by,

$$T_{SO} = - < f_2 i_1|V|i_1 i_2 > G^+(\epsilon_{i_1} - \epsilon_{f_1} + \epsilon_{P_i} - \epsilon_{P_f}) < f_1|v|i_1 > \ . (7.40)$$

Some authors [Carter and Kelly, 1981] sensibly refer to GSC as initial state correlation and the sum of TS1 and SO as final state correlation. In MBPT the term shake (e.g. SO) often refers to the first order contribution, whereas in other contexts (e.g. where the shake amplitude is $< f'|i >$) shake corresponds to the sum of all MBPT SO diagrams.

For transitions out of the initial state of N electrons the lowest order MBPT terms are terms with a total of N interactions with either the projectile via V or the electrons via v.

Atomic electronic states. Consider an atomic target containing N electrons. Transition rates, cross sections and other properties calculated with approximate theories are often sensitive to the choice of electronic states chosen for the atomic targets. One convenient set of states are the uncorrelated independent electron states developed by Kelly [1969]. These basis states are eigenfunctions of the single electron Hamiltonian,

$$H_T|s_j >= \left(\frac{1}{2}\nabla_j^2 - Z_T/r_j + v_j(r)\right)|s_j > \qquad j = 1, 2, ..., N \ , \quad (7.41)$$

where the mean field potential $v_j(r)$ is chosen to approximate the electrostatic interactions with the other electrons. The eigenstates of the N electron atom are a Hartree Fock wavefunction formed from these states including exchange. Then,

$$v = \sum_{i>j}^{N} \frac{1}{|\vec{r}_i - \vec{r}_j|} - \sum_j v_j(r_j) \tag{7.42}$$

defines the correlation potential v.

Using the one electron states $|s_j> (1 \leq j \leq N)$ generated for the initial ground state Kelly first introduced the V^N potential defined by,

$$< m|v_j(r)|n >=< m|V^N|n >\equiv \sum_{j=1}^{N} [< mi|v|ni > - < mi|v|in >] \;, \tag{7.43}$$

where

$$< ab|v|cd >= \int\int < a(r)b(r')|\frac{1}{|\vec{r} - \vec{r}'|}|c(r)d(r') > d\vec{r}d\vec{r}\,' \;.$$

A difficulty with the V^N potential is that all excited states have continuum energies because the occupied orbitals in the ground state completely screen the nuclear charge, Z_T. Consequently the MBPT expansion usually converges slowly for the V^N potential. Kelly found faster convergence using the V^{N-1} potential defined by,

$$< m|V^{N-1}|n >\equiv \sum_{j=1}^{N-1} [< mi|v|ni > - < mi|v|in >] \;. \tag{7.44}$$

The excited orbitals computed using the V^{N-1} potential include states with negative energies since the V^{N-1} potential at large distances corresponds to a singly charged ion. Further discussion of these V^N and V^{N-1} potentials is given by Starace [1982].

Scattering mechanisms. It is sometimes useful to discuss various observable properties of the interactions of atomic matter in terms of physical mechanisms. However, agreeing on the precise meaning of these mechanisms is not always easy. The MBPT amplitudes are frequently used to describe mechanisms for various scattering processes [Starace, 1982; Andersen *et al.*, 1987; McGuire, 1991; Hino *et al.*, 1993]. The MBPT description has various advantages. The conceptual meaning is intuitively clear because the diagrams lend to visualization. Translation from diagrams to calculations may be done in a well defined way. Other theoretical methods may usually be described in terms of MBPT diagrams. The MBPT method is used in much of physics and some of chemistry so that the meaning is clear to people in different disciplines. Difficulty can arise, however, when one tries to make a one to one correspondence with a quantum amplitude and an observable (sometimes classical) process. Also, since the MBPT amplitudes are not gauge invariant they have no consistent meaning as mechanisms unless the gauge is specified (Cf. section 9.3.1).

7.4 Observations

1 Z^3 terms

For collisions of atoms with incident projectiles of charge Z at a high collision velocity v perturbation series is often convergent. The first two terms are then adequate for two electron transitions, namely,

$$a \simeq a^1 + a^2 = C_1(Z/i\mathrm{v}) + C_2(Z/\mathrm{v})^2 , \tag{7.45}$$

where C_1 and C_2 are in general complex coefficients independent of Z and v. Transition probabilities and cross sections vary as $|a|^2$. Thus, if both C_1 and C_2 are non-zero, then a Z^3 term representing interference between first and second order amplitudes is possible. In this case transition probabilities and cross sections for particles of opposite charge are different.

If $|i>$ and $|f>$ are real in (7.13), then C_1 is purely imaginary and one may set $C_1 = -ic_1$ where c_1 is real. In the second order coefficient there are both real and imaginary terms, i.e. $C_2 = \bar{c}_2 - ic_2$ where c_2 and $\bar{c}_2$ are real. It may be shown (section 7.1.2) that c_1 arises from an electron-electron interaction and c_2 arises from time ordering effects (i.e. $T \neq T_{av} = 1$ in (7.15)). Eq(7.45) may now be expressed as,

$$\begin{aligned} a &= -ic_1(Z/\mathrm{v}) - (\bar{c}_2 - ic_2)(Z/\mathrm{v})^2 \\ &= -i(c_1 - c_2 Z/\mathrm{v})(Z/\mathrm{v}) - \bar{c}_2(Z/\mathrm{v})^2 . \end{aligned} \tag{7.46}$$

The term $\bar{c}_2$, which has no time ordering, may carry some electron correlations and does include the lowest order independent electron approximation, where a reduces to a simple product of first order probability amplitudes.

Under these conditions, probabilities for two electron transitions in second order perturbation theory have the general form,

$$\begin{aligned} P &= |a|^2 \simeq |a^1 + a^2|^2 \\ &= |-i(c_1 - c_2 Z/\mathrm{v})(Z/\mathrm{v}) - \bar{c}_2(Z/\mathrm{v})^2|^2 \\ &= |(c_1 - c_2 Z/\mathrm{v})(Z/\mathrm{v})|^2 + |\bar{c}_2(Z/\mathrm{v})^2|^2 \\ &= c_1^2(Z/\mathrm{v})^2 - 2c_1c_2(Z/\mathrm{v})^3 + (c_2^2 + \bar{c}_2^2)(Z/\mathrm{v})^4 . \end{aligned} \tag{7.47}$$

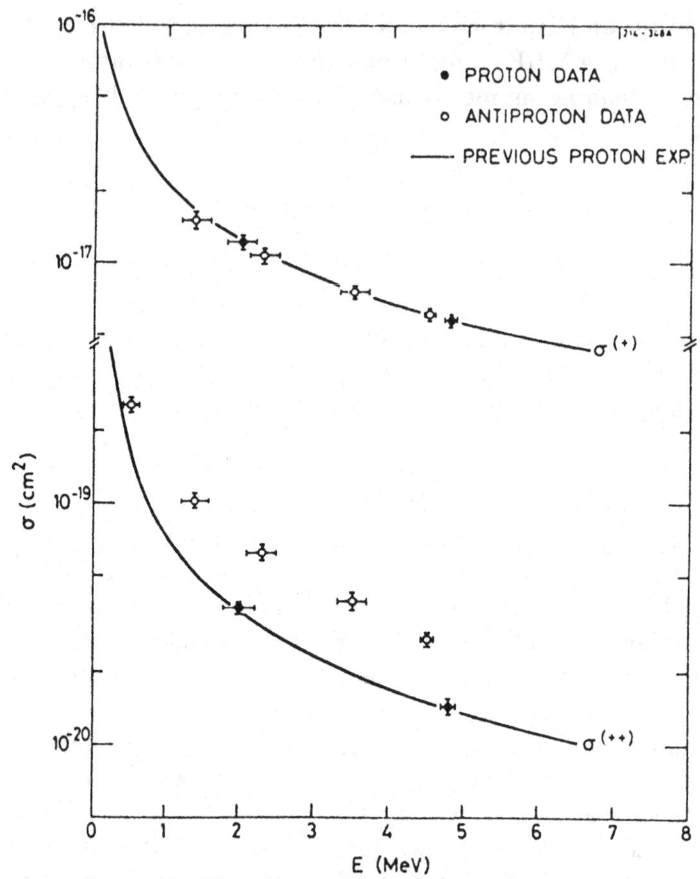

Fig. 7.5. Observed cross sections for both single and double ionization of helium by impact of protons and anti-protons [Andersen *et al.*, 1986, 1987]. Only data are shown in this figure.

The Z^3 term is opposite sign for projectiles of opposite charge. If the the electron-electron effects in the Z^3 terms include only shake, then these effects may be included in an uncorrelated formulation as illustrated in section 7.1.1.

Total cross sections σ are found by integrating over the impact parameter $\vec{b}$ of the projectile. This integration leads to a $ln(\text{v})$ term in the first order term. The difference in double excitation by particles of opposite charge is given by,

$$\sigma(-) - \sigma(+) = 4C_{12}Z^3 \ . \tag{7.48}$$

where $4C_{12}$ comes from the integration of the opposite $2c_1c_2/v^3$ terms in (7.41). The C_{12} term is non-zero only if effects due to both spatial correlation in c_1 and time ordering in c_2 are present. This result may

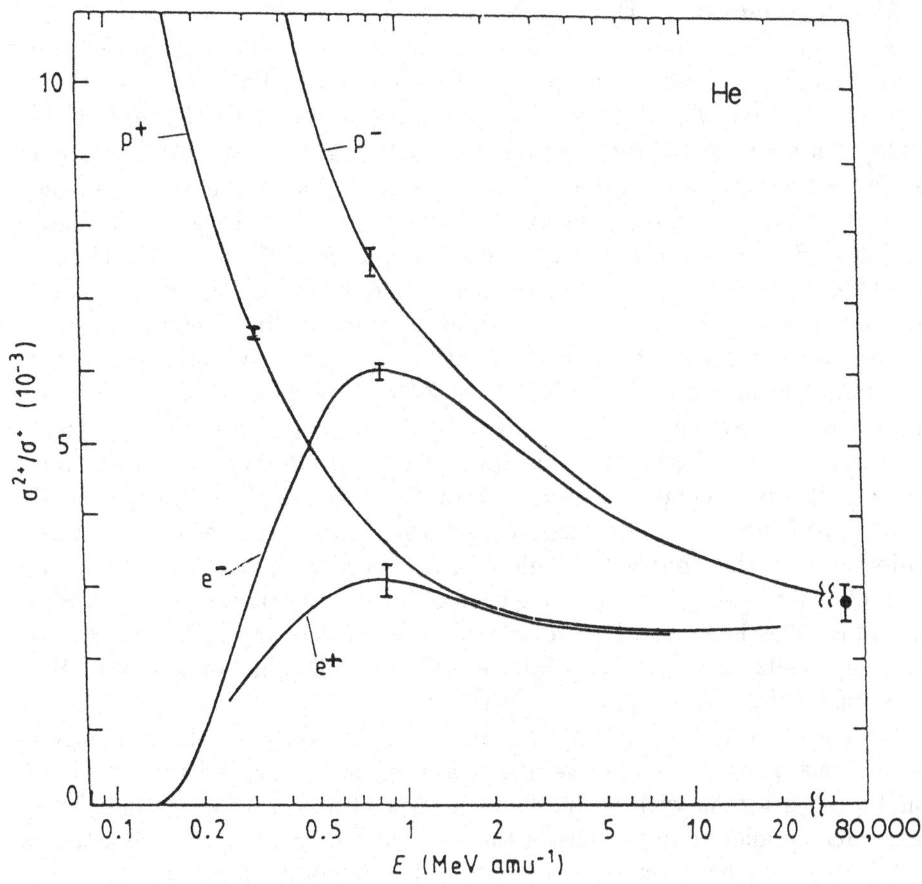

Fig. 7.6. Observed ratios of double to single ionization in helium by protons (p^+), anti-protons (p^-), electrons (e^-), and positrons (e^+) of the same incident velocity [Charlton *et al.*, 1988]. The 80 GeV/amu point is from Müller *et al.* [1983a].

not apply to capture or ionization since ϕ_i and ϕ_f are assumed to be real. Also coupling to a continuum background is ignored. Some of these corrections have been addressed by Godunov *et al.* [1996].

2 Cross sections and ratios

Both cross sections and cross section ratios for multiple electron transitions for impact by charged particles have been observed [McGuire, 1991; Knudsen and Reading, 1992]. Two electron transitions have been the most extensively studied. Many of these studies have been in helium [McGuire *et al.*, 1995] and a number have been in H_2 as well. A brief catalogue of observations of few electron transitions is given below.

Multiple ionization. Double ionization has been the most widely observed two electron transition since this process is less difficult to observe than other two electron transitions [Haugen *et al.*, 1982; Müller *et al.*, 1983a; Charlton *et al.*, 1988; Knudsen and Reading, 1992; Ullrich *et al.*, 1993]. An example of cross sections for both single and double ionization by impact of protons and by anti-protons is shown in figure 7.5. At the energies shown the single ionization cross sections are in good agreement with first Born calculations and scale as the square of the projectile charge Z. At lower energies there are significant differences between protons and anti-protons for single ionization [Knudsen and Reading, 1992].

The ratio of double to single ionization in helium by various charged projectiles is shown in figure 7.6. The Z^3 differences are due to the dynamics of electron correlation and time ordering, but the specific physical mechanisms are unclear at this time [McGuire, 1991]. The forced impulse method calculations of Ford and Reading [1994] for protons and anti-protons are in excellent agreement with these observations. Similar differences in the double to single cross section ratio in H_2 between impact of protons and electrons have been observed by Edwards *et al.* [1988]. Z^3 differences have also been observed in neon and argon [Knudsen and Reading, 1992] and in CH_4 [Malhi *et al.*, 1987]. Double ionization of H^- is discussed by Melchert *et al.* [1991].

Gray and MacAdam [1994] have observed Z^3 effects in single electron transitions using both positive and negative singly charged ions incident on Rydberg atoms with electrons in a level of $n \sim 25$. Since these ions are much smaller than the size of the orbit of the Rydberg electron, these projectiles may be effectively regarded as heavy point charges in this case.

Differential studies of both double ionization cross sections and ratios to single ionization have been observed. Kamber *et al.* [1988] have effectively measured the ratio as a function of the energy transferred by the projectile by detecting events in which a binary encounter between the projectile and a target electron occurs. This differential cross section is related to photoionization in the Bethe limit (Cf. section 10.2). A peak in the angular distribution of scattered projectiles for this ratio was observed by Giese and Horsdal [1988] and subsequently analyzed in terms of a double binary encounter event [Vegh, 1989; Olson *et al.*, 1989a]. Single and double ionization cross sections triply differential in the angle and energy of an emitted electron and the defection angle of the projectile have been observed by Skogvall and Schweitz [1990]. Both total [Charlton *et al.*, 1988] and various differential [Lahmam-Bennani and Duguet, 1992, 1996; Brion, 1995; Marji *et al.*, 1995; Duguet *et al.*, 1991; Lahmam-Bennani *et al.*, 1989] cross sections have also been observed for double ionization by incident electrons. Triple ionization of CO_2 has been considered by Handke *et al.* [1996].

Cross sections for multiple ionization of atoms and molecules by highly charged ions have been observed for many years [McGuire, 1991]. Scaling for cross sections for multiple ionization by electron impact has been developed by Fisher *et al.* [1995].

Total cross sections for multiple ionization of ions by electron impact have been observed by Hofmann *et al.* [1993].

The high energy limit of the ratio of double to single ionization approaches a value independent of the charge or energy of the projectile as indicated in figure 7.6. As discussed in the section above, this limiting value of the ratio depends on both the dynamics of electron correlation and on time ordering. This ratio has been measured in helium with different charged projectiles by various groups [Müller *et al.*, 1983a; Knudsen *et al.*, 1984; Ullrich *et al.*, 1993]. The accepted value is 0.26% with error bars that depend on the incident projectile. Ullrich *et al.* quote a minimum error of about 5% in the case of proton impact.

As a rule of thumb multiple electron transition cross sections tend to be dominated by ionization at high velocity and by electron capture at low velocity, especially for highly charged projectiles.

Multiple transfer. Some data for double electron transfer in helium have been available for many years [Fogel *et al.*, 1959; Nikolaev *et al.*, 1961; Schryber, 1967; Toburen *et al.*, 1968; Gilbody, 1994]. Multiple electron capture is discussed by Schlachter *et al.* [1990]. Since the cross section falls rapidly with energy, few data are available at very high energies. Calculations at intermediate and high velocities have been done by [Tokesi and Hock, 1996; Fritsch and Lin, 1991; Theisen and McGuire, 1979] Multiple capture at low velocity is described in the simple over-the-barrier-model (section 2.6.3) by Niehaus [1986].

Double excitation. Observations of double excitation of helium have been available for many years as auto ionizing resonances in electron emission spectra [Rudd, 1956; Stolterfoht, 1971; Stolterfoht *et al.*, 1972; Bruch *et al.*, 1984; Pedersen and Hvelplund, 1989; Giese *et al.*, 1990; Schulz *et al.*, 1995]. However, in this case it is not possible to separate double excitation from single ionization in helium, even for total cross sections. The doubly excited states of helium are above the threshold for single ionization and the doubly excited resonance state may decay into the single ionization continuum via an Auger transition with one of the excited electrons going into the continuum and the other falling back to the ground state. Since the final state is the same as for single ionization, interference between these two amplitudes occurs. This interference may be characterized in the angular distributions of the emitted Auger electrons by parameters introduced by Fano [1961] or by the equivalent parameters of Shore [1967] (Cf. section 3.5). Since the single ionization background continuum varies with energy in the case of helium, this interference does not decouple in the total cross sections [McGuire, 1991, p.286]. Double excitation in proton-helium collisions has been reviewed by Schulz [1995].

Calculations of double excitation to the $n = 2$ level of helium have been done using the above method by Straton *et al.* [1992] and also by Godunov *et al.* [1992, 1995]. The later authors include the interference between the double excitation resonances and the corresponding single ionization continuum to which these states decay. Excellent agreement with observations of Pedersen and Hvelplund [1989] is found. Similar calculations have been done by Martin and Salin [1995,1996] and also by Morishita *et al.* [1994] who find excellent agreement with observations of Bordenave-Montesquieu *et al.* [1995] and of Giese *et al.* [1990] for cross sections differential in the ejected electron distributions.

Some atoms have doubly excited states lying below the ionization limit, so that double excitation is clearly distinct from any ionization continuum. Godunov and Schipakov [1993] have calculated double excitation of such a state in beryllium produced by impact of protons and antiprotons.

Ionization-excitation. Several observations of excitation with ionization have been made in helium [Bruch *et al.*, 1993: Fülling *et al.*, 1992; Schartner, 1990; Pedersen and Folkmann, 1990; Hippler, 1988; Bruch *et al.*, 1979], in neon [Bruch *et al.*, 1979] and in H_2 [Edwards, 1990]. In both helium and H_2 there is a factor of two difference between electrons and protons at speeds around 2 MeV/amu in the observed ratio of excitation ionization to single ionization. This Z^3 effect is similar to that observed in the double ionization of helium in both helium and H_2.

The observed high energy limit of the ratio of ionization-excitation (with excitation to $n = 2$) to single ionization tends toward a value of about 0.1%.

Transfer ionization. Total cross sections for transfer with ionization in helium and ratios of these cross sections to single electron transfer have been observed over a broad range of energies [Hvelplund *et al.*, 1980; Horsdal-Pedersen and Larsen, 1985; McGuire *et al.*, 1988; Kristensen and Horsdal, 1990; Datz *et al.*, 1990] for various projectiles. At lower energies transfer ionization is indistinguishable from double electron transfer followed by auto ionization.

The high energy ratio of total cross sections for transfer ionization to single electron transfer tends to a value about 2%, although it has been suggested that this ratio may increase at sufficiently high energies due to shakeover (Cf. section 6.3.1) [McGuire *et al.*, 1988].

Evidence for a double ridge structure due to second order Thomas scat-

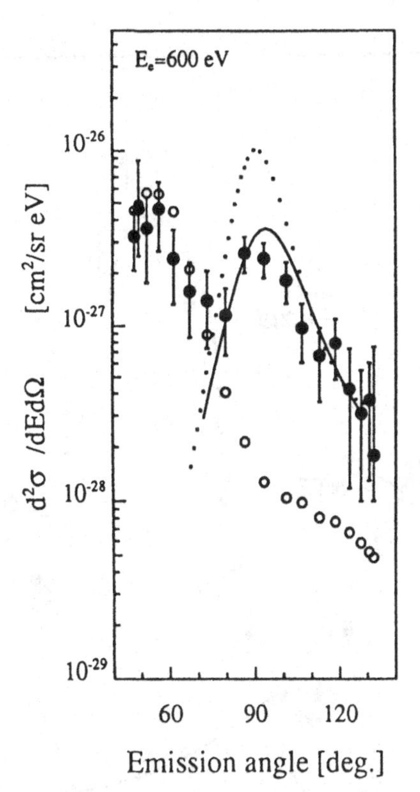

Fig. 7.7. Transfer ionization: $p^+ + \mathrm{He} \to \mathrm{H} + \mathrm{He}^{-2} + e^-$. Observations at 1 MeV as a function of the angle of emission of a 600 eV electron are from Palinkas *et al.* [1989]. The solid and dotted lines [Briggs and Taulbjerg, 1979; Ishihara and McGuire, 1988] are theoretical results which describe second order Thomas scattering where the first electron rescatters from the second (emitted) electron before the first electron is transferred. The theory corresponds to a cut along the broad ridge through the sharp ridge in the figure at the right from McGuire *et al.* [1989].

tering has been observed by Palinkas *et al.* [1989] as illustrated in figure 7.7. This Thomas mechanism occurs when one electron scatters from the second electron in order to satisfy kinematic conditions necessary for capture [Briggs and Taulbjerg, 1978; McGuire *et al.*, 1989, 1996a]. Here the locus of the sharp ridge in figure 7.7 is determined by the constraints of conservation of overall energy and momentum, while the locus of the other broader ridge is determined by conservation of energy in the intermediate state‡. Thus, this feature is due directly to the dynamics of the electron-

‡ In the classical limit both ridges are infinitely steep. This ironically suggests that the function of quantum mechanics in this case is to insure that nature is continuous (i.e. not too lumpy).

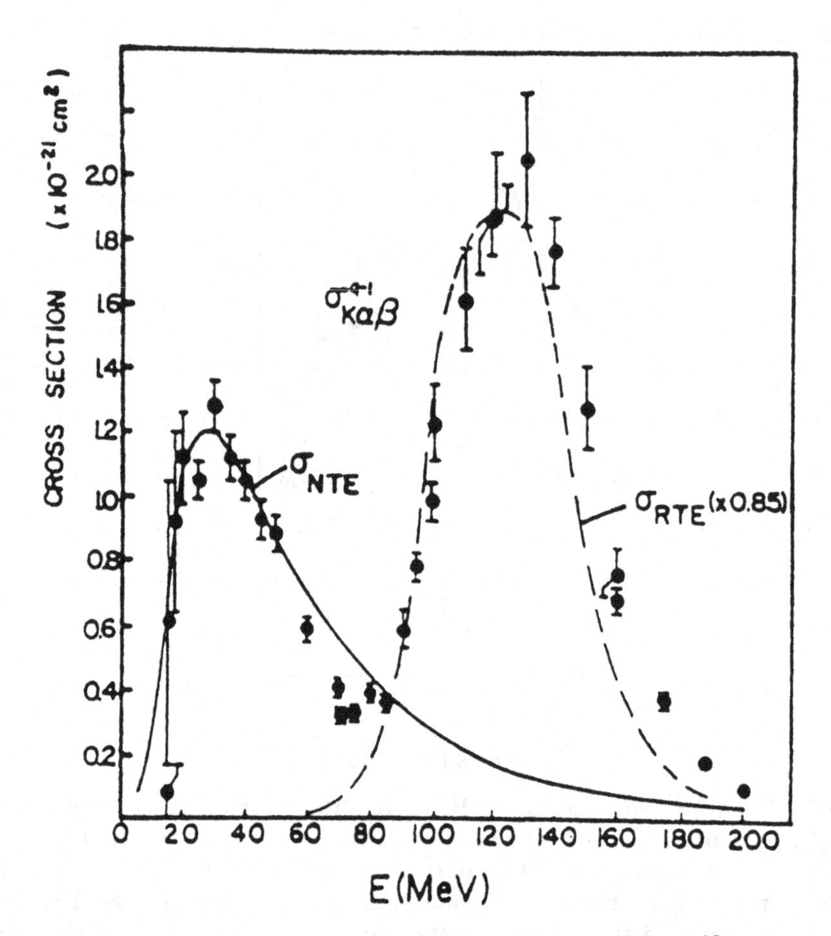

Fig. 7.8. Observed cross section for transfer-excitation in Si^{+13} + He [Tanis, 1990]. The dashed curve is a non-resonant transfer-excitation (NTE) calculation. The solid curve is a resonant transfer-excitation (RTE) calculation. Both calculations are normalized to the data.

electron interaction. It is expected that this feature will dominate the differential cross section at higher projectile energy.

The ratio of transfer ionization to single transfer was observed for p^++He as a function of the scattering angle of the projectile [Horsdal *et al.*, 1986]. A pronounced peak was measured at a scattering angle of 0.5 mrad. A diffraction model of Gayet and Salin [1990] gives a plausible agreement.

Transfer excitation. The two-electron transition in which one target electron is transferred and a second projectile electron is excited has been observed in a variety of collision partners [Tanis, 1990; Graham, 1990;

Schuch *et al.*, 1990; Mokler *et al.*, 1989]. In 1982 Tanis *et al.* discovered a resonance in the total cross section. This resonance was interpreted as an inverse Auger process (or dielectronic recombination) where the projectile electron is excited by an interaction with the captured target electron. This process occurs when the kinetic energy of the projectile electron matches the transition energy of the excited electron so that this is a resonance condition on the collision velocity. Hence, this process is called 'resonant transfer excitation' (RTE). A non-resonant process in which transfer and excitation occur due to independent interactions of the two electrons with the projectile (transfer) and target (transfer) nuclei was observed using X-rays by Pepmiller *et al.* [1983] and by more highly resolved Auger electrons by Swenson *et al.* [1986]. This 'non-resonant transfer excitation' (NTE) corresponds to the independent electron approximation. Thus, RTE depends on the electron-electron interaction and NTE does not. An observed cross section for these processes are shown in figure 7.8. Other processes leading to transfer-excitation have also been proposed [Hahn, 1990].

A general review of multiple electron transitions in ion-atom collisions is given by McDaniel *et al.* [1993].

Some of the known ratios of two to one electron transition cross sections in the high velocity limit for various projectiles producing various transitions in helium are tabulated in section 6.3.3.

Note on Z^3 effects, correlation and interference. The Z^3 effects discussed here require both electron correlation and time-ordering. As a simple rule correlation is important when $1/\text{v}_{orbit}$ is larger than both $Z_T^{eff}/\text{v}_{orbit}$ or Z/v, where v_{orbit} is the orbit speed of the active electron(s). Z^3 effects may occur classically as discussed in section 7.1.2. In this sense Z^3 effects do not require quantum interference between first and second order terms in Z. Interference generally requires that the systems be of an extent in both space and time that is neither too big nor too small. Correlation, on the other hand, may occur in systems of arbitrary extent in space and time.

8

Projectiles carrying electrons

In previous chapters interactions with structureless point charge projectiles have been considered. There are many interactions, however, which involve at least two atomic centers with one or more electrons on each center. In such cases the projectile is not well localized and there is a need to integrate over the non localized electron cloud density of the projectile. Evaluation of cross sections and transition rates for such processes requires a method for dealing with at least four interacting bodies. If multiple electron transitions occur on any of the atomic centers, then some form of even higher order many body theory is required. In general such a many body description is difficult.

In this chapter the probability amplitude for a transition of a target electron caused by a charged projectile carrying an electron is formulated. This probability amplitude may be used for transitions of multiple target electrons if the correlation interaction between the target electrons is neglected. Unless the projectile is simply considered as an effective point projectile with a charge Z^{eff}, the interaction between the target and the projectile electrons may not be ignored. Since this interaction is between electrons on two different atomic centers, the effects of this interaction have been referred to as two center correlation effects (Cf. section 6.2.4). In first order perturbation theory [Bates and Griffing, 1953; McGuire *et al.*, 1981; Montenegro *et al.*, 1994; Wang *et al.*, 1995b] the effect of the projectile electron may be described in terms of an effective projectile charge $Z^{eff}(q)$ which varies with the momentum transfer q (or corresponding impact parameter b) of the interaction. One effect of the two center electron-electron interaction is to produce a transition on the projectile as well as the target. Consequently, in this chapter electron transitions on different atomic centers are considered.

The emphasis in this chapter will be on conceptually simple first order methods. Higher order methods for the screening contribution have been fruitfully developed by Wang *et al.* [1991] and by Jakubassa-Amundsen [1992], while an impulse like model for the antiscreening contribution was developed by Zouros *et al.* [Zouros *et al.*, 1993; Montenegro and Zouros, 1994]. The theory of interactions between atoms and projectiles carrying electrons has been recently reviewed by Montenegro *et al.* [1994].

8.1 Formulation

Consider a collision system consisting of a projectile carrying a single electron incident on a single-electron target. For simplicity, assume both the projectile and the target electrons are hydrogenic with effective nuclear charges of Z and Z_T respectively. Furthermore, one may use the semi-classical approximation in which the internuclear motion is assumed to be classical with a given classical trajectory specified by $\vec{R}(t)$. In general $\vec{R}(t)$ is arbitrary and may be calculated using the net potential between the two atomic centers (as discussed in the appendix to chapter 3). In evaluation of electronic transition probabilities, the replacement of the actual trajectory by a straight line trajectory $\vec{R}(t) = \vec{b} + \vec{v}t$ is usually well justified because of the small electron-nucleon momentum ratio at high v.

The total Hamiltonian of the system is,

$$H = H_P + H_T - \frac{Z}{|\vec{R} - \vec{r}_T|} - \frac{Z_T}{|\vec{R} + \vec{r}_P|} + \frac{1}{r_{12}} + \frac{ZZ_T}{R} , \tag{8.1}$$

where H_P (H_T) is the Hamiltonian of the projectile (target) and $\vec{r}_P$ ($\vec{r}_T$) the coordinates of the projectile (target) electron with respect to its parent nucleus, and $r_{12} = |\vec{R} + \vec{r}_P - \vec{r}_T|$ is the distance between the electrons. With the use of the classical trajectory, one may drop the internuclear potential ZZ_T/R from the total Hamiltonian. Then the the total Hamiltonian H may be replaced by the electronic Hamiltonian H_{el} given (as described in the appendix of chapter 3) by,

$$H_{el} = H_P + H_T + V, \qquad V = -\frac{Z}{|\vec{R} - \vec{r}_T|} - \frac{Z_T}{|\vec{R} + \vec{r}_P|} + \frac{1}{r_{12}} . \tag{8.2}$$

Here $-\frac{Z}{|\vec{R}-\vec{r}_T|}$ is the interaction between the projectile nucleus and the target electron, $-\frac{Z_T}{|\vec{R}+\vec{r}_P|}$ the interaction between the target nucleus and the projectile electron, and $1/r_{12}$ the interaction between the projectile electron and the target electron. This electronic Hamiltonian H_{el} may be used to find the full electronic wavefunction $\psi_i(t)$ satisfying

$$i\frac{\partial \psi_i}{\partial t} = H_{el}\psi_i , \tag{8.3}$$

as given by (3.40).

The exact transition probability amplitude describing the change of states $i \rightarrow f$ under the influence of the interaction V is (Cf. (2.1), (2.17) and (2.22)),

$$a_{f_T,f_P}(\vec{b}) = \lim_{t\rightarrow\infty} \langle f|\psi_i(t)\rangle = -i\int dt \langle f|V|\psi_i\rangle ,$$

where $\psi_i(t)$ is the exact wavefunction of (8.3) at time t propagated from the initial state $|i\rangle$ of the system at $t = -\infty$. Here i and f denote the electronic states on both the target and the projectile.

The cross section for the transition of an arbitrary number of electrons on either atomic center is given by,

$$\sigma = \int d\vec{b} |a_{f_T,f_P}(\vec{b})|^2 , \tag{8.4}$$

where $|a_{f_T,f_P}(\vec{b})|^2$ is the probability that the transition $i \rightarrow f$ has occurred.

1 Basic ideas

As an example, consider a collision between a projectile and a target in which the target loses an electron, as illustrated in figure 8.1. The electron loss is governed by two different and competing mechanisms which are related to the state of the projectile after the collision. In the first (single transition) mechanism the projectile electron remains inactive in its initial state. Loss of the target electron may be caused either by by a projectile nucleus - target electron interaction or by a projectile electron - target electron interaction. These two amplitudes add coherently and interfere destructively (Cf. (8.6)). The effect is to reduce (or 'screen') the strength of the charge of the projectile nucleus and the cross section for electron loss. This is called 'screening'. In the second (double transition) mechanism , loss of the target electron may be caused entirely by the projectile electron - target electron interaction with the projectile electron also undergoing a transition. The cross sections for different excited projectile final states add incoherently so that the contribution from this second mechanism is always positive. It cannot reduce the cross section. This is called 'antiscreening'. This picture holds in first order perturbation theory.

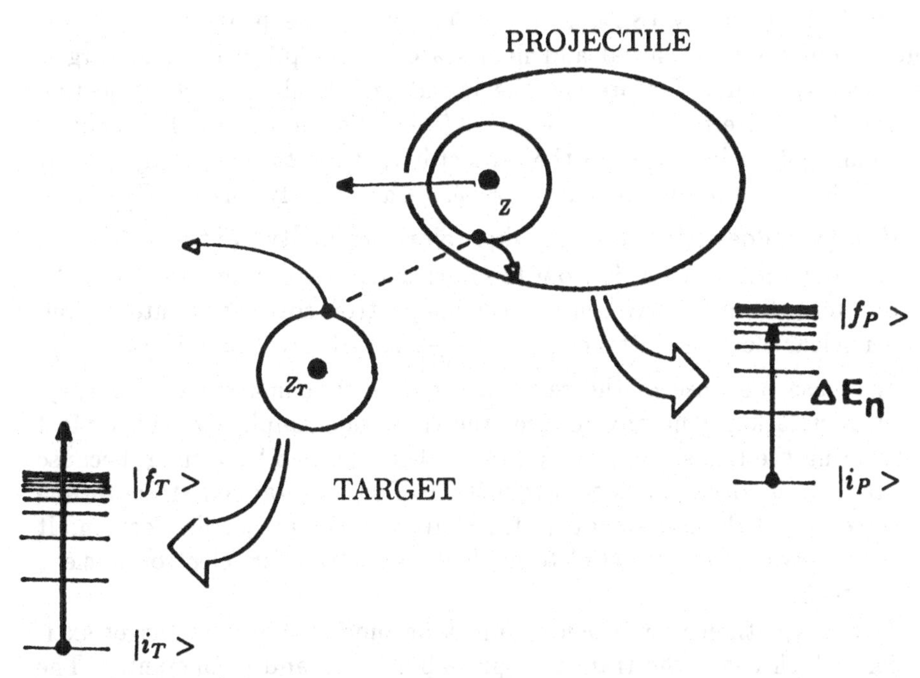

Fig. 8.1. Diagrammatic representation of first order electron loss mechanisms. In this example there is an active electron on the target which goes from state $|i_T>$ to $|f_T> \neq |i_T>$. When this occurs, the projectile can either remain in the ground state $|f_P> = |i_P>$ (screening) or be excited to $|f_P> = \neq |i_P>$ (antiscreening) [adapted from Montenegro *et al.*, 1994].

8.2 The first Born approximation

For now the target is regarded as having a single active electron. Extension to collisions in which more than one transition occurs in the target may be done using the independent electron approximation (Cf. section 8.2.3). Within the first Born approximation, the exact wavefunction $\psi_i(t)$ is approximated by its zeroth order unperturbed initial state wavefunction $|i\rangle$. The transition amplitude in the first Born approximation as a function of impact parameter for target excitation is given by,

$$a_{f_T,f_P}(\vec{b}) = -i \int dt \langle f|V|i\rangle \tag{8.5}$$

$$= i \int dt e^{i\Delta E t} \left[\delta_{f_P i_P} \langle f_T | \frac{Z}{|\vec{R} - \vec{r}_T|} | i_T \rangle - \langle f | \frac{1}{r_{12}} | i \rangle \right] ,$$

where V is given by (8.2), Z is the charge of the projectile and $i_{P,T}$ and $f_{P,T}$ denote the initial and final states of the projectile and target, respectively, and i, f represent the initial and final product projectile-target states. Here $\Delta E = \varepsilon_f^P + \varepsilon_f^T - \varepsilon_i^P - \varepsilon_i^T$ is the energy difference of the final and initial states of the projectile and the target. In arriving at (8.6) it has been assumed that $f_T \neq i_T$. Consequently, the $-\frac{Z_T}{|\vec{R}+\vec{r}_P|}$ term in V of (8.2) does not contribute due to orthogonality of $|f_T\rangle$ and $|i_T\rangle$.

The first term in (8.5) is from the nuclear-electron interaction and the second term from the two center electron-electron correlation interaction. The nuclear-electron interaction $-\frac{Z}{|\vec{R}-\vec{r}_T|}$ contributes only if $f_P = i_P$. In this case, because of the relative negative sign in front of the $1/r_{12}$ term, correlation generally reduces the transition amplitude. This effect of lowering the transition probability has been termed 'screening' because the reduction corresponds to a reduction of the interaction strength due to screening of the nuclear projectile charge by the projectile electron. It is also known as the 'elastic' term, 'coherent' term or 'electron-nucleus (e-N)' term.

If $f_P \neq i_P$, then more open channels become available to target excitation, which cause the transition probabilities to add incoherently. The effect of enhancing the transition rate is called 'antiscreening' because the cross section is increased rather than decreased. It is also known as the 'inelastic' term, 'incoherent' term or 'electron-electron (e-e)' term. It immediately becomes clear from (8.5) that for antiscreening term (i.e., $f_P \neq i_P$), the transition amplitude is identically zero unless the correlation term $1/r_{12}$ is included. This furnishes a direct link between correlation and antiscreening.

Further reduction of $a_{f_T, f_P}(\vec{b})$ can be carried out using Bethe's integral which expresses the Coulomb potential in momentum space,

$$\frac{1}{R} = \frac{1}{2\pi^2} \int d\vec{q} \frac{e^{-i\vec{q}\cdot\vec{R}}}{q^2} .$$

Substitution into (8.5) of $1/|\vec{R} - \vec{r}_T|$ and $1/r_{12}$ by their momentum representation quickly yields,

$$a_{f_T, f_P}(\vec{b}) = i \int dt e^{i\Delta E t} \frac{1}{2\pi^2} \int d\vec{q} \frac{e^{-i\vec{q}\cdot\vec{R}}}{q^2} \left[\delta_{f_P i_P} Z - F_{f_P i_P}(-\vec{q}) \right] F_{f_T i_T}(\vec{q}) \tag{8.6}$$

where the atomic form factor F is defined as,

$$F_{f_{P,T}i_{P,T}} = \langle f_{P,T}|e^{i\vec{q}\cdot\vec{r}}|i_{P,T}\rangle \ . \tag{8.7}$$

This atomic form factor is proportional to the generalized oscillator strength (Cf. the appendix to chapter 9). The integration over time in (8.6) may now be carried out explicitly,

$$\int dt e^{i\Delta Et} e^{-i\vec{q}\cdot\vec{R}} = e^{-i\vec{q}\cdot\vec{b}} \int dt e^{i(\Delta E - \vec{q}\cdot\vec{v})t} = e^{-i\vec{q}\cdot\vec{b}} 2\pi\delta(\vec{q}\cdot\vec{\mathrm{v}} - \Delta E) \,,$$

to give

$$a_{f_T,f_P}(\vec{b}) = \frac{i}{\pi} \int d\vec{q}\delta(\vec{q}\cdot\vec{\mathrm{v}} - \Delta E) \frac{e^{-i\vec{q}\cdot\vec{b}}}{q^2} \left[\delta_{f_P i_P} Z - F_{f_P i_P}(-\vec{q})\right] F_{f_T i_T}(\vec{q}) \tag{8.8}$$

The Dirac δ function in (8.8) restricts the component of $\vec{q}$ along beam axis $\hat{\mathrm{v}}$ (defined as the z-axis) such that $q_z \equiv q_{\parallel} = \Delta E/\mathrm{v}$. Introducing the transverse and parallel momentum transfers $\vec{q}_{\perp}$ and $q_{\parallel}$ such that,

$$\vec{q} = \vec{q}_{\perp} + q_{\parallel}\hat{\mathrm{v}} \ , \quad \vec{q}_{\perp}\cdot\hat{\mathrm{v}} = 0 \ , \quad q_{\parallel} = \frac{\Delta E}{\mathrm{v}} \ ,$$

into (8.8) and carrying out the integration over q_z one has,

$$a_{f_T,f_P}(\vec{b}) = \frac{i}{\pi\mathrm{v}} \int d\vec{q}_{\perp} \frac{e^{-i\vec{q}_{\perp}\cdot\vec{b}}}{q^2} \left[\delta_{f_P i_P} Z - F_{f_P i_P}(-\vec{q})\right] F_{f_T i_T}(\vec{q}) \ . \tag{8.9}$$

This result holds for both screening ($f_p = i_p$) and antiscreening ($f_p \neq i_p$). This is the usual first Born transition amplitude in the impact parameter approach. It is expressed here in terms of momentum transfer $\vec{q}$. Note that a finite minimum momentum transfer $q_{\parallel}$ is required for given transition with a transition energy ΔE.

1 *The effective projectile charge*

Taking the Fourier transform of (8.8), one obtains the scattering amplitude in first order perturbation theory for a transition in an atomic target scattering from a projectile carrying an electron, namely,

$$T_{f_T,f_P}(\vec{q}) = Z^{eff}_{f_P i_P}(\vec{q}) T_{f_T i_T}(\vec{q}) \ , \tag{8.10}$$

where

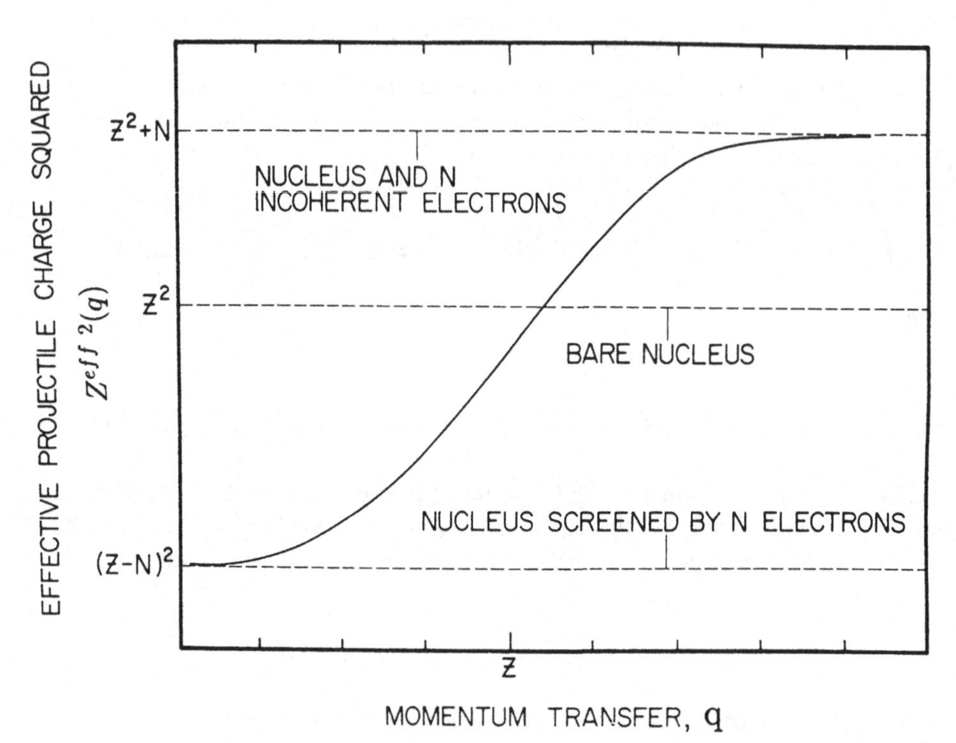

Fig. 8.2. The square of the effective projectile charge $Z^{eff\ 2}$ as a function of the momentum transfer q. For small q corresponding to large impact parameters, the projectile charge Z is fully screened by the N electrons on the projectile and the electron-electron contribution adds coherently to the nuclear contribution. In the large q limit the charge Z and the N electrons scatter incoherently.

$$T_{f_T i_T}(\vec{q}) = \frac{4\pi i}{\mathrm{v}} \frac{F_{f_T i_T}(\vec{q})}{q^2} \tag{8.11}$$

is independent of the projectile and,

$$Z^{eff}_{f_P i_P}(\vec{q}) = \frac{1}{(2\pi)^2} \left[\delta_{f_P i_P} Z - F_{f_P i_P}(-\vec{q})\right] \tag{8.12}$$

is independent of the target and acts as an effective projectile charge which varies with the momentum transfer $\vec{q}$.

The effective projectile charge $Z^{eff}_{f_P i_P}(\vec{q})$ has physically interesting limits for both small and large values of q. These limits follow from the nature of the atomic form factor $F_{f_P i_P} = \langle f_P | e^{i\vec{q}\cdot\vec{r}} | i_P \rangle$ from (8.7). For large q this form factor goes to zero and for small q the atomic form factor reduces to $\langle f_P|i_P\rangle = \delta_{f_P i_P}$.

A simple form for $Z^{eff}_{f_P i_P}(\vec{q})^2$ may be obtained by summing over the final states of the projectile. Using the definition of the atomic form factor for a projectile with N electrons, one has,

$$\begin{aligned} Z_P^{eff\ 2}(\vec{q}) &= \sum_{f_P} Z^{eff}_{f_P i_P}(\vec{q})^2 = \sum_{f_P} | < f_P|Z - \sum_j^N e^{i\vec{q}\cdot\vec{r}_j}|i_P > |^2 \qquad (8.13) \\ &= | < i_P|Z - \sum_j^N e^{i\vec{q}\cdot\vec{r}_j}|i_P > |^2 + \sum_{f_P \neq i_P} | < i_P|\sum_j^N e^{i\vec{q}\cdot\vec{r}_j}|i_P > |^2 \\ &= |Z - < i_p|\sum_j^N e^{i\vec{q}\cdot\vec{r}_j}|i_P > + |N - < i_P|\sum_j^N e^{i\vec{q}\cdot\vec{r}_j}|i_P > |^2 , \end{aligned}$$

where the closure relation,

$$\begin{aligned} &\sum_{f_T} | < f_P|\sum_j^N e^{i\vec{q}\cdot\vec{r}_j}|i_P > |^2 \qquad (8.14) \\ &= < i_P|\sum_j^N e^{-i\vec{q}\cdot\vec{r}_j} \sum_{f_P} |f_P >< f_P|\sum_j^N e^{i\vec{q}\cdot\vec{r}_j}|i_P > \\ &= < i_P|N|i_P >= N , \end{aligned}$$

has been used. The first term in the second line of (8.14) comes from elastic scattering of the projectile and is never greater than Z^2. This is the screening term. The second antiscreening term comes from inelastic scattering of the projectile. It comes from the interaction between electrons on the two different centers and is a two center correlation effect. Application of this closure relation is approximate because the minimum momentum transfer q_{min} varies with the final state of the projectile $|f_P>$. So for $q \simeq q_{min}$ closure does not apply unless a common value of q_{min} is chosen for all $|f_P>$. Since q_{min} depends on the projectile velocity v, not all terms may contribute to the sum in (8.14) for small q at low v [Anholt, 1986; Montenegro *et al.*, 1994]. Using the small and large q limits of the form factor, it follows from (8.14) that

$$Z_P^{eff\ 2}(0) = (Z - N)^2 \leq Z_P^{eff\ 2}(q) \leq Z^2 + N^2 = Z_P^{eff\ 2}(\infty)$$

as illustrated in figure 8.2.

In figure 8.2 as $q \to 0$, the scattering cross section becomes coherent. Then, for scattering from N particles the observable intensity varies as N^2. Small q corresponds to large impact parameters, i.e. scattering from large distances. One may observe an effect such as this on a foggy day: it is clear nearby, but fog (or clouds) are visible further away. The range of visibility depends strongly on the density of the particles. At small distances (large q), the scattering is incoherent and varies linearly with N and is thus weaker. Baym notes, 'This effect is why we can see clouds which are composed of dense coherently scattering droplets, even though we cannot see the water vapor' (i.e. individual particles). The scattering from the electron cloud on a projectile is similar to scattering of light from mist. The incoherent scattering ($\sim N$) is nearby and usually hardly noticeable. The coherent scattering ($\sim N^2$) is stronger at large distances and in some instances is so dense as to entirely screen visibility of distant objects.

The measurement problem and a classical limit. A long standing problem in physics is the measurement problem: where and how does the reduction of quantum mechanics to classical physics occur? The classical limit of quantum mechanics is sometimes associated with the disappearance of interference terms (as in the Schrödinger cat paradox where interference between 'living' states and 'dead' states is possible quantum mechanically. When does quantum interference disappear? That is the paradox (i.e., the measurement problem.). A reduction from interfering quantum amplitudes to an incoherent classical limit on the atomic level is contained in $Z^{eff\ 2}$ illustrated in figure 8.2. Here at small momentum transfer q the projectile charge is fully screened and $Z^{eff\ 2} = (Z - N)^2$, which is proportional to N^2. For large q the amplitudes become incoherent as the $\exp\{i\vec{q} \cdot \vec{r}_j\}$ interference terms oscillate rapidly and disappear so that $Z^{eff\ 2} = Z^2 + N$ corresponding to incoherent scattering by the particles in the projectile, linear in N. This is a limit that could have been found from classical wave methods without quantum mechanics. While the measurement problem in its full form includes a variety of classical limits ($\hbar = 0, n = \infty$, distinguishable states, onset of entropy, decoupling of a detector from observation, etc.) the change of $Z^{eff\ 2}$ from coherent screening to incoherent scattering of individual particles is one of the simplest examples of the reduction of quantum mechanics to classical physics. Additional technical detail is given by Goldberger and Watson [1964, pp. 734-6].

2 Virtual impact parameter method

From the Fourier transform of the scattering amplitude of (8.9) one obtains the probability amplitude a_{f_T,f_P} for a transition from $|i_T>$ to $|f_T>$ on the target and from $|i_P>$ to $|f_P>$ on the projectile, namely,

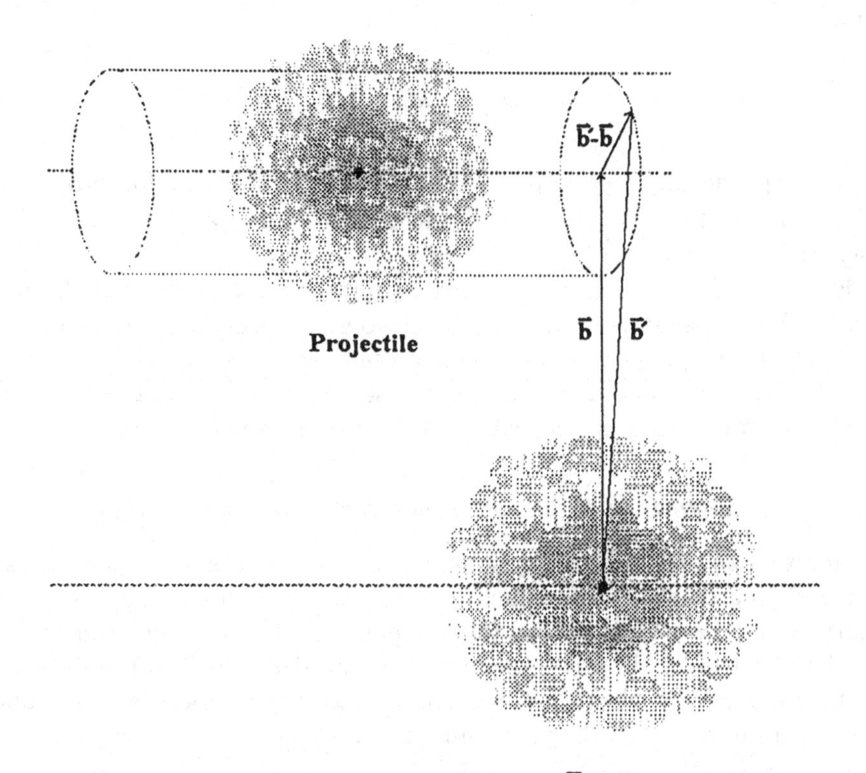

Fig. 8.3. The probability amplitude is found by convoluting over impact parameter b the probability per unit projectile charge per impact parameter for a transition on the target over the projectile electron cloud charge distribution. Here $\vec{b}$ is the impact parameter of the projectile nucleus and $\vec{b}'$ is the virtual impact parameter of the electron cloud [Wang *et al.*, 1995b].

$$a_{f_T,f_P}(\vec{b}) = \int d\vec{q}_\perp e^{-i\vec{q}_\perp\cdot\vec{b}} Z^{eff}_{f_Pi_P}(\vec{q}) T_{f_Ti_T}(\vec{q}) \,. \tag{8.15}$$

One may now apply the general convolution (or Faltung) theorem for the Fourier transform, namely,

$$\int d\vec{q}_\perp e^{-i\vec{q}_\perp\cdot\vec{b}} A(\vec{q}_\perp) B(\vec{q}_\perp) = \int d\vec{b}' \tilde{A}(\vec{b}' - \vec{b}) \tilde{B}(\vec{b}') \,, \tag{8.16}$$

where A and B are arbitrary functions and $\tilde{A}$ and $\tilde{B}$ their respective Fourier transforms. Consequently, one has the central expression,

$$a_{f_T,f_P}(\vec{b}) = \int d\vec{b}' \varsigma^{eff}(\vec{b}' - \vec{b}) a^T_{fi}(\vec{b}') \,, \tag{8.17}$$

where,

$$\varsigma^{eff}(\vec{b}) = \int d\vec{q}_\perp e^{i\vec{q}_\perp \cdot \vec{b}} Z^{eff}_{f_P i_P}(\vec{q}) \tag{8.18}$$

is a charge density per impact parameter $\vec{b}$. Note that for projectiles with a point charge $Z^{eff}_{f_P i_P}(\vec{q}) = Z$ and $\varsigma^{eff}(\vec{b} - \vec{b}') = Z\delta(\vec{b} - \vec{b}')$, which is physically sensible .

Eq (8.17) describes the transition amplitude as a product of the projectile charge per impact parameter convoluted with the transition probability per unit projectile charge convoluted over $\vec{b}'$ as illustrated in figure 8.3. Unlike the impact parameter $\vec{b}$ it is seldom the case that the variable $\vec{b}'$ is observable, so it is referred to as a 'virtual' impact parameter.

3 Application to the independent electron approximation

In addition to providing a physical picture of a collision of an atomic target with a projectile with a non-localized electron cloud, this first order method may be used to extend the independent electron approximation of section 4.1 to include projectiles which carry electrons into the collision. In the independent electron approximation for bare projectiles the probability amplitude a_{f_T} in the independent electron approximation reduces to a product of single electron probabilities, i.e., $a_{f_T} = \prod_j a^j_{f_T}$ where $\prod_j a^j_{f_T}$ are the probability amplitudes for the independent (or uncorrelated) electrons. This may be used in (8.4) to evaluate total cross sections. Now one may consider projectiles carrying one active electron which may interact with the target electrons. Then for multiple electron transitions in the independent electron approximation, the probability amplitude of (8.17) becomes,

$$a_{f_T, f_P}(\vec{b}) = \prod_j a^j_{f_T, f_P}(\vec{b}) \ ,$$

where the $a^j_{f_T, f_P}(\vec{b})$ is defined for each electron by Eq (8.17). Here two center correlation (with both screening and antiscreening) for one projectile electron is included. The target electrons are still independent of one another, but not of the electron on the projectile. Extension to a projectile with N_P independent electrons is straightforward.

8.3 Examples

To calculate the transition amplitudes from Eq (8.17), knowledge of both $\varsigma^{eff}(\vec{b}' - \vec{b})$ and $a^T_{fi}(\vec{b}')$ is required. Evaluation of ς^{eff} is done below.

Evaluation of a^T_{fi} is straightforward using the techniques of chapter 2. Consider excitation from $1s$ to $1s$ to $2s$ (dipole forbidden) and $2p+$ (dipole allowed) by a projectile of unit charge with $\hat{v}$ as the quantization axis. The transition to $2p0$ is ignored compared to the dominant $2p+$ transition at large velocity v.

Using the well-known atomic form factor in q-space [McDowell and Coleman, 1970] the first order scattering amplitudes for these two states are for $1s \to 2s$,

$$T_{2s,1s}(\vec{q}) = \frac{i4\pi}{\mathrm{v}} \frac{4\sqrt{2}Z_T^4}{[(\frac{3}{2}Z_T)^2 + q^2]^3} , \tag{8.19}$$

and for $1s \to 2p+$,

$$T_{2p+,1s}(\vec{q}) = \frac{i4\pi}{\mathrm{v}} \frac{6iZ_T^5 q_\perp}{q^2[(\frac{3}{2}Z_T)^2 + q^2]^3} \exp(i\varphi_{\vec{q}_\perp}) . \tag{8.20}$$

The target excitation amplitudes in b-space may be obtained by using the Fourier transforms of the above expressions. Doing this one has for $1s \to 2s$,

$$a^T_{2s,1s}(\vec{b}) = \frac{i\sqrt{2}Z_T^4}{\mathrm{v}} \frac{b^2 K_2(\sqrt{(\frac{3}{2}Z_T)^2 + q_\parallel^2}\, b)}{[(\frac{3}{2}Z_T)^2 + q_\parallel^2]^3} , \tag{8.21}$$

and for $1s \to 2p+$,

$$\begin{aligned} a^T_{2p+,1s}(\vec{b}) &= -\frac{i12Z_T^5}{\mathrm{v}\alpha^6} e^{i\varphi} \Big[|q_\parallel| K_1(|q_\parallel| b) - \beta K_1(\beta b) \\ &- \frac{\alpha^2 b}{2} K_0(\beta b) - \frac{\alpha^4 b^2}{8\beta} K_1(\beta b) \Big] , \end{aligned} \tag{8.22}$$

where φ is the azimuthal angle of $\vec{b}$, K_i the modified Bessel function of the second kind, $\alpha = 3Z_T/2$ and $\beta = \sqrt{\alpha^2 + q_\parallel^2}$.

1 *Screening terms*

The screening terms correspond to elastic scattering of the projectile. The electron-electron interaction reduces the effective projectile charge in this case.

Surface charge density for screening. The effective charge in q-space for screening may be obtained from (8.12) by setting $f_P = i_P = 1s$, which after the Fourier transform (8.18) yields the effective surface charge density for screening term,

$$\begin{aligned} \varsigma_{sc}(\vec{b}' - \vec{b}) &= Z\delta(\vec{b}' - \vec{b}) - \frac{4Z^4}{\pi} \frac{|\vec{b}' - \vec{b}| K_1(\sqrt{4Z^2 + q_{\|}^2}|\vec{b}' - \vec{b}|)}{\sqrt{4Z^2 + q_{\|}^2}} \\ &\equiv \varsigma_{bare}(\vec{b}' - \vec{b}) - \varsigma_{e-e}(\vec{b}' - \vec{b}) , \end{aligned} \tag{8.23}$$

where the total charge density has been divided into the bare projectile component ς_{bare} and the electron charge cloud part ς_{e-e}. Eq(8.23) shows three key features: i) for a point-particle such as the projectile nucleus, the charge density $\varsigma_{bare}(\vec{b}' - \vec{b})$ is localized as a Dirac delta function as expected; ii) ς_{e-e} and ς_{bare} are of opposite sign so that $\varsigma_{sc} < \varsigma_{bare}$; and iii) the electronic charge cloud $\varsigma_{e-e}(\vec{b}' - \vec{b})$ is 'fuzzy', i.e., smeared out over a range of $\vec{b}'$ around $\vec{b}$.

Gaussian-like surfaces. Following directly from (8.7), (8.12), and (8.18), the electronic charge density of (8.23) may be alternatively expressed as,

$$\begin{aligned} \varsigma_{e-e}(\vec{b}' - \vec{b}) &= \int d\vec{q}_\perp e^{i\vec{q}_\perp \cdot (\vec{b}' - \vec{b})} \frac{1}{(2\pi)^2} \langle 1s | e^{-i\vec{q}_\perp \cdot \vec{r}} | 1s \rangle \\ &= \langle 1s | \frac{1}{(2\pi)^2} \int d\vec{q}_\perp e^{i\vec{q}_\perp \cdot (\vec{b}' - \vec{b} - \vec{r}_\perp)} e^{-iq_{\|} z} | 1s \rangle \\ &= \langle 1s | \delta(\vec{b}' - \vec{b} - \vec{r}_\perp) e^{-iq_{\|} z} | 1s \rangle . \end{aligned} \tag{8.24}$$

This expression is connected to the static surface charge density which can be obtained by integrating over z of the volume charge density as,

$$\begin{aligned} \varsigma_{static}(\vec{b}' - \vec{b}) &= \int dz |\phi_{1s}(\vec{r})|^2_{\vec{r}_\perp = \vec{b}' - \vec{b}} = \int d\vec{r} \delta(\vec{b}' - \vec{b} - \vec{r}_\perp) |\phi_{1s}(\vec{r})|^2 \\ &= \langle 1s | \delta(\vec{b}' - \vec{b} - \vec{r}_\perp) | 1s \rangle . \end{aligned} \tag{8.25}$$

As pictured in figure 8.3, ς_{static} may be viewed as a surface charge density accumulated on a Gaussian-like cylindrical surface along z direction of a volume charge density. Similarly, ς_{e-e} is an effective surface charge density differing from the static charge density only by a modulating factor $e^{-iq_{\|} z}$. This factor is related to the fact that a minimum momentum transfer is required, which limits the spatial extent along z which can contribute to the transition. In the limit of large impact speed v, the minimum momentum transfer becomes small $\sim \mathrm{v}^{-1}$ and ς_{e-e} approaches ς_{static} within $O(\mathrm{v}^{-2})$. That is, within the accuracy of the first order perturbation theory,

$$\varsigma_{e-e}(\vec{b}' - \vec{b}) \simeq \varsigma_{static}(\vec{b}' - \vec{b}) \,. \tag{8.26}$$

Thus, ς_{e-e} is a static surface charge density. This surface charge density ς_{e-e} contributes to the transition probability amplitude at a fixed value of $\vec{b}$ and $\vec{b}'$ and a given time t. The target probability amplitude per unit charge $a^T_{fi}(\vec{b}')$ is weighted with ς_{static}, which is the static projectile charge density per $(\vec{b}' - \vec{b})$ that accumulates over a path $z = vt$ during the collision. The total probability amplitude is convoluted over all virtual impact parameters $\vec{b}'$ as illustrated in figure 8.3.

$1s - 2s$ *target excitation.* The convolution process is illustrated for a $1s \rightarrow 2s$ transition in figures 8.4a and 8.4b, for which the transition probability amplitude can be evaluated from (8.17) and (8.21), namely,

$$\begin{aligned} a_{2s,sc}(\vec{b}) &= \int d\vec{b}' \varsigma_{sc}(\vec{b}' - \vec{b}) a^T_{2s,1s}(\vec{b}') \\ &= \int d\vec{b}' \left[Z\delta(\vec{b}' - \vec{b}) - \varsigma_{e-e}(\vec{b}' - \vec{b}) \right] a^T_{2s,1s}(\vec{b}') \,. \end{aligned} \tag{8.27}$$

Explicit evaluation of the (8.27) yields a simple analytic result

$$\begin{aligned} a_{2s,sc}(\vec{b}) &= Z \frac{i\sqrt{2} Z_T^4}{v} \frac{b^2 K_2(\beta b)}{\beta^2} - \frac{i 2^{\frac{15}{2}} Z^4 Z_T^4}{v \lambda^8} \left[\frac{\lambda^2}{2} \frac{b K_1(\gamma b)}{\gamma} - 3K_0(\gamma b) + \right. \\ & \left. 3K_0(\beta b) + \lambda^2 \frac{b K_1(\beta b)}{\beta} + \frac{\lambda^4}{8} \frac{b^2 K_2(\beta b)}{\beta^2} \right] \,, \end{aligned} \tag{8.28}$$

where $\gamma = \sqrt{4Z^2 + q_{\|}^2}$ and $\lambda^2 = \alpha^2 - 4Z^2$ (α and β are given earlier following (8.22)).

The solid curve in figure 8.4a is due to the bare projectile nucleus and corresponds to the convolution of the first term in (8.27). The surface charge density of the nucleus is simply $Z\delta(\vec{b}' - \vec{b})$, represented by the dashed curve in figure 8.4a. Convolution of the this surface charge density with the target excitation amplitude (8.21) yields the first term in (8.28).

The convolution from the second term in (8.27) results in the contribution due to the electronic screening (solid curve in figure 8.4b) and corresponds to the bracketed term in (8.28). The broadening of the transition amplitude due to a broad electronic cloud (figure 8.4b) is shown clearly by comparing with the transition amplitude due to the bare nucleus (figure 8.4a). The width of $\varsigma_{e-e}(\vec{b}' - \vec{b})$ is seen to be approximately the size of the electron cloud of the projectile as would be expected.

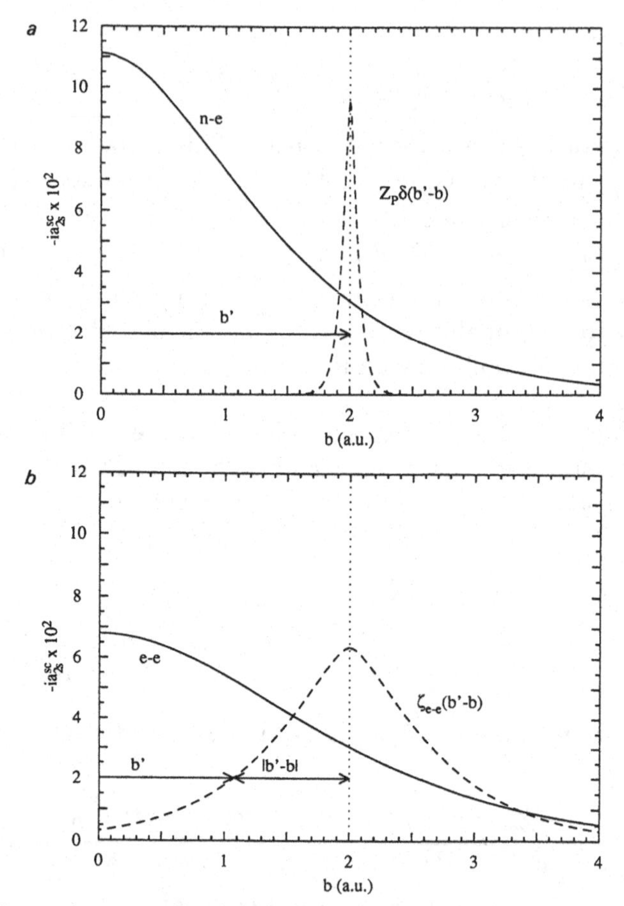

Fig. 8.4. Screening contribution for target excitation $1s \to 2s$ in H + H collisions at $v = 5$ au. (a) The bare nuclear contribution. The solid curve shows the transition amplitude, and the dashed curve shows the point charge density ς_{bare}. (b) The electronic contribution. The solid solid curve shows the transition amplitude, and the dashed curve shows the electronic charge density ς_{e-e} [Wang *et al.*, 1995b].

2 Antiscreening terms

The antiscreening term corresponds to excitation of the projectile electron by the target electron. Proper inclusion of all the channels in antiscreening is relatively difficult even within the first Born approximation.

Summing over the excited states of the projectile. Although no analytic form for the antiscreening terms in (8.17) is known, some simplification can be made by considering the transition probability $|a_{f_T,f_P}(\vec{b})|^2$ instead of the transition amplitude $a_{f_T,f_P}(\vec{b})$. The transition probability for antiscreening is an incoherent sum over all final states except the initial state, namely,

$$|a_{f_T,anti}(\vec{b})|^2 = \sum_{f_P \neq i_P} |a_{2s,f_P}(\vec{b})|^2 = \int\int d\vec{b}' d\vec{b}'' \times \sum_{f_P \neq i_P} a^T_{fi}(\vec{b}') a^{T*}_{fi}(\vec{b}'') \varsigma^{eff}_{f_P,1s}(\vec{b} - \vec{b}') \varsigma^{eff}_{f_P,1s*}(\vec{b} - \vec{b}'') . \quad (8.29)$$

In order to apply closure to (8.29) one may use the alternative expression for $\varsigma^{eff}(\vec{b})$ in the form of (8.24), namely,

$$\varsigma^{eff}_{f_P,1s}(\vec{b}) = \langle f_P | \delta(\vec{b} - \vec{r}_\perp) e^{-iq_\| z} | 1s \rangle , \quad (8.30)$$

where $f_P \neq 1s$. Substituting (8.30) into (8.29) one obtains,

$$|a_{f_T,anti}(\vec{b})|^2 = \int\int d\vec{b}' d\vec{b}'' \sum_{f_P \neq i_P} a^T_{fi}(\vec{b}') a^{T*}_{fi}(\vec{b}'') \times \langle 1s | \delta(\vec{b} - \vec{b}'' - \vec{r}_\perp) e^{iq_\| z} | f_P \rangle \times \langle f_P | \delta(\vec{b} - \vec{b}' - \vec{r}_\perp) e^{-iq_\| z} | 1s \rangle . \quad (8.31)$$

This expression is the antiscreening transition probability summed over all final excited states of the projectile. It is exact within the first order perturbation theory. Next, (8.31) is simplified using the closure approximation.

Closure approximation. In the closure approximation [Anholt, 1986], it is assumed that the target excitation amplitude, which is an implicit function of $q_\|$ (hence f_P through $q_\| = \Delta E/\mathrm{v}$ and $\Delta E = \varepsilon^P_f + \varepsilon^T_f - \varepsilon^P_i - \varepsilon^T_i$), may be approximated at some average $\bar{q}_\|$. Denoting this by $\bar{a}^T_{fi}$ and factoring both of the terms outside the sum in (8.31), one has,

$$|a_{f_T,anti}(\vec{b})|^2 = \int\int d\vec{b}' d\vec{b}'' \bar{a}^T_{fi}(\vec{b}') \bar{a}^{T*}_{fi}(\vec{b}'') \sum_{f_P \neq i_P} \langle 1s | \delta(\vec{b} - \vec{b}'' - \vec{r}_\perp) e^{i\bar{q}_\| z} | f_P \rangle \times \langle f_P | \delta(\vec{b} - \vec{b}' - \vec{r}_\perp) e^{-i\bar{q}_\| z} | 1s \rangle . \quad (8.32)$$

Using the exact closure property of a complete basis set,

$$\sum_{f_P \neq i_P} |f_P\rangle\langle f_P| = 1 - |i_P\rangle\langle i_P| ,$$

one obtains from (8.32),

$$\begin{aligned}
|a_{f_T,anti}(\vec{b})|^2 &= \int\int d\vec{b}' d\vec{b}'' \bar{a}^T_{fi}(\vec{b}')\bar{a}^{T*}_{fi}(\vec{b}'') \qquad (8.33)\\
&\times\langle 1s|\delta(\vec{b}-\vec{b}''-\vec{r}_\perp)e^{i\bar{q}_\| z}(1-|1s\rangle\langle 1s|)\\
&\times\ \delta(\vec{b}-\vec{b}'-\vec{r}_\perp)e^{-i\bar{q}_\| z}|1s\rangle\\
&= \int\int d\vec{b}' d\vec{b}'' \bar{a}^T_{fi}(\vec{b}')\bar{a}^{T*}_{fi}(\vec{b}'')\\
&\times\ \Big[\delta(\vec{b}'-\vec{b}'')\langle 1s|\delta(\vec{b}-\vec{b}'-\vec{r}_\perp)|1s\rangle\\
&-\ \langle 1s|\delta(\vec{b}-\vec{b}''-\vec{r}_\perp)e^{i\bar{q}_\| z}|1s\rangle\langle 1s|\delta(\vec{b}-\vec{b}'-\vec{r}_\perp)e^{-i\bar{q}_\| z}|1s\rangle\Big]\\
&= \int d\vec{b}' \varsigma_{static}(\vec{b}'-\vec{b})\left|\bar{a}^T_{fi}(\vec{b}')\right|^2 - \left|\int d\vec{b}' \bar{\varsigma}_{e-e}(\vec{b}'-\vec{b})\bar{a}^T_{fi}(\vec{b}')\right|^2
\end{aligned}$$

where $\bar{\varsigma}_{e-e}$ is understood to be evaluated at an average $\bar{q}_\|$ just like $\bar{a}^T_{fi}$.

Eq (8.33) is the final result for antiscreening after using the closure approximation. The validity of this approximation has been well studied and various schemes have been suggested [Montenegro *et al.*, 1994]. One simple value of $\bar{q}_\|$ is,

$$\bar{q}_\| = \frac{\varepsilon^T_f - \varepsilon^T_i - \varepsilon^P_i}{\mathrm{v}}, \qquad (8.34)$$

in which it is assumed that the mean energy of the projectile final state is zero. This may be reasonable in view of the bound and continuum contributing states, as well as the observation that most contribution to antiscreening comes from low-lying continuum states of the projectile. Other choices of $\bar{q}_\|$, however, may also be used [Day, 1981].

Threshold for antiscreening. In the limit of full antiscreening, the projectile electrons and the projectile nucleus are fully decoupled as noted above. In this case one might expect that the antiscreening contribution is zero unless the projectile has enough speed so that a projectile electron has sufficient kinetic energy to excite the target by itself. This defines a minimum threshold velocity for the antiscreening contribution, as first suggested by Hagmann [1986] and by Anholt [1986]. Partial theoretical justification for this antiscreening threshold was given by Montenegro and Meyerhof [1991], whose results suggest that the threshold is somewhat gradual.

8.4 Impulse approximation

The impulse approximation is based on a perturbation expansion which includes higher order effects of the projectile interaction V on the target electron.

The impulse approximation has been used to describe a variety of interactions ranging from nuclear collisions to various atomic collisions [Goldberger and Watson, 1964]. This approximation is particularly well suited to collisions in which the particles undergoing a transition (e.g. the target electrons here) may be considered as 'quasi-free', uncorrelated particles that interact strongly with the perturbing collision partner (e.g., the entire projectile).

Starting from a full many body theory, the many body operators are replaced by two body operators in the impulse approximation [McDowell and Coleman, 1970, section 6.8; Coleman, 1968]. This simplifies the many body problem considerably. On the other hand, the two body operators are the full two body scattering operators (e.g. T matrices), so that there is a significant improvement over the first Born approximation in which an effective interaction V_{eff} appears in first order only. In the impulse approximation the active electrons are regarded as uncorrelated, effectively free electrons, and the cross section is determined by calculating the transition for one or more quasi-free target electrons weighted by a momentum distribution that is determined by the undisturbed initial state target wavefunctions. The impulse approximation is valid if the collision time is much shorter than the orbiting time $1/I_{target}$ (in atomic units) of the target electrons, where I_{target} is their binding energy [Mott and Massey, 1965, section XII.5.2]. A less restrictive criterion is that I_T be much smaller than the kinetic energy of the projectile.

The term 'impulse approximation' can lead to confusion because this term is used to denote different physical and mathematical approximations by different authors. Moreover, in standard treatments [Goldberger and Watson, McDowell and Coleman, and herein] projectile and target are used in the conventional sense, but in some other cases [Montenegro *et al*, 1994, and many of the other references of this chapter] the target and projectile are interchanged because it is the projectile in which an electron is observed to occur. Also, the projectile in the standard notation is usually taken to be structureless, such as a bare ion or a simple electron. In this chapter it is the structure of the projectile that is of interest.

Since one-center correlation is of some interest as are N-electron effects, consideration below is first given to target excitation by a target with N correlated electrons interacting with a projectile via a single effective potential V_{eff}. This is 'quasi-elastic scattering' [Goldberger and Watson,

1964] which includes many body effects on one center (the target here), but does not address higher order effects due to the detailed structure of the projectile. Next, simpler systems with only one target electron are considered where the impulse approximation is defined for target excitation via an interaction V_{eff} with a structureless projectile. In a still simpler case the cross section in the impulse approximation is described by the cross section for a target electron interacting freely with V_{eff} averaged over a bound state momentum distribution for the target electron. Finally, some attempts to go beyond the usual impulse approximation and include the structure of the projectile are considered.

Quasi-elastic scattering. The full many electron problem is difficult to solve exactly. If the electron-electron correlation between the projectile electrons is retained, one obtains a widely used approximation that is intermediate between the exact many electron solution and the impulse approximation. This is called 'quasi-elastic scattering' [Goldberger and Watson, 1964, section 11.2].

Consider a projectile with N identical correlated electrons interacting with a target whose detailed internal structure is rigid. Quasi-elastic scattering occurs when a small excitation energy is imparted to the target by the projectile but electron correlation in the projectile is retained. For a target with N identical, correlated electrons, the differential cross section for target excitation (with a sum of elastic plus inelastic scattering of the projectile) may be expressed in the quasi-elastic scattering approximation as [Goldberger and Watson, (200) in 11.2],

$$d\sigma/d\Omega = \left\{N + N(N-1)\left[|F(\vec{q})|^2 + \frac{1}{\mathcal{V}}C(\vec{q})\right]\right\}d\sigma_{Two-Body}/d\Omega\,, \quad (8.35)$$

where $\vec{q}$ is the momentum transfer of the projectile to the target, $F(\vec{q})$ is the target one-electron form factor corresponding to (8.6), with $F(\vec{q}) = F_{f_{P,T}i_{P,T}} = F_{i_T i_T}$ in (8.7). $\mathcal{V}$ is the volume of the target, $C(\vec{q})$ is a multi-electron target correlation function and $d\sigma_{Two-Body}/d\Omega$ is the differential cross section for scattering of the projectile from a single free target electron into a solid angle $d\Omega$.

Some conceptual physics is evident in (8.7). Incoherent scattering from the N projectile electrons occurs at $q >> 1/r_{\text{interaction}}$ where the last two terms above go to zero and $d\sigma/d\Omega$ becomes linear in N. At the other limit, where q becomes small, elastic scattering of the target dominates, which is coherent so that the cross section varies as N^2 [since $F(0) = 1$ and $C(0) = 0$ in the equation above]. The transition from a coherent to an incoherent cross section for a projectile with N electrons is also discussed at the end of section 8.2.1 above. In the impulse approximation, $\frac{1}{\mathcal{V}}C(\vec{q})$

is omitted, i.e., multi-electron projectile correlation is neglected, and the target electrons are treated independently, but the two-body cross section is folded with density distributions of the active target electron.

The impulse approximation. Now consider the impulse approximation including the two-center electron-electron interaction, but keeping the projectile rigid (i.e. frozen in its ground state), which is equivalent to retaining only the screening term. Here only a single target electron is included (the extension to multiple target electrons is briefly discussed below). The full perturbing interaction from (8.2) is $V = -\frac{Z}{|\vec{R}-\vec{r}_T|} - \frac{Z_T}{|\vec{R}+\vec{r}_P|} + \frac{1}{r_{12}}$ of (8.2), where the last term is the interaction of the projectile electron with the target which does not contribute in the first Born approximation. With this full interaction V the exact transition operator T may be expressed as [McDowell and Coleman, 1970, chapter 6, section 8.1.; Wang *et al.*, 1992],

$$T = V \sum_{n=0}^{\infty} (G_0^+ V)^n \, , \tag{8.36}$$

where higher order scattering (e.g. with the target nucleus) is fully included. Here G_0^+ is the Green function describing the unperturbed propagation of the target and projectile electrons. One now neglects the electron-electron interactions within the projectile (or approximates them by a mean field potential). To obtain the impulse approximation for a system in which the interaction V_{eff} is a sum of interactions, one neglects in T the $\frac{Z_T}{|\vec{R}+\vec{r}_P|}$ interaction of the projectile electron with the target nucleus (which to first order does not affect the transition of the target electron) and replaces V by [Wang *et al.*, 1992],

$$V_{eff} = -\frac{Z}{|\vec{R} - \vec{r}_T|} + \frac{1}{r_{12}} \, . \tag{8.37}$$

If $Z_T/\mathrm{v} << 1$, it is not expected that the interaction of the target charge Z_T with the projectile electron is important in the transition of the target electron. This last idea is the essence of the impulse approximation. Only the projectile perturbs the target electron in this approximation. The transition operator is now given by,

$$T_{IA} = V_{eff} \sum_{n=0}^{\infty} (g_0^+ \, V_{eff})^n \, , \tag{8.38}$$

where g_0^+ is the one-electron free particle Green function. The $n = 0$ term corresponds to the PWBA approximation discussed in section 8.2.

In the impulse approximation, the cross section for the transition of a target electron caused by the interaction V_{eff} is sometimes further approximated in terms of the two-body cross section from the two-body matrix $T_{\rm IA}$ defined by (8.38), namely,

$$d\sigma_{IA}/d\Omega = \int d^3q \; < i_t(\vec{q})|i_T(\vec{q}) > \; d\sigma_{Two-Body}/d\Omega \, , \tag{8.39}$$

where $\vec{q}$ is the target electron momentum, $< i_T(\vec{q})|$ is the momentum representation of state of the target electron in its initial state and $d\sigma_{Two-Body}/d\Omega$ is the exact cross section for scattering of an unbound target electron with the projectile. If the projectile were a single charged particle, then $d\sigma_{Two-Body}/d\Omega$ would be evaluated from the off-shell two-body Coulomb T matrix for scattering of the target electron by the charged (projectile) particle [Wang *et al.*, 1991]. This expression is commonly used [Goldberger and Watson, (60b) in 11.1]. Eq(8.39) is intuitively understandable since it describes a two-body cross section folded with a momentum spectrum of the active bound target electron.

Wang *et al.* [1992] point out that (8.39) is valid only if the target emits a fast electron in the backward direction. Otherwise the momentum space density distribution $< i_T(\vec{q})|i_T(\vec{q}) >$ of the ground state wavefunction should be replaced [McDowell and Coleman, 1970] by the inelastic target form factor $F'(\vec{q}) =< f_T|e^{-i\vec{q}\cdot\vec{r}}|i_T >$. It is only when the final state is approximated by an outgoing plane wave that (8.39) is valid [Wang *et al.*, 1991].

If there is more than one target electron, $< i_T(\vec{q})|i_T(\vec{q}) >$ is replaced by $\sum_{j=1}^{n} < f_T|e^{-i\vec{q}\cdot\vec{r}_j}|i_T >$ and one may recover the n dependence of (8.35) with $C = 0$. The relation of the impulse approximation to second Born theories is discussed by Jakubassa-Amundsen [1992] who also describes how to include the antiscreening term (inelastic projectile scattering) in the impulse approximation. An impact-parameter formulation of the impulse approximation has been given by Briggs [1977].

In practice the impulse approximation does not usually include the internal structure of the projectile. Hence the impulse approximation is not often used to treat the antiscreening terms that arise from excitation or loss of a projectile electron.

Related models. Based on work of Brandt [1983], Zouros *et al.* [1989, 1993] have used (8.39) as the basis for an intuitively appealing model expressing the four-body cross section (projectile nucleus and electron, colliding with the target nucleus and target electron) in terms of the three-body cross section (projectile nucleus and electron, colliding with the target electron). Zouros *et al.* use the latter cross section at a given relative velocity $\vec{v}$ between the target electron and the projectile, and weight it by the Compton profile of the target electron. If $\vec{p}_0$ is the momentum of the target electron in the target frame, then its momentum in the projectile frame is $\vec{p} = \vec{p}_0 + \vec{v}$. In many cases $v >> p_0$ so that $p = (v^2 + 2vp_{0z} + p_0^2)^{1/2} \approx (v^2 + 2vp_{0z})^{1/2}$. Then, in analogy with (8.39),

$$\sigma_{Four-Body} = \int_{-\infty}^{\infty} dp_{0z}\ J_0(p_{0z})\ \sigma_{Three-Body}(v^2 + 2vp_{0z})\ . \qquad (8.40)$$

Here, $J_0(p_{0z}) = \int\int dp_{0x} dp_{0y} < i_T(\vec{p}_0 | i_T(\vec{p}_0 >$ is the Compton profile which is the probability to find a target electron with a momentum component p_{0z}. Here $< i_T(\vec{p}_0|$ is the momentum representation of the ground state of the target electron. The model of Zouros *et al.* is somewhat different than the usual impulse approximation (Cf. Montenegro *et al.* [1994]).

8.5 Observed results

It is well established experimentally that a projectile with N electrons of a net charge $+Q = Z - N$ acts like a point particle of charge $+Q$ when the size of the projectile is much smaller than the size of the target. For example, in collisions of U^{+90} + He the uranium ion (which has a nuclear charge of +92) acts like a point projectile of charge +90. In this screening limit, the tightly held electrons simply screen the charge of the nucleus.

When the projectile is much larger than the target, the first Born approximation predicts that the projectile nucleus and the projectile electrons act as independent particles. This limit corresponds to the free collision model of Bohr [1948]. An observation illustrating this antiscreening limit is shown in figure 8.5. In this experiment of Wang *et al.* [1986] excited hydrogen H* in a high Rydberg state collides with a much smaller argon atom. The data observed for H* is equal to the sum of data for p^+ and for e^- colliding with argon.

Data illustrating both screening and antiscreening are shown in figure 8.6. In these cases from Huelskötter *et al.* [1991] the one electron target ion O^{+7} loses its electron in collisions with both H_2 and helium. There is a threshold for antiscreening at a velocity where the H_2 or helium electron has enough energy itself to cause transitions in the O^{+7} target. Below this threshold the antiscreening contribution is small, while above the antiscreening threshold, both the screening and the antiscreening contributions

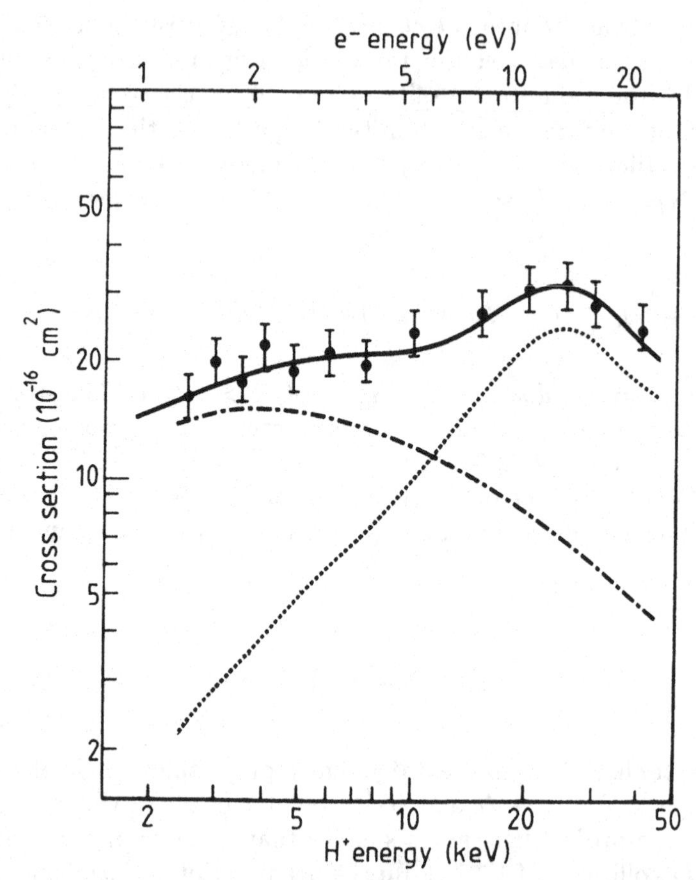

Fig. 8.5. Cross section for scattering of H* from argon [Wang *et al.*, 1986]. The dash-dot line is data for scattering of e^- from argon and the dotted line is for p^+ from argon. The solid curve is the sum of the e^- and p^+ results and is in agreement with the H* data. This corresponds to the incoherent (i.e. antiscreening) limit of figure 8.2 where the H* nucleus and the H* electron scatter independently.

are significant. The effects of projectile electrons have been observed* in a variety of systems [Richard, 1990; Montenegro *et al.*, 1994].

Observations testing the impact parameter dependence of the screening and antiscreening effects have been carried out by Montenegro *et al.* [1993] and by Wu *et al.* [1995]. Wu *et al.* have also observed triple ionization events in which double ionization occurs on one center and single ionization occurs on the other atomic center.

Effects of projectile electrons on ejected target electrons in binary encounters have been discussed by Wang *et al.* [1993].

* In many of these cases the role of target and projectile are reversed.

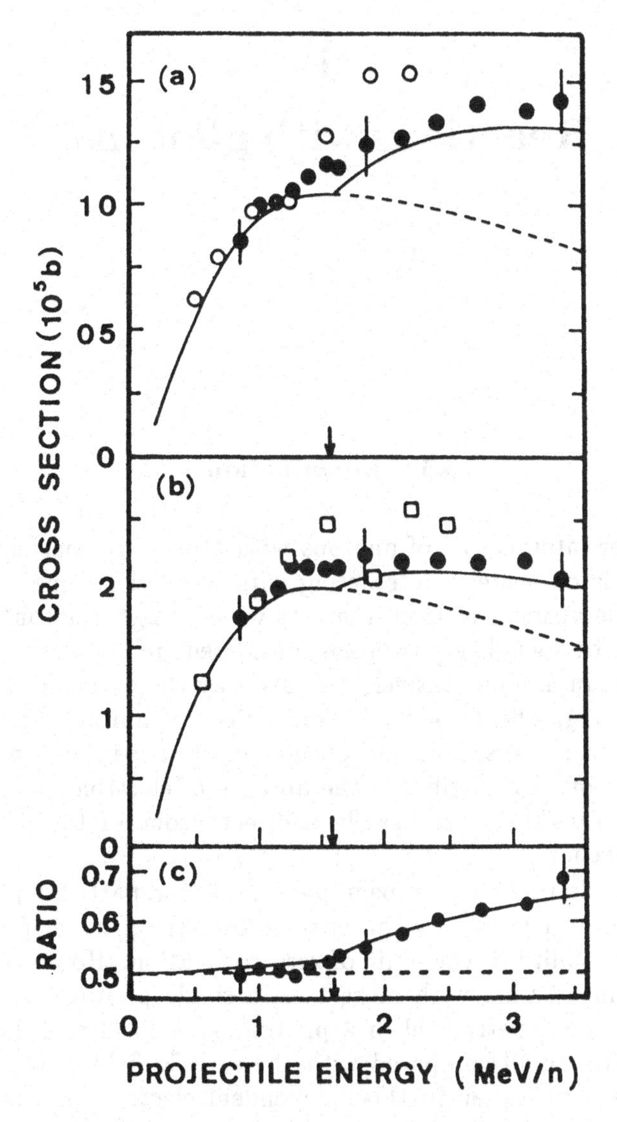

Fig. 8.6. Cross sections for electron loss in O^{+7} in collisions with a) H_2 and (b) helium as a function of the collision velocity. The arrows indicate the location of the threshold for ionization for free electrons. The dashed curves give the continuation of the calculated screening contributions above the antiscreening threshold. The ratio of H_2 to He cross sections is given in (c). The solid curve is the first Born calculation with antiscreening and the dashed curve is the same calculation with only the screening term [Huelskötter *et al.*, 1991].

9
Reactions with photons

9.1 Introduction

In this chapter interactions of photons with atoms are considered. Here the emphasis is on systems interacting with weak electromagnetic fields so that a single atomic electron interacts with a single photon*. Initially interactions with a single electron are considered. In this case the photon tends to probe in a comparatively delicate way the details of the atomic wavefunction (e.g. effects of static correlation in multi-electron atoms). Later two electron transitions are considered. Because these two electron transitions are often negligible in the absence of electron correlation the two electron transitions are usually a direct probe of the dynamics of electron correlation.

In previous chapters the impact parameter (or particle) picture has been used wherever possible in order to recover the product form for the transition probability in the limit of zero correlation. However, here the likelihood of interacting with more than a single photon is quite small since the electromagnetic field of a photon, even for strong laser fields, is almost always small compared with the electric field provided by the target nucleus. Consequently, this independent electron limit is not often useful. Also, photon wavepackets are usually much larger in size than an atom. Consequently the wave picture is used where the electric and magnetic fields of the photon are considered to be plane waves. Transformation to the particle picture may be done using the usual Fourier transform from the scattering amplitude $f(\vec{q})$ to the probability amplitude $a(\vec{b})$ (Cf.

---

* The photon is thought to be composed of oscillating electric and magnetic fields. There is no corresponding model for the electron which we mathematically describe as a conceptually simple matter wave, although the electron appears to be more complex than the photon (e.g. the magnetic moment and the classical radius of the electron suggest internal structure as yet unknown).

section 3.3.3). In this case, however, as in the prior chapter, the impact parameter $\vec{b}$ is not well localized and is not physically observable, while the momentum transfer $\vec{q}$ is well defined.

9.2 The Hamiltonian

The Hamiltonian for a single electron in the presence of an atomic nucleus of charge Z_T is,

$$H_T = \frac{1}{2}p^2 + V_{e,Z_T} = -\frac{1}{2}\nabla^2 - Z_T/r \ . \tag{9.1}$$

In the plane wave limit the incident photon is described [Gasiorowicz, 1996, (21-39); Starace, 1982; Heitler, 1954, p.56] as an oscillating electromagnetic field described by a vector potential,

$$\vec{A} = \sqrt{2\pi c^2/\omega\mathcal{V}}\vec{\lambda}e^{i(\vec{k}\cdot\vec{r}-\omega t)} \ , \tag{9.2}$$

where $\vec{\lambda}$ is the direction of polarization (i.e. direction of the electric field vector) of the photon, $\vec{k} = \omega/c\hat{k}$ is the momentum of the photon perpendicular to $\vec{\lambda}$, $\mathcal{V}$ is the spatial volume and ω is the energy of the photon (in atomic units). Here the Coulomb gauge is used [Jackson, 1975] where $\nabla \cdot \vec{A} = 0$ and $V(\vec{r}) = \int \rho(\vec{r} - \vec{r}\,')d\vec{r}\,'$ with $\rho(\vec{r} - \vec{r}\,')$ representing the density of electric charge. Here the radiation source is far away and thus one may take $\rho = 0$. In the Coulomb gauge the photon has no longitudinal components of $\vec{A}$.

For an atom with a single electron of charge $e = -|e| = -1$ (in atomic units) in the field of a photon one has [Heitler, 1954],

$$\begin{aligned} H &= \frac{1}{2}(\vec{p} + \vec{A}/c)^2 - Z_T/r = \alpha\vec{p}\cdot\vec{A} + \frac{1}{2}\alpha^2A^2 + H_T \\ &\equiv V + H_T \ , \end{aligned} \tag{9.3}$$

where $\alpha = 1/c = 1/137$. Eq(9.3) is frequently used to describe the interaction of light with matter. Here $V = \vec{p}\cdot\vec{A}/c + \frac{1}{2}A^2/c^2$ is the interaction potential of a target atom with an incident photon. At this point all multipole terms are retained in the $e^{i\vec{k}\cdot\vec{r}}$ term in $\vec{A}$.

The canonical momentum here is $\vec{P} = \vec{p} - \vec{A}/c$ so that (9.3) is of the form $H = \frac{1}{2}P^2 + V(Q)$ where from (6.27) one has $\dot{Q} = \frac{\partial H}{\partial P} = P$ and $\dot{P} = -\frac{\partial H}{\partial Q} = -\frac{\partial V}{\partial Q}$. It is the canonical momentum $\vec{Q}$ that carries the creation and annihilation operators which create and annihilate the photons and not the 'mechanical' momentum $\vec{p}$ nor the 'quiver' momentum $\vec{A}/c$

corresponding to the oscillating electromagnetic field [Heitler, chapter 2; Aberg and Tulkki, 1985]. It is the $\vec{p}\cdot\vec{A}$ term that couples the mechanical momentum of the electron with the quiver momentum of the photon. Aberg [1996] points out that that $\vec{p}\cdot\vec{A}$ can contribute to (γ,γ') reactions. The role of the creation and annihilation operators is most simply given in a relativistic formulation where the creation and annihilation operators are linear in $H_{int} = -\gamma_\mu A^\mu$. In such a relativistic formulation the A^2 term in (9.3) arises from the second order term in $-\gamma_\mu A^\mu$ using a closure approximation appropriate for non-relativistic energy differences in intermediate states small compared to the electron rest mass energy $m_e c^2 = \alpha^{-2}$.

1 Dipole approximation

The dipole approximation in (9.2) is $e^{i\vec{k}\cdot\vec{r}} \approx 1$, so that,

$$\vec{A} = \sqrt{2\pi c^2/\omega V}\vec{\lambda}e^{i(\vec{k}\cdot\vec{r}-\omega t)} \simeq \sqrt{2\pi c^2/\omega V}\vec{\lambda}e^{-i\omega t} \,. \tag{9.4}$$

This approximation is valid when $kr \simeq \omega/c << 1$, where $r \simeq r_{atom} \simeq 1$ (in atomic units). The parameter, kr, exceeds unity for photon energies above 3.7 keV with r equal the Bohr radius. Now $\vec{\lambda}$ is taken to be the direction of the z axis.

2 L, V and A forms of the dipole operator

From (9.3) it is evident that in calculating properties of matter interacting with photons one may often wish to evaluate the matrix element,

$$< f|\vec{p}\cdot\vec{A}|i> \sim < f|\vec{p}\cdot\vec{\lambda}|i> = i < f|p_z|i> = -i < f|\frac{d}{dz}|i> -i \equiv d^V \tag{9.5}$$

This is called the 'velocity' (V) form of the dipole operator because $-i\frac{d}{dz}$ corresponds to the velocity of the electron in the atomic target. It is also the time derivative of the length form $d^L =< f|z|i>$ as is next shown.

To define the 'length' (L) form of the dipole matrix element, the following identity may be used,

$$\begin{aligned}[z, H_T] &= [z,(p^2/2 - Z_T/r)] = [z,p^2/2] = -\tfrac{1}{2}[z,\nabla^2] \\ &= -\tfrac{1}{2}[z\tfrac{d^2}{dz^2} - \tfrac{d^2}{dz^2}z] = -\tfrac{d}{dz} \,.\end{aligned} \tag{9.6}$$

The length form of the dipole matrix element is now defined by the relation,

$$d^V = \frac{\partial}{\partial t} d^L = [d^L, H_T] = -[< f|z|i >, H_T] \; ,$$

so that,

$$d^L = - < f|z|i > \tag{9.7}$$

In the length form the dipole operator is simply the length z.

In a similar fashion the 'acceleration' (A) form of the dipole operator may be defined,

$$\begin{aligned} d^A = \frac{\partial}{\partial t} d^V &= [d^V, H_T] =< f|[\frac{d}{dz}, (p^2/2 - Z_T/r)]|i > \\ &= - < f|[\frac{d^2}{dz^2}, Z_T/r]|i > =< f|zZ_T/r^3|i > \; . \end{aligned} \tag{9.8}$$

The 'acceleration' (or A) form corresponds to second time derivative of the length form since $d^A = [[d^L, H_T], H_T]$ acts as a second time derivative of $d^L = z$. Faisal [1987] and Giusti-Suzor and Zoller [1987] point out that in this gauge the electron is transformed to a frame in which the electron is freely oscillating in the electric field of the photon.

Noting that, for any operator d,

$$\begin{aligned} < f|[H_T, d]|i > &=< f|H_T d - dH_T|i >= -(E_f - E_i) < f|d|i > \\ &= \omega < f|d|i >= - < f|[d, H_T]|i > \; , \end{aligned} \tag{9.9}$$

so that the corresponding L, V and A matrix elements are related by,

$$\begin{aligned} -\omega d^L &= d^V = -1/\omega d^A \\ d^a &= \partial d^V/\partial t = \partial^2 d^L/\partial t^2 \; . \end{aligned} \tag{9.10}$$

Consequently, for one electron atoms the L, V and A forms of the dipole matrix element will give identical cross sections. However, the matrix elements are different. The length form depends most strongly on contributions at large distances and the acceleration form is sensitive to contributions close to the nucleus and eliminates long range effects as is apparent from (9.5), (9.7) and (9.8).

L, V and A gauge transformations. The transformations between L, V and A forms of the dipole operator may be related to electrodynamic gauge transformations [Grant, 1974]. The relation between electromagnetic and quantum gauge transformations is given in the appendix to this chapter. The velocity form of the dipole operator corresponds to the

Coulomb gauge. In the long wavelength (i.e. dipole) limit, the gauge transformation from the velocity form to the length form corresponds [Aberg and Tulkki, 1985] to,

$$\vec{A}^V = \vec{A}^L - \nabla\eta \qquad \phi^V = \phi^L + \frac{\partial\eta}{\partial t}, \tag{9.11}$$

where $\eta = \vec{r}\cdot\vec{A}^V(t)$. Here $\phi^V = 0$, $\nabla\cdot\vec{A}^V = 0$ and $[\vec{p},\vec{A}^V] = 0$ for the transverse radiation fields in the Coulomb gauge. This defines the velocity form of the vector and scalar potentials, $\vec{A} = \vec{A}^V$ and $\phi = \phi^V = 0$. Consequently, $\vec{E} = -\frac{1}{c}\frac{\partial A}{\partial t}$, where $\vec{E}$ is the electric field of the photon. This gauge transformation corresponds to,

$$\frac{1}{c}\vec{p}\cdot\vec{A}^V - \Phi^V \;\rightarrow\; -\Phi^L = -\frac{1}{c}\vec{r}\cdot\frac{\partial\vec{A}^L}{\partial t} = \vec{r}\cdot\vec{E}, \tag{9.12}$$

as is easily verified, with an ω dependence consistent with (9.11) in the dipole limit.

The transformation from velocity to acceleration form of the dipole operator corresponds to the gauge transformation $\eta = \frac{1}{c}\int^t dt'\vec{A}^V(t')\cdot\vec{p}$. Using (9.51) below it is straightforward to show [Wang, 1994] that in the dipole limit $\frac{1}{c}\vec{A}^V\cdot\vec{p}$ is transformed to $-\frac{Z_T}{c}\int^t dt'\vec{A}^V(t')\cdot\frac{\vec{r}}{r^3}$ where $V = -Z_T/r$.

In the case of the interactions of charged particles (including electron-electron interactions in multi-electrons targets) it is conventional to use the Coulomb gauge with,

$$\nabla\cdot\vec{A} = 0, \qquad V = \int\frac{\rho(\vec{r}\,')}{|\vec{r}-\vec{r}\,'|}d\vec{r}\,'. \tag{9.13}$$

For photons all three gauges, corresponding to the L, V and A forms of the dipole operator are commonly used. In principle, to be consistent in the choice of gauge, if one works in a gauge other than the Coulomb gauge (corresponding to the V form of the dipole operator), then one might wish to transform the V of the Coulomb potential in (9.13) above and introduce the corresponding vector potential, $\vec{A}$, for interactions with the charged particles. In practice, this is not usually done†. This freedom to choose different gauges for different interactions is fortunate since otherwise it could be difficult to use unusual non-Coulomb potentials to

† In transforming $\frac{1}{c}\vec{p}\cdot\vec{A} - \Phi + V$, it is sufficient to neglect the influence of the gauge transformation on V if V is considered to be a part of $-\Phi + V$ since the $\frac{\partial\eta}{\partial t}$ may be associated with Φ and not V. An example is given in section 9.3.1 for double photoionization.

evaluate wavefunctions for systems of charged particles in a problem where the non-Coulomb acceleration gauge, for example, were used to describe the interaction of photons with matter.

The results of this subsection have been developed for one electron atoms. They may usually be applied to multi-electron atoms if $\frac{1}{|\vec{r}_i-\vec{r}_j|}$ electron-electron terms and exchange terms are approximated by a central potential $v(r)$. However, for approximate calculations using non-local $1/|\vec{r}_i-\vec{r}_j|$ potentials there are often differences in the L, V and A forms of cross sections for single ionization by photo annihilation [Starace, 1982]. These differences in L, V and A calculations are due to multi-electron effects and are used as a measure of the error in these calculations.

9.3 Photoionization (Photo effect)

Photoionization occurs when a photon is annihilated (photoannihilation) in the process of ionizing an atomic electron. This is the photo effect first explained by Einstein and is denoted by 'PE' to distinguish it from Compton scattering in which the incident photon is inelastically scattered and not annihilated. Photoannihilation by a free electron is forbidden by conservation of energy and momentum. The cross section for photoionization may be expressed in terms of the $\alpha\vec{p}\cdot\vec{A}+\frac{1}{2}\alpha^2A^2$ term in (9.3) which represents the interaction of radiation with matter. The $\frac{1}{2}\alpha^2A^2$ term is usually dropped for photoionization because it is a factor of α smaller than the leading term and also because it corresponds to a two photon process. (The $\frac{1}{2}\alpha^2A^2$ term is important for lowest order Compton scattering, however.).

Using standard first order perturbation theory [Landau and Liftshitz, section 42], one has the transition rate into a final state $|f>=|\vec{k}>$, for an electron in a final continuum state with momentum, $\vec{k}$,

$$dW = 2\pi\alpha\left|< f|\vec{p}\cdot\vec{A}|i>\right|^2\delta(\epsilon_k+I-\omega)k^2dkd\Omega\,, \tag{9.14}$$

where I is the binding energy of the electron and $d\Omega$ is a solid angle about the direction of the ejected electron $\hat{k}$ with kinetic energy $\epsilon_f=\epsilon_k$. The final state wavefunctions are orthonormal so that $< f'|f>=<\vec{k}'|\vec{k}>=\delta(\vec{k}'-\vec{k})$. Dividing this transition rate by the incident photon density, $c/\mathcal{V}$, one has the usual differential scattering cross section for photoionization in the velocity form of the dipole matrix element,

$$\frac{d\sigma_{PE}^{+}}{d\Omega} = \frac{4\pi^2}{\omega c}k\int|<\vec{k}|\vec{\lambda}\cdot\vec{p}|i>|^2 \tag{9.15}$$

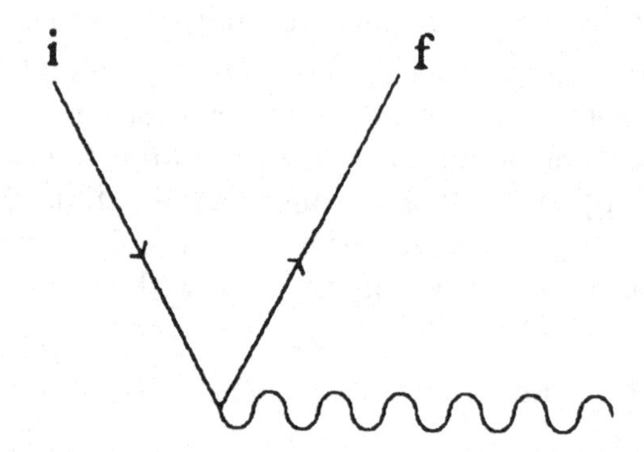

Fig. 9.1. The MBPT diagram for single ionization in photoannihilation.

$$= \frac{4\pi^2}{\omega c} k d^V ,$$

where $\vec{\lambda} \cdot \vec{p} = \frac{d}{dz}$ is the velocity form of the dipole operator and the ejected electron energy is $\epsilon_k = \omega - I$. This expression may be used for an atom with N electrons by replacing $\vec{p}$ by $\vec{p}_{total} = \sum_{i=1}^{N} \vec{p}_i$.

Single ionization. The many body perturbation theory (MBPT) diagram for single ionization is shown in figure 9.1. The length and acceleration forms of the dipole operator may be used in place of the velocity form in accord with (9.11).

Total cross sections may be obtained from the equation above by integrating over $d\Omega = sin\theta d\theta d\phi$ of the ejected electron. For electrons initially in the $1s$ atomic state this leads to the convenient expression for large ω [Heitler, 1954, p. 207],

$$\sigma_{PE}^{+} = 2^5 \frac{\alpha^3}{Z_T^2} \left(\frac{cI}{\omega} \right)^{7/2} \phi_T , \tag{9.16}$$

where $\phi_T = 8\pi r_0^2/3$ is the classical cross section for Thomson scattering of a free electron by a photon (Cf. the appendix of this chapter) and σ^+ is the cross section per atomic electron (i.e. half the total cross section for atoms with two $1s$ electrons). The rapid $\omega^{-7/2}$ decrease of the photoionization cross section with photon energy reflects the fact that photoannihilation by a free electron is forbidden.

Double ionization. Evaluation of the cross section for double ionization is similar to that for single ionization, except that the final state must be

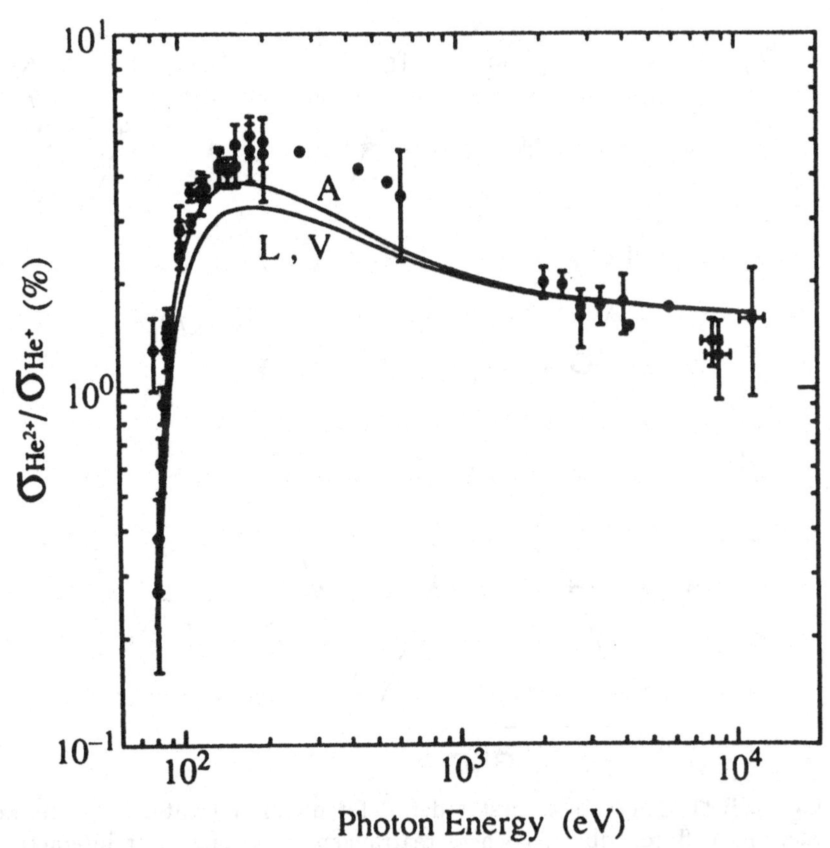

Fig. 9.2. Ratio of double to single ionization cross sections in helium versus photon energy. The calculations are the L, V and A forms of the dipole matrix element used in the MBPT calculations of Hino *et al.* [1993] and the data are due to Levin *et al.* [1991,1993], Carlson [1967], Schmidt *et al.* [1976], Wight and Van der Weil [1976], Holland *et al.* [1979], Wehlitz *et al.* [1991] and Samson *et al.* [1993].

represented by two electrons in the continuum with momenta $\vec{k}_1$ and $\vec{k}_2$. For double ionization,

$$\frac{d\sigma_{PE}^{++}}{d\Omega_1} = \frac{4\pi^2}{\omega c} k| < \vec{k}|\vec{\lambda} \cdot \vec{p}|i > |^2 k_2^2 dk_2 d\Omega_2 \,. \tag{9.17}$$

For the total double ionization cross section one must integrate over the energies and angles of both electrons subject to the constraint of conservation of energy, $\epsilon_{k1} + \epsilon_{k2} = \omega + I$.

Calculations of single and double ionization of helium based on MBPT [Hino *et al.*, 1993] have used to evaluate the ratio $R_{PE} = \sigma_{PE}^{+}/\sigma_{PE}^{++}$. This ratio of double to single ionization cross sections is shown in figure 9.2.

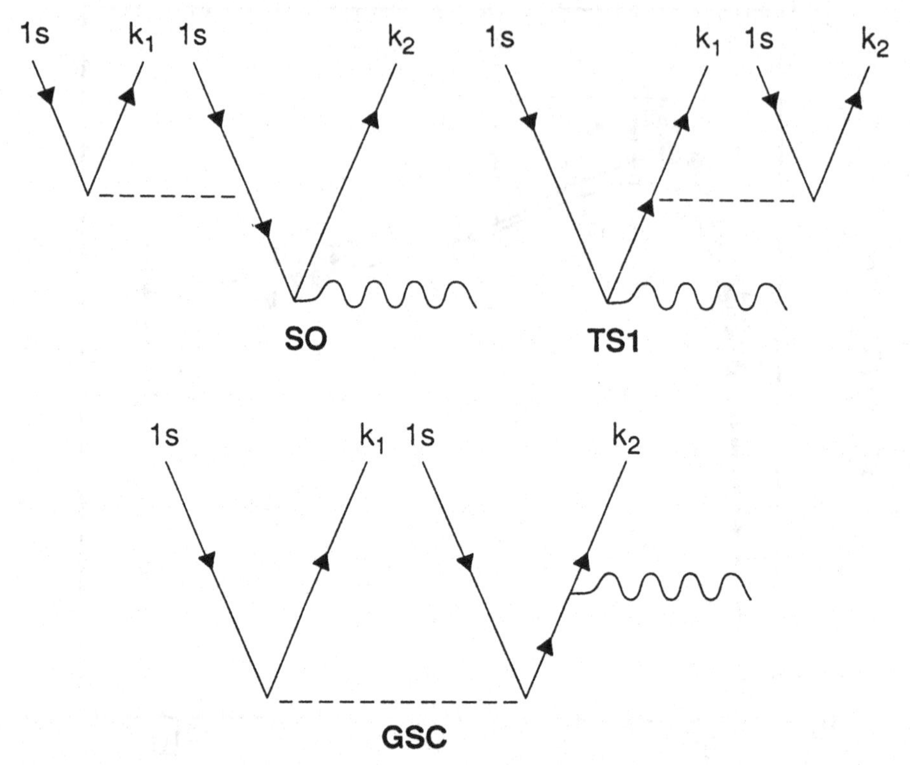

Fig. 9.3. MBPT diagrams to first order $\vec{A}$ for double ionization. SO is the first order shakeoff contribution where rearrangement occurs after interaction with the photon due to a change in screening of the electrons. TS1 is a two step process where one electron excited by the photon in the first step interacts with the second electron on the way out of the collision. GSC is the lowest order contribution to ground state correlation where the effects of the electron-electron interaction before interaction with the photon are included.

There is no difference between the V and L forms of the MBPT calculation. The difference with the A form becomes small at high energies.

1 Gauge dependence

Gauge dependence of individual MBPT amplitudes. Although the L, V and A forms of the dipole operator all give the same cross sections in an exact calculation, the individual MBPT amplitudes vary significantly with L, V and A forms (or gauges) of the dipole operator for double ionization. For single ionization there is only one lowest order MBPT diagram. Consequently, for single ionization the L, V and A gauges all give the same result if a local potential is used in accord with (9.11) above.

The gauge dependence of MBPT diagrams becomes evident for double ionization where there is now more than one MBPT diagram in lowest order. The three MBPT diagrams for double ionization to first order in the interaction with the photon and first order in the correlation potential v are shown in figure 9.3. The T matrix for these diagrams with a dipole operator, $d^{\alpha}(\alpha = L, V or A)$ is given as discussed in section 7.3 by,

$$T^{\alpha} = T^{\alpha}(SO) + T^{\alpha}(TS1) + T^{\alpha}(GSC) \tag{9.18}$$

where,

$$T(SO) = - < k_2 1s|v|1s1s > (\epsilon_{1s} - \epsilon_{k_1} + \omega + i\eta)^{-1} < k_1|d^{\alpha}|1s > \tag{9.19}$$

$$T(TS1) = \sum_k < k_1 k_2|v|k1s > (\epsilon_{1s} + \omega - \epsilon_{k_1} + i\eta)^{-1} < k|d^{\alpha}|1s > \tag{9.20}$$

and

$$T(GSC) = \sum_k < k_1|d^{\alpha}|k > (2\epsilon_{1s} - \epsilon_{k_2} - \epsilon_k + i\eta)^{-1} < kk_2|v|1s1s > \ . \tag{9.21}$$

Each of these amplitudes is different in the L, V and A gauges since the dipole matrix elements of d^{α} differ for $\alpha = L, VandA$, as is apparent from (9.7) - (9.8). This difference is specifically illustrated in (9.23) - (9.25) below.

Calculations of the ratio of double to single ionization cross sections for individual as well as summed MBPT diagrams are shown for the L gauge in figure 9.4, for the V gauge in figure 9.5 and for the A gauge in figure 9.6. In these figures SO, TS1 and GSC correspond (9.19), (9.20) and (9.21) which in turn correspond to the MBPT diagrams shown in figure 9.3. The Total curve uses the sum of SO, TS1 and GSC amplitudes as indicated by (9.18). At energies between 100 and 2000 eV the TS1 amplitude dominates. Above about 2000 eV the importance of individual MBPT terms depends on the form (or gauge) of the dipole operator. In the L form of the dipole operator, the largest contribution comes from TS1 at high energies and SO is relatively small. In the A form, TS1 is negligible and SO is the largest term. This behavior was first predicted by Dalgarno and Sadeghpour [1992].

Gauge invariance of the sum of lowest order MBPT amplitudes. Ishihara showed analytically [Hino *et al.*, 1993] that the individual MBPT terms are indeed different in the L, V and A forms and also that the sum

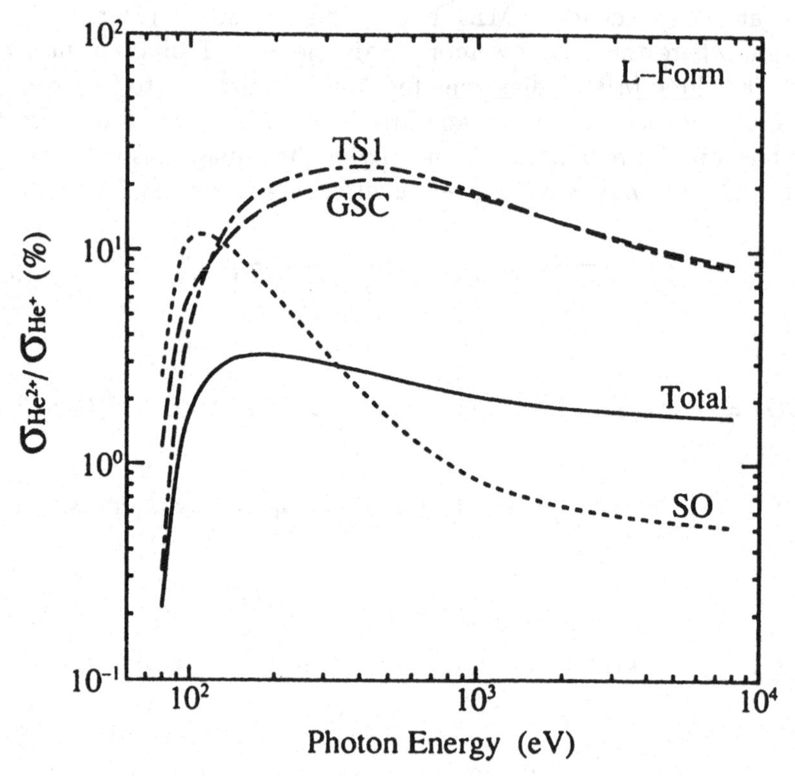

Fig. 9.4. Length form of the MBPT calculation of the ratio of double to single ionization in helium.

of all lowest order MBPT amplitudes is (nearly) invariant at high photon energies. The analysis is straightforward. Since the correlation potential v is chosen to be the local V^{N-1} potential of (7.44), one may use the relation,

$$d^V = [d^L, h] \,, \tag{9.22}$$

where h is the single electron Hamiltonian used to define the basis states. Using this relation in (9.19) -(9.21) one quickly has

$$T^V(TS1) + \omega T^L(TS1) = \sum_k < k_1 k_2 |v| k1s >< k|d^L|1s > \,. \tag{9.23}$$

Similarly,

$$T^V(SO) + \omega T^L(SO) = - < k_2 1s|v|1s1s >< k_1|d^L|1s > \,, \tag{9.24}$$

and

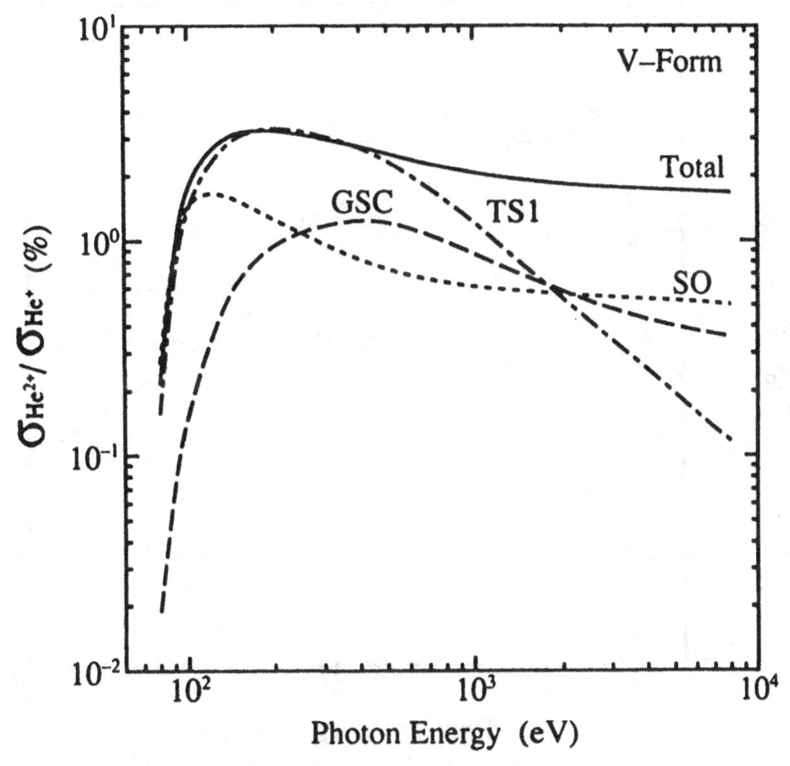

Fig. 9.5. Velocity form of the MBPT calculation of the ratio of double to single ionization in helium.

$$T^V(GSC) + \omega T^L(GSC) = -\sum_k < k_1|d^L|k >< kk_2|v|1s1s > \quad . \quad (9.25)$$

The differences are not small, particularly in TS1 and GSC as seen in figures 9.4 - 9.6. The V and A terms may be similarly related. Thus, individual MBPT terms are clearly dependent on the form (or gauge) of the dipole operator. This means that MBPT amplitudes have no consistent meaning as mechanisms for double ionization unless one defines the form (or gauge) of the dipole interaction. This is not unexpected since changing the interaction changes the matrix elements and the resulting observables in an approximate calculation. A common gauge in atomic physics is the Coulomb gauge [Jackson] where the interaction between two charged particles is Z_1Z_2/r. The Coulomb gauge corresponds to the velocity form of the dipole operator [Grant, 1974].

This gauge problem leads to a difficulty with the concept of correlation as well. Usually the SO amplitude is not considered as correlation (Cf.

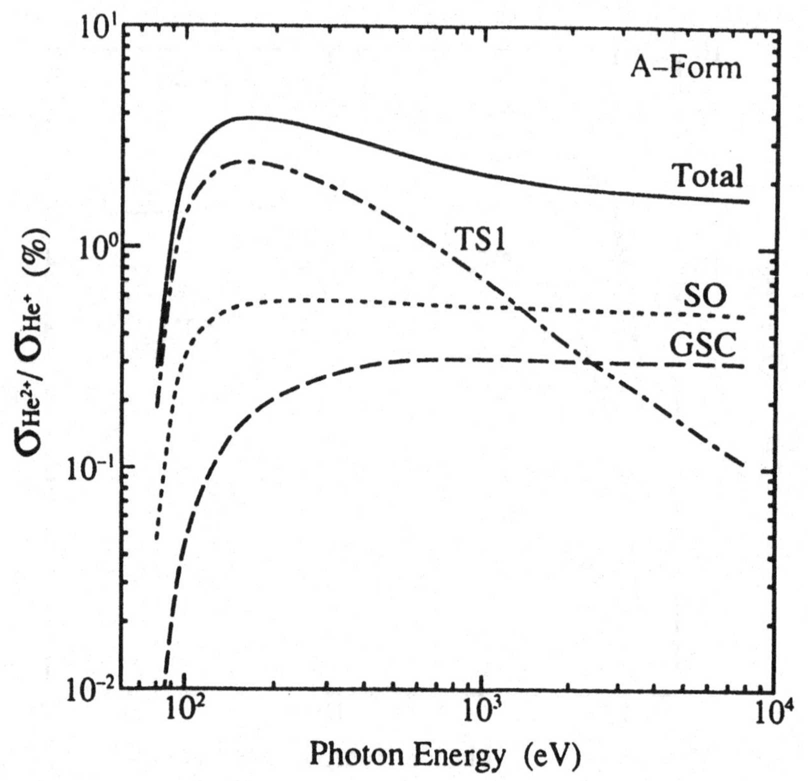

Fig. 9.6. Acceleration form of the MBPT calculation of the ratio of double to single ionization in helium.

section 7.1.1). Since the sum of SO, GSC and TS1 is gauge invariant (shown below), the correlated GSC and TS1 amplitudes vary with the L, V and A gauges. Thus if MBPT amplitudes are used to define correlation, the concept of correlation varies with gauge. This difficulty is avoided with other methods of defining shakeoff (e.g., $P_S = | < f_2'|i_2 > |^2$ in (6.20) [which corresponds to an infinite sum of MBPT shake terms]).

In order to show that the sum of all first order MBPT terms is the same, it is necessary to show that the terms on the right hand side of (9.23) - (9.25) sum to zero. To show this one may apply the closure relation,

$$\sum_{k \neq 1s} |k >< k| = 1 - |1s >< 1s| \,, \qquad (9.26)$$

to the right hand side of (9.25) to obtain,

$$-\sum_k < k_1|d^L|k >< kk_2|v|1s1s >= \quad - \quad < k_1|d^L v_{k_2 1s}|1s > \qquad (9.27)$$

$$+ \quad < k_2|d^L|1s >< 1sk_2|v|1s1s > ,$$

where,

$$v_{k_21s} = v_{k_21s}(\vec{r}) = \int < k_2(\vec{r}\,')|\frac{\vec{r}\,'}{|\vec{r}-\vec{r}\,'|}|1s(\vec{r}\,') > d\vec{r}\,' . \tag{9.28}$$

The second term of (9.28) cancels the right hand side of (9.24). The first term of (9.28) cancels the right hand side of (9.23) because,

$$\begin{aligned} &\sum_k < k_1k_2|v|k1s >< k|d^L|1s > \\ &=< k_1|v_{k_21s}d^L|1s > - < k_1k_2|v|k1s >< 1s|d^L|1s > \\ &=< k_1|d^Lv_{k_21s}|1s > , \end{aligned} \tag{9.29}$$

using $[v_{k_21s}, d^L] = 0$ and $< 1s|d^L|1s >= 0$. Thus, all terms on the right hand side of (9.23) - (9.25) sum to zero, so that,

$$\begin{aligned} T^V &= T^V(TS1) + T^V(SO + T^V(GSC) \\ &= -\omega\left(T^L(TS1) + T^L(SO + T^L(GSC)\right) = -\omega T^L . \end{aligned} \tag{9.30}$$

There is no difference in the L and the V form of the double ionization cross section if v is a local potential.

For the A form one may define,

$$\tilde{d}^A = [d^V, h] = d^A + \frac{d}{dz}v_{1s1s} , \tag{9.31}$$

where,

$$v_{1s1s} = -\frac{2}{r} + \int < 1s(\vec{r}\,')|\frac{1}{|\vec{r}-\vec{r}\,'|}|1s(\vec{r}\,') > d\vec{r}\,' , \tag{9.32}$$

which is the V^{N-1} MBPT potential of section 7.2. Using (9.31) one may obtain expressions similar to (9.23) - (9.25) with d^L replaced by d^V and d^V by $\tilde{d}^A$. However, since $d^V = \frac{d}{dz}$ does not commute with v_{k_21s} one has,

$$\begin{aligned} T^V &= T^V(TS1) + T^V(SO) + T^V(GSC) \\ &= -\frac{1}{\omega}\left(\tilde{T}^A(TS1) + \tilde{T}^A(SO) + \tilde{T}^A(GSC)\right. \\ &\quad \left. - < k_1|[v_{k_21s}, d^V]|1s >\right) , \end{aligned} \tag{9.33}$$

where,

$$\tilde{T}^A = T^A + \Delta T \,, \tag{9.34}$$

with ΔT obtained using $\frac{d}{dz} v_{1s1s}$ in place of d^A in the defining equation of T^A. In this case there is a difference in the V and A form, due to the noncommutivity of $v_{k_2 1s}$ and d^V and also the difference between d^A and $\tilde{T}^A$. In figure 9.3 it appears that the difference between the V and A forms becomes small at high photon energy.

Thus, observable double ionization cross sections are invariant under transformations between L, V and A forms of the dipole limit of the $\mathcal{O} = \vec{p} \cdot \vec{A}$ in photoannihilation. A similar invariance between L, V and A forms for the $\mathcal{O} = A^2$ term for double ionization via Compton scattering has been explicitly demonstrated by Hino [Bergstrom *et al.*, 1995]. In both of these cases this invariance between L, V and A forms of the operator corresponds to a generalized continuity equation as noted in the appendix to this chapter .

Extension of gauge invariance to higher orders in MBPT. This gauge invariance of summed MBPT terms may be extended to higher orders in MBPT [Wang *et al.*, 1994]. Consider the expansion of the exact quantum amplitude, a^G, for some physical observable (such as a transition probability) where the gauge is denoted by G, namely,

$$a^G(Z) = a_0^G + a_1^G Z + a_2^G Z^2 + \ldots + a_j^G Z^j + \ldots \,, \tag{9.35}$$

where Z is some expansion parameter. The nth order term is $a_n^G Z^n$.

Gauge invariance demands for amplitudes evaluated in different gauges G and G' that,

$$|a^G|^2 = |a^{G'}|^2 \,. \tag{9.36}$$

Thus,

$$a^G = e^{i\eta} a^{G'} \,. \tag{9.37}$$

For the gauge transformations corresponding to electrodynamic gauge transformations, including L, V and A gauges, η affects only the vector potential $\vec{A}$ and the scalar potential ϕ. Thus, the phase η is assumed to be independent of the expansion parameter, Z. Then,

$$\begin{aligned} a^G(Z) &= a_0^G + a_1^G Z + a_2^G Z^2 + \ldots + a_j^G Z^j + \ldots \\ &= e^{i\eta} a^{G'} = e^{i\eta} \left(a_0^{G'} + a_1^{G'} Z + a_2^{G'} Z^2 + \ldots + a_j^{G'} Z^j + \ldots \right) . \end{aligned} \tag{9.38}$$

Since Z may vary arbitrarily, the coefficients are equal, i.e.,

$$a_j^G = e^{i\eta} a_j^{G'} \ . \tag{9.39}$$

Thus, gauge invariance is expected to hold in each order in the MBPT expansion.

For expansions in both an interaction V with an incident projectile and the correlation potential v gauge invariance holds for the sum of all MBPT diagrams nth order in V and mth order in v for each n and m. This result has also been noted in other cases of interactions of photons with atoms [Drake, 1995]. It is not clear whether this result holds for a multi-photon expansion since η may depend on $\vec{A}$.

9.4 Compton and Raman scattering

Compton and Raman scattering‡ occur when an incident photon scatters inelastically. Compton scattering is usually first order in the operator A^2 and Raman scattering is usually second order in $\vec{p} \cdot \vec{A}$.

1 Compton scattering

The classical cross section for scattering of a photon by a free electron is Thomson cross section [Heitler, 1954, p.34],

$$d\phi_T/d\Omega = \frac{\omega_f}{\omega_i} r_0^2 |\vec{\lambda}_i \cdot \vec{\lambda}_f|^2 \ , \tag{9.40}$$

where $\vec{\lambda}_i$ and $\vec{\lambda}_f$ are the directions of the polarization of the initial and final photon and $r_0 = \alpha^2 = 2.818 \times 10^{-13}$ cm is the classical radius of the electron. Averaging over the directions of polarization of the photons gives a factor of $\frac{1}{2}(1 + cos^2\theta)$ where θ is the angle between the incident and the outgoing photons.

Quantum mechanically, since Compton scattering involves two photons, the lowest order matrix element comes from the A^2 term in (9.3)§. The Compton scattering cross section for inelastic scattering from an arbitrary initial state, $|i>$, to an arbitrary final state, $|f>$, may be expressed [Baym, 1974, Eq(13-132)] in first order as,

$$d\sigma_C/d\Omega = \frac{\omega_f}{\omega_i} r_0^2 |\vec{\lambda}_i \cdot \vec{\lambda}_f < f| e^{i(\vec{k}-\vec{k}')\cdot\vec{r}} |i> |^2 \ , \tag{9.41}$$

‡ Various names and forms for Compton scattering are listed in the appendix to this chapter.

§ For resonance Compton or resonance Raman scattering which occurs at photon energies close to the resonance energy, the $(\vec{p} \cdot \vec{A})^2$ term is dominant. This is ignored here.

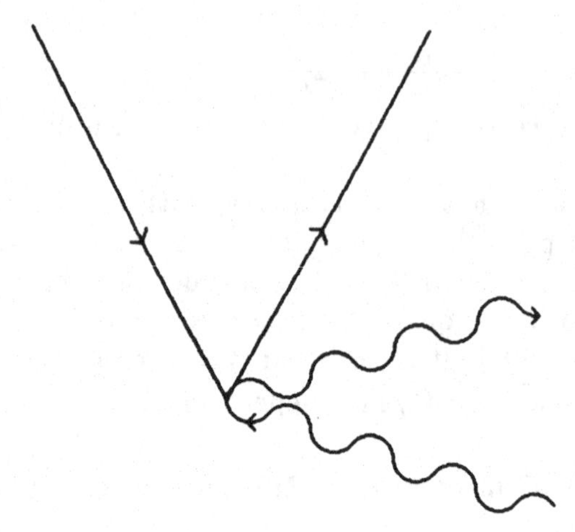

Fig. 9.7. MBPT diagram for single ionization via Compton scattering.

where $\vec{k}$ and $\vec{k}'$ are the initial and final momenta of the photon.

For single ionization with $|f>=|\vec{k}_e>=|\epsilon_f,\Omega_e>$ where ϵ_f and Ω_e are the final energy and solid angle of the continuum electron. Then, after averaging over the polarization of the photons the Compton scattering cross section doubly differential in the energy transfer (ΔE) - momentum transfer (q) plane becomes [Heitler, 1954; Burgdörfer *et al.*, 1994],

$$\left(\frac{d^2\sigma}{d\Delta E dq^2}\right)_C = \frac{\pi r_0^2}{2k^2}\left(1+\left(1-\frac{q^2}{2k^2}\right)^2\right) \cdot F_I(\Delta E, q^2)\,, \qquad (9.42)$$

where $\Delta E = \epsilon_f - \epsilon_i = \epsilon_k - I$ is the change in energy of the electron and $q^2 = (\vec{k}-\vec{k}')^2 = k^2 + k'^2 - 2kk'\cos\theta \simeq 2k^2(1-\cos\theta$ so that $\left(1+\left(1-\frac{q^2}{2k^2}\right)^2\right) = 1 + cos^2\theta)$ and with $\epsilon_f = \epsilon_k = \Delta E + \epsilon_i$,

$$F_I(\Delta E, q^2) = \int d\Omega_e \sqrt{2(\Delta E + \epsilon_i)}|\langle f|\sum_{j=1}^{N} e^{i\vec{q}\cdot\vec{r}_j}|i\rangle|^2\,. \qquad (9.43)$$

$F_I(\Delta E, q^2)$ is the inelastic transition form factor (or atomic form factor proportional to the generalized oscillator strength [Aberg and Tulkki, 1985]) for the N-electron atom integrated over all emission angles of the emitted electron and weighted with the density of continuum final states. The atomic initial state $|i>$ with binding energy ϵ_i is assumed to be isotropic but arbitrary otherwise. Consequently, F_I depends only on the magnitude of the momentum and energy transfer. The dependence of F_I

on the energy transfer ΔE is implicit through the energy of the accessed final state $\Delta E = \epsilon_f - \epsilon_i$.

Eqs(9.41) and (9.42) hold for arbitrary final states, including transitions of multiple electrons. In the case of multiple ionization, F_I is understood to include an integral over emission angles of all electrons and the proper density of states. The importance of Compton scattering in double ionization by keV photons was recognized by Samson, Greene and Bartlett [1993] who realized that Compton scattering is independent of the photon energy, ω, at high photon energies, while photoionization varies as $\omega^{-7/2}$. The first calculation was done by Andersson and Burgdörfer [1993].

MBPT calculations of Compton scattering have been carried out by Hino *et al.* [1994]. The lowest order MBPT diagrams for Compton scattering are the same as for photoionization except that the single photon vertex is changed to a double photon vertex as in figure 9.7 for single ionization.

2 Compton scattering and photoionization

Ionization by Compton scattering and photoionization are different processes. In photoionization the photon is annihilated, while in Compton scattering the photon is inelastically scattered and not annihilated. Using a generalized shake probability (section 6.3.2), it may be shown [Andersson and Burgdörfer, 1994; Suric *et al.*, 1994] that for Compton scattering,

$$R_C = 1 - \sum_B \int d^3r_1 \left| \int \phi_B^*(r_2)\Phi_i(r_1, r_2)d^3r_2 \right|^2 . \qquad (9.44)$$

where the summation runs over bound states. This may be compared to the expression from Dalgarno and Sadeghpour [1992] for photoionization in the acceleration gauge that,

$$R_{PE} = 1 - \frac{\sum_B |\int \phi_B^*(r_2)\Phi_i(r_2, r_1 = 0)d^3r_2|^2}{\int |\Phi_i(r_2, r_1 = 0)|^2 d^3r_2} . \qquad (9.45)$$

In the simple shake limit $\Phi_i = \phi_1'\phi_2'$ (without exchange), then both ratios reduce to the same simple shake ratio, which for ionization (not including bound states) gives 0.71% if $s = 5/16$, corresponding to a variational hydrogenic wavefunction with s chosen to minimize the binding energy. For more complete wavefunctions, however, Eqs(9.44) and (9.45) differ because photoionization depends on the value of Φ at $r_1 = 0$ while for Compton scattering there are contributions from all values of r_1. This point may also be understood from the generalized shake probability $P(\vec{q} - \vec{k})$ corresponding to (6.26) which depends on the momentum transferred to the residual target ion $(\vec{q}-\vec{k})$, where $\vec{q}$ is the momentum transferred by the projectile and $\vec{k}$ is the momentum of the outgoing electron(s). In Compton scattering $(\vec{q}-\vec{k}) \simeq 0 << 1$ corresponding the the binary ridge. This probes small momentum components of the initial state wavefunction at large distances from the target nucleus. In photoannihilation at high but non-relativistic photon energies $\omega = qc > 1$ with $q << k$ one has $(\vec{q}-\vec{k}) \simeq \vec{k} >> 1$ which probes distances close to the nucleus. Charged particle scattering covers both regions.

Dalgarno [1994] has found a simple functional form that fits values of R_{PE} for various two electron atoms and ions, namely,

$$R_{PE} = \frac{0.09}{Z^2}(1 - \frac{1}{3Z}) . \tag{9.46}$$

3 Raman scattering

Raman scattering (sometimes called second order resonant Compton scattering) is a two photon process (one in and one out) which is evaluated using the $\vec{p}\cdot\vec{A}$ term in second order perturbation theory. Resonances can occur when the energy difference between the incoming and outgoing photons is equal to a transition energy for the target.

Consider inelastic (γ_i, γ_f) scattering of a photon by an atom. The interaction potential is $V = \vec{p}\cdot\vec{A}/c + A^2/c^2$ in the Coulomb gauge. For weak fields such a two photon process is associated with terms quadratic in $\vec{A}$. The matrix element for such a transition [Aberg and Tulkki, 1985; Heitler, 1954] is of the form,

$$< f|\mathcal{O}'|i > = \frac{1}{c^2} < f|A^2 + \sum_I \vec{p}\cdot\vec{A}\frac{1}{E - E_I + i\eta}\vec{p}\cdot\vec{A}|i > . \tag{9.47}$$

This leads to a cross section of the form,

$$\begin{aligned} \frac{d^2\sigma}{d\Omega d\omega_f} &= \frac{\omega_f}{\omega_i} r_0^2 | < f|e^{i(\vec{k}_i-\vec{k}_f)\cdot\vec{r}}\vec{\lambda}_f\cdot\vec{\lambda}_i|f > \\ &+ \sum_I < f|e^{-i\vec{k}_f\cdot\vec{r}}\vec{\lambda}_f\cdot\vec{p}|I > \frac{1}{E - \mathcal{E}_I} < I|e^{i\vec{k}_f\cdot\vec{r}}\vec{\lambda}_i\cdot\vec{p}|i > |^2 \\ &\times \delta(E_i - E_f) \end{aligned} \tag{9.48}$$

where $\mathcal{E}_I$ are complex energies with widths $i\frac{\Gamma}{2}$. Here $\vec{\lambda}_{i,f}$ represents the polarization direction of the initial and final photons of momentum $\vec{k}_i$ and $\vec{k}_f$ respectively. The sum $\sum_I$ over intermediate states implicitly contains both absorption and emission terms which are explicit in Heitler and

Aberg and Tulkki. For final continuum states the cross section is per momentum $\vec{k}$ of each of the outgoing electrons. Eq(9.48) is often called the Kramers-Heisenberg equation [Heitler, 1954].

The first term in the amplitude of (9.48) above corresponds to Compton scattering as described in section 9.4.1 above. The second term contains resonances when the energy denominators are large. This corresponds to resonant Raman scattering. Both terms lead from the same initial to the same final state and add coherently. In some cases near resonances the second term may dominate over the first and may be treated alone. In other cases at high (but not relativistic) photon energies the leading A^2 term dominates Compton scattering.

The non-trivial transformation from the simple $\frac{1}{E-E_I+i\eta}$ propagator of (9.47) to the more useful $\frac{1}{E-\mathcal{E}_I}$ propagator of (9.48) requires interaction between discrete and continuum states. The complex energy $\mathcal{E}_I$ contains an energy shift Δ and a width $i\Gamma$ as described by Aberg and Tulkki. In the simplest case $\Delta = 0$ and $i\Gamma$ reduces to the width of a simple isolated resonance at $E = E_{res}$. This simple Green function $\frac{1}{E-E_{res}+i\frac{1}{2}\Gamma}$ representing a Lorenzian energy profile corresponds to propagation in time given by $e^{i(E-E_{res})t}e^{-\frac{1}{2}\Gamma}$ which decays with a half life of $1/\Gamma$.

Various Raman terms are discussed by Aberg [1996] including resonant Raman transitions. Raman emission by X-rays, resonant inelastic X-ray scattering, resonant Raman scattering, radiative and radiationless resonant Raman scattering and radiative and non-radiative resonant inelastic scattering. Application of Raman scattering in solids with conduction bands is discussed by Ma [1996] and by Cara [1996]. In some applications the coherent sum over intermediate states in (9.48) is approximated by a simpler incoherent sum over resonant states. These are often called 'one step' and 'two step' processes respectively. Non-dipole contributions are discussed by Langhoff [1996]. For off resonant scattering (9.48) may be simplified by using a closure approximation using an average intermediate energy. An extended analysis in the case of strong fields created by highly charged ions including distortions of the strong field has been attempted by Godunov *et al.* [1996].

Since the lowest order terms for Raman scattering are second order, second order methods including time ordering (Cf. section 7.1.2) are generally applicable [Heller *et al.*, 1982; Lee and Heller, 1979]. Further theoretical detail is given by Aberg and Tulkki [1985].

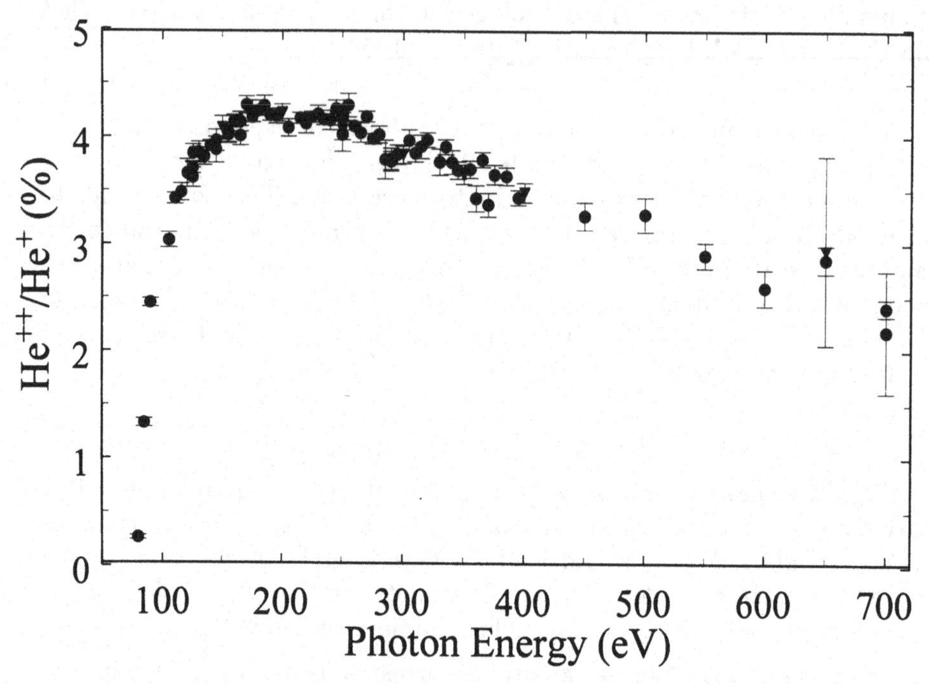

Fig. 9.8. Observed ratio of double to single ionization in helium by photon impact [Levin *et al.*, 1996].

9.5 Observations

There have been many observations of interactions of photons with atoms involving transition of a single electron [McDaniel and Mansky, 1994; Caldwell and Krause, 1996; Crasemann, 1996; Bell and Kingston, 1994]. Single photoionization has been reviewed by a number of authors for both atoms [Starace, 1996; Amusia, 1996] and molecules [Hitchcock, 1995; Hatano, 1995]. Atomic and molecular physics with synchrotron radiation have been reviewed by Crasemann [1992, 1996]. In quantum optics interactions of lasers with atoms [Lucatorto *et al.*, 1996] and molecules [White, 1996] have been observed as have multiphoton and strong field processes [Kulander, 1996; Karule, 1990]. There have been fewer observations of multiple electron transitions and some of these are discussed below.

Photoannihilation. Various experiments have been done for both single and double ionization of atoms via annihilation of an incident photon [Lablanquie, 1992]. From the late seventies until the present the threshold region has been extensively investigated [Schmidt *et al.*, 1976; Wight and

Van der Weil, 1976; Holland *et al.*, 1979; Wuilleumier, 1982; Wehlitz *et al.*, 1991; Hall *et al.*, 1993] and agreement [Kossmann *et al.*, 1988; Lablanquie *et al.*, 1990] was found with the theory of Wannier [1953]. More recently in the high energy limit, measurements of the ratio of double to single photoionization at several photon energies, from 2 to 12 keV have been reported to be 1.6 (± 0.2)% [Levin *et al.*, 1993, 1995; Spielberger *et al.*, 1995; McGuire *et al.*, 1995] consistent with a calculated asymptotic value of 1.66% [Byron and Joachain, 1967; Aberg, 1976; Carter and Kelly, 1981; Ishihara *et al.*, 1991; Dalgarno and Sadeghpour, 1992] Some of these data are shown in figure 9.8. At high photon energies ω both the single and the double ionization cross sections vary as $\omega^{-7/2}$. The rapid decrease with increasing ω occurs because the process is quasi-forbidden. Annihilation of a photon with a free electron is kinematically forbidden, i.e., it violates conservation of energy and momentum.

Differential cross sections are in general a more sensitive test of physical mechanisms than total cross sections where differences and structures tend to average out. Effects of electron correlation in differential studies of double ionization of atoms by a single photon have been considered by Ceraulo *et al.* [1994] who compare theory with experiment for sixfold differential cross sections in helium. Angular distributions of double ionization of helium have been reviewed by Huetz [1996].

Compton scattering. At high photon energies ω the Compton scattering cross section is independent of ω and is related to the Thomson scattering cross section for inelastic scattering of a photon from a free electron (Cf. section 9.6.2, or Heitler, section V.22). Thus, at sufficiently high energies the Compton cross section is expected to dominate over the photoannihilation cross section [Samson *et al.*, 1993; Andersson and Burgdörfer, 1993]. Data for the transition from photoannihilation to Compton scattering has been given by Azuma *et al.* [1995]. The classical threshold for Compton scattering is easily computed from conservation of energy and momentum. Observations of the ratio of double to single ionization via Compton scattering at energies above several keV have been reported by Samson *et al.* [1994] and Levin *et al.* [1995]. Observations of recoil ions for both photoannihilation and Compton scattering have been done by Spielberger *et al.* [1995].

Raman scattering. A variety of observations of Raman scattering have been reported for vibrational and rotational transitions in molecules [Heller *et al.*, 1982]. More recently Raman scattering in solids has been observed [Ederer and McGuire, 1996]. However, few if any observations of few or many electron transitions have been reported.

Biexcitons. An example of a two photon, two electron transition is the biexciton observed in solid state matter [Gale and Mysyrowicz, 1974]. It has been observed in optical transitions in semiconductors at low temperatures . The biexciton is created by two photon absorption from the ground state of a crystal with a full upper valence band and an empty conduction band. It is the excitonic equivalent of a hydrogen (or positronium) molecule. It consists of two electrons in the conduction band tied by an electron exchange interaction with the two holes left behind in the valence band.

The two photon transition which creates the biexciton has a giant cross section for the following reason. Instead of having one electron interacting simultaneously with two photons (as in the more usual two photon transition), here two electrons interact simultaneously each with its own photon. The only constraint is that the mean separation between the two electrons is of the order of the 'molecular' dimension, typically 10 - 100 atomic units [Hanamura, 1973]. This large spatial delocalization enhances the transition cross section by several orders of magnitude. Biexcitons give rize to strong non-linear optical effects. Also, since the biexciton is the lowest quantum electronic excitation in most semiconductors and since biexcitons consist of an even number of fermions (two electrons and two holes), they can form a Bose Einstein condensate at sufficiently high particle densities and sufficiently low temperatures ¶.

N-photon processes. Multi-photon processes [Kulander, 1996; Faisal *et al.*, 1995; Schafer *et al.*, 1995; Potvliege and Shakeshaft, 1992; Gavrila, 1992; Eberly *et al.*, 1991; Faisal, 1987; Lambropoulos, 1976] occur in strong electric fields which may be produced by high intensity lasers. These processes include above threshold ionization (ATI), multi-ionization, Stark shifted induced resonances, higher order harmonic generation, and barrier suppression ionization [Potvliege and Shakeshaft, 1992]. In ATI ionization by a single photon is classically forbidden for photons whose energy ω is less than the ionization threshold I of the atom. In this case ionization may occur via the absorption of N photons where $N\omega > I$. Single and multiple ionization of atoms under these conditions has been observed [Corkum, 1993; Helm, 1992; Rottke *et al.*, 1990] and analyzed using methods similar to those described in sections 6.2.3 and 9.3. Various analyses have been developed including the Floquet formalism [Potvliege and Shakeshaft, 1990; Chu and Cooper, 1985] for a time-periodic Hamil-

¶ Experimental evidence of Bose Einstein condensation of biexcitons was reported by Chase *et al.* in 1983. Bose Einstein condensation was later reported for single excitons in Cu_2O, a special case where biexcitons are not stable [Mysyrowicz, 1980, Mysyrowicz *et al.*, 1990]. Anomalous transport properties pointing to electronic superfluidity in Cu_2O have also been observed [Fortin *et al.*, 1993, Mysyrowicz *et al.*, 1996].

tonian, numerical integration of the time dependent Schrödinger equation [Esry and Greene, 1994; Schafer *et al.*, 1993; Kulander, 1987a,b] and use of Keldysh-Volkov states [Faisal, 1973; Reiss, 1980] (Cf. section 2.6.1). At optical wavelengths most multiple ionization occurs via the sequential ionization of successive charge states [Gavrila, 1992]. Non-sequential double [Fittinghoff *et al.*, 1992; Walker *et al.*, 1994; Dietrich *et al.*, 1994; Walker, 1995] and triple [Augst *et al.*, 1995] ionization has been observed with sub-picosecond laser pulses. The theoretical explanation for this is given in terms of the above TS1 mechanism (i.e. electron-electron interaction in the final state) by Becker and Faisal [1996]. The uncorrelated limit with strong laser fields was not yet been done as of 1995.

Quantum effects in squeezed light (exhibiting a partial violation of the Uncertainty principle) and Fock states are effects not included in the classical description of light. The effects of these non-classical phenomena on radiative transitions may be observed in two photon transitions [Georgiades *et al.*, 1995].

Laser assisted electron-atom excitation experiments are reviewed by Newell [1992] for transitions of a single electron. Examples of laser assisted ion-atom experiments are given by De Paola [1991]. Laser-induced recombination in merged electron and proton beams is reviewed by Schramm *et al.* [1992].

9.6 Appendix

1 *Gauge transformations*

Here two types of gauge transformations are considered: electrodynamic and quantum mechanical. Both gauge transformations leave physical observables unchanged.‖

Electrodynamic gauge transformations. An electrodynamic gauge transformation corresponds to a simultaneous change of the vector potential $\vec{A}$ and the scalar potential ϕ such that,

$$\vec{A} \to \vec{A} + \nabla\eta\,, \qquad \phi \to -\frac{1}{c}\frac{\partial\eta}{\partial t}\,. \tag{9.49}$$

‖ Gauge invariance appears to be related to the continuity equation, corresponding to a conserved quantity. For example in the case of transformations between L, V and A gauges described above, the gauge transformation may be written in the form $[\mathcal{O}, h] = -\vec{\nabla}\cdot(\vec{\nabla}\mathcal{O})$ which is of the form $\frac{\partial\mathcal{O}}{\partial t} + \vec{\nabla}\cdot\vec{j} = \partial_\mu j^\mu = 0$ where in this case $\mathcal{O}$ is the dipole operator and h is a single particle Hamiltonian.

This electrodynamic gauge transformation [Jackson, 1975] does not change Maxwell's four equations or the physical quantities such as $\vec{E}$ and $\vec{B}$ derived from Maxwell's equations. Electrodynamic gauge invariance arises from conservation of electrical charge. Two commonly used electrodynamic gauges are the Coulomb and the Lorentz gauges. In the Coulomb gauge defined by $\nabla \cdot \vec{A} = 0$ the uncoupled scalar and vector potentials are defined by the Poisson equation with charges and transverse currents as source terms, which is physically appealing. For example, the Coulomb potential is defined in the Coulomb gauge by $\phi(\vec{r},t) = \int \rho(\vec{r}',t)/|\vec{r}-\vec{r}'|d\vec{r}'$ where ρ is the electric charge density. The Lorentz gauge defined by $\nabla^2\eta - \frac{1}{c^2}\frac{\partial^2\eta}{\partial t^2} = 0$ gives equations that are relativistically covariant, but with $\vec{A}$ and ϕ coupled by the Lorentz condition $\nabla \cdot \vec{A} + \frac{1}{c}\frac{\partial\phi}{\partial t} = 0$.

For all gauges, $\nabla^2\vec{A} - \frac{1}{c^2}\frac{\partial^2\vec{A}}{\partial t^2} = -\frac{4\pi}{c}\vec{J}$ and $\nabla^2\phi - \frac{1}{c^2}\frac{\partial^2\phi}{\partial t^2} = -4\pi\rho$ which may be written relativistically as, $\Box A^\mu = -4\pi j^\mu$. The Lorentz condition, which may be written, $\partial_\mu A^\mu = 0$, is a form of the continuity equation which forces the gauge transformations to satisfy $\Box\eta = 0$ which leaves invariant the electric charge and current, j^μ, which is the physical origin of the electromagnetic fields. Thus, electrodynamic gauge invariance corresponds to conservation of electric charge.

Quantum gauge transformations. A quantum gauge transformation changes the wavefunction, ψ, by an overall phase, η, namely, $\psi \to \psi' = e^{-i\eta}\psi$. Physical observables are unchanged by an overall quantum gauge transformation**. Quantum gauge invariance arises from conservation of matter non-relativistically. However, the Hamiltonian is changed by the addition of a potential-like term††. If,

$$H\psi = i\frac{\partial\psi}{\partial t}, \tag{9.50}$$

then under the gauge transformation $\psi \to \psi' = e^{-i\eta}\psi$ with $H'\psi' = i\frac{\partial\psi'}{\partial t}$ it is obvious that H' becomes,

$$H' = e^{-i\eta}\ H\ e^{i\eta} + \frac{\partial\eta}{\partial t}, \tag{9.51}$$

** Berry's phase and the Aharonov-Bohm effect are examples of quantum phases which can give observable effects [Milonni and Singh, 1990]. At low temperatures the phase due to $\vec{A}$ may be quantized so that superconductivity corresponds to correlated states of pairs of electrons [Gasiorowicz, 1996, p.230].

†† The transformation $V \to V - \bar{V}$ is a quantum gauge transformation that changes the zero point energy of the Hamiltonian. The zero point energy should not affect physical observables which depend only on the change in energy in principle. In this case $\eta = -i\int^t \bar{V}(t')dt'$.

as pointed out by Wang [1994].

Relation between em and qm gauge transformations. In both the electrodynamic (em) and the quantum mechanical (qm) gauge transformations a term is added to the scalar potential while physical observables remain unchanged. The Schrödinger Equation for a particle of mass m and charge $+q$ in an electromagnetic field described by $\vec{A}$ and ϕ with an additional potential energy V is given by,

$$i\frac{\partial\psi}{\partial t} = \left(\frac{1}{2m}(\vec{p} - \frac{q}{c}\vec{A})^2 + q\phi + V\right)\psi \,. \tag{9.52}$$

The form of this equation remains the same under the electromagnetic gauge transformation of (9.49) if the quantum gauge transformation $\psi \rightarrow \psi' = e^{-iq\eta}\psi$ is simultaneously applied since the 'extra' $\frac{\partial\eta}{\partial t}$ terms then cancel. Thus, while electrodynamic and quantum gauge transformations are somewhat different, they are also related [Schiff, 1968].

Gauge transformations with non-local potentials V are discussed by Starace [1982]; relativistic gauge transformations are discussed by Grant [1974]. Some specific gauge transformations are discussed in section 9.2.2.

2 Two electrons in a magnetic field

The problem of two electrons interacting via their $1/r_{12}$ Coulomb potential in a constant magnetic field may be solved by a simple transformation of coordinates [Taut, 1994]. Following Burke, the Hamiltonian for two electrons in a magnetic field is given by,

$$H = \sum_{i=1,2}\left\{\frac{1}{2}\left(\vec{p}_i + \frac{1}{c}\vec{A}(\vec{r}_i)\right)^2\right\} + \frac{1}{|\vec{r}_2 - \vec{r}_1|} \,. \tag{9.53}$$

Using $\vec{A}(\vec{r}_i) = \frac{1}{2}(\vec{B}\times\vec{r}_i)$ with $\vec{B}$ constant and perpendicular to the $(\vec{r}_1, \vec{r}_2)$ plane, one obtains,

$$H = \sum_{i=1,2}\left\{-\frac{1}{2}\nabla^2_{r_i} + \frac{1}{2}kr_i^2\right\} + \frac{1}{|\vec{r}_2 - \vec{r}_1|} \tag{9.54}$$

where $k = B^2/4c^2$. This is of the form of a Hamiltonian for two interacting electrons connected to an infinite mass by springs with the same spring constant k.

Analytic solutions may be found by using the simple transformation

$\vec{R} = (\vec{r}_1 + \vec{r}_2)/2$ and $\vec{u} = \vec{r}_2 - \vec{r}_1$, whence,

$$\begin{aligned} H &= -\frac{1}{4}\nabla_R^2 + kR^2 - \nabla_u^2 + \frac{1}{4}ku^2 + \frac{1}{u} \\ &= H_R + H_u \,. \end{aligned} \tag{9.55}$$

The wavefunction separates into a product of a three-dimensional oscillator in $\vec{R}$ (of mass 2 and spring constant $2k$) and a simple equation in $\vec{u}$:

$$\Psi(\vec{r}_1, \vec{r}_2) = \left(\frac{2\omega}{\pi}\right)^{3/4} \exp\left(-\omega R^2\right) \phi_0(\vec{u}) \tag{9.56}$$

where $\omega = \sqrt{k}$ is the Larmor frequency. Now $\phi_0(\vec{u})$ satisfies,

$$(-\nabla_u^2 + \frac{1}{4}ku^2 + 1/u)\phi_0(\vec{u}) = \epsilon\phi_0(\vec{u}), \tag{9.57}$$

where the total energy $E = 3\omega/2 + \epsilon$. Equation (9.57) can easily be solved numerically [Laufer and Krieger, 1986; Burke, 1996], but can also be solved analytically for certain discrete values of k.

To obtain an analytic solution, expand $\phi_0(\vec{u})$ as a power series in u times the Gaussian decay due to the oscillator potential:

$$\phi_0(\vec{u}) = Y_{lm}(\Omega_u) \sum_{j=1}^{N} c_j u^j \exp(-u^2/2u_o^2), \tag{9.58}$$

where $u_o = \sqrt{2/\omega}$ is the length scale of the oscillator in the absence of the Coulomb repulsion, and Ω_u denotes the direction of $\vec{u}$. Insertion of this form into Eq. (9.57) yields a double recursive series for the coefficients c_j which terminates at finite j only for certain values of k [Taut, 1994]. For $l = 0$, the first few values are $k = \infty$ (the independent electron limit), $k = 1/4$, and $k = 1/100$, with energies $\epsilon = 3\omega/2, 5/4, 7/20$, for $N = 0, 1, 2$, respectively. For these values of k, the wavefunction may be written analytically e.g., for $k = 1/4$ [Kais *et al.*, 1989],

$$\phi_0(u) = \frac{(1 + u/2)\exp(-u^2/8)}{\sqrt{4\pi(5\sqrt{\pi} + 8)}}. \tag{9.59}$$

This 'Hooke's law atom' has been used to study the Coulomb cusp condition on the ground state two electron wavefunction at $\vec{r}_1 = \vec{r}_2$ [Burke *et al.*, 1994], the 'pair' Wigner crystal in solid state physics [Taut, 1994a], as a test of density functional approximations to the ground-state energy of electronic systems [Filippi *et al.*, 1994] and as a pedagogic tool for illustrating concepts of conditional probability densities [Burke *et al.*, 1995].

(This subsection was contributed by K. Burke.)

3 Two electrons in an electromagnetic field

The Hamiltonian for two particles of equal mass m and charge Z ($Z = -1$ for electrons), in an electromagnetic field described by the vector potential $\vec{A}$ is [Faisal, 1994],

$$H = \sum_{i=1,2} \left\{ \frac{1}{2m} \left(\vec{p_i} - \frac{Z}{c}\vec{A}(t) \right)^2 \right\} + \frac{Z^2}{|\vec{r}_2 - \vec{r}_1|} \tag{9.60}$$

where

$$\vec{A}(t) = \vec{\lambda} A_0(t) cos(\omega t + \delta) \tag{9.61}$$

describes the vector potential in the dipole approximation with a polarization $\vec{\lambda}$ and a pulse envelope $A_0(t)$.

In the absence of the Coulomb interaction (i.e. $Z = 0$), the above equation is separable and the solutions may be written as the product of the Volkov Keldysh states (Cf. section 2.6.1), namely,

$$\psi_0(\vec{r_i}, \vec{r}_2.t) = \phi_{k_1}(\vec{r}_1, t)\phi_{k_2}(\vec{r}_2, t) \tag{9.62}$$

where the continuum states of momentum $\vec{k}_1$ and $\vec{k}_2$ are,

$$\phi_{k_i}(\vec{r_i}, t) = \frac{1}{(2\pi)^{3/2}} e^{i[\vec{k}_i \cdot \vec{r}_i - \frac{1}{2m}k^2 t + \vec{k}_i \cdot \vec{\alpha}_0(t) - \beta(t)]} , \tag{9.63}$$

with

$$\vec{\alpha}_0(t) = \frac{Z}{mc} \int^t d\tau \vec{A}(\tau) , \qquad \beta(t) = \frac{Z^2}{2mc} \int^t d\tau A^2(\tau) . \tag{9.64}$$

The solution with $Z \neq 0$ in (9.60) may be found [Bergou *et al.*, 1981] using relative coordinates in both position and momentum, namely,

$$\vec{R} = (\vec{r}_1 + \vec{r}_2)/2 , \quad \vec{r} = \vec{r}_1 - \vec{r}_2 , \qquad \vec{P} = \vec{p_i} + \vec{p}_2 , \quad \vec{p} = (\vec{p}_1 - \vec{p}_2)/2 . \tag{9.65}$$

Using $M = 2m$ and $\mu = m/2$, then (9.60) becomes,

$$H = \frac{P^2}{2M} + \frac{p^2}{2\mu} + \frac{Z^2}{r} - \frac{Z}{mc}\vec{P} \cdot \vec{A}(t) + \frac{Z^2}{mc^2} A^2(t) \tag{9.66}$$

which is separable.

The explicit solution of (9.66) is

$$\psi^-(\vec{R},\vec{r},t) = \psi_C^-(\vec{r})\frac{1}{(2\pi)^{3/2}}e^{i\vec{K}\cdot\vec{R}}e^{i[-\frac{1}{2}(K^2/2M+k^2/2\mu)t+\vec{K}\cdot\vec{\alpha}_0(t)-2\beta(t)]}\ , \tag{9.67}$$

with $\vec{K} = \vec{k}_1 + \vec{k}_2$ and $\vec{k} = \vec{k}_1 - \vec{k}_2$. Here $\psi_C^-(\vec{r})$ is the standard repulsive Coulomb two body wavefunction with an incoming boundary condition, namely,

$$\psi_C^-(\vec{r}) = exp(-\pi\eta_k/2)\Gamma(1-i\eta_k)e_1^{i\vec{k}\cdot\vec{r}}F_1(i\eta_k, 1; i(\vec{k}\cdot\vec{r} - kr))\ , \tag{9.68}$$

with $\eta_k = Z^2\mu/k$. The solution of (9.67) may be verified by direct substitution into $i\frac{\partial\psi}{\partial t} = H\psi$.

4 Mathematical scattering factors

These factors are given for single electron targets. A seminal paper for many electron atoms is that of Waller and Hartree [1929] which gives the basic sum rules in a clear fashion. There is extensive discussion in Goldberger and Watson [1964, chapter 11, e.g., Eq(200)].

1. Atomic form factor (elastic scattering factor, scatter form factor): $< i|\exp i\vec{q}\cdot\vec{r}|i >$.
2. Total inelastic scattering factor (or Compton scattering factor): $1-|< i|\exp i\vec{q}\cdot\vec{r}|i >|^2$. These are tabulated by Cromer and Mann [1967] for various atoms.
3. Incoherent scattering function: $|< f|\exp i\vec{q}\cdot\vec{r}|i >|^2$.
4. Structure factor: This gives the interference for coherent scattering from N identical centers separated by R_j. $N + \sum_i^N \sum_{k>i}^N cos(\delta_i - \delta_k)$ where $\delta_j = \vec{q}\cdot\vec{R}_j$. This leads to Bragg scattering.
5. Generalized oscillator strength: $\Delta E/q^2|< f|\exp i\vec{q}\cdot\vec{r}|i >|^2$. The oscillator strength is the dipole limit of this quantity. These quantities are independent of the mass and charge of the projectile. These obey sum rules.

5 Expansion parameters

Various expansion parameters can be used in describing interactions of photons with atoms, namely,

1. $Z_T\alpha = Z_T/137.04$ (fine structure constant). Used to characterize the strength of effects of quantum electrodynamics (QED).

2. $\hbar\omega/mc^2$. Measures the strength of relativistic effects.

3. kr. Size of the atomic distance r compared to the wavelength of the photon. If this is small, then the multipole expansion is valid.

4. $Ze^2/\hbar v$. Massey parameter used to determine the strength of an interaction of a particle of charge Z moving at velocity v.

5. $I/\hbar\omega$. Binding energy I of the atomic electron in comparison to the energy of the photon.

6. N. Photon number which measures how many photons are involved in an interaction.

Except where noted all of these parameters are taken to be small in this chapter.

6 Names of processes

1. Thomson scattering. This is scattering of a free electron by a photon (Cf. Heitler pp. 34-35). $d\sigma_T/d\Omega = 1/2r_0^2(1+cos^2\theta)$ $\sigma = 8\pi/3r_0^2 = 6.65 \times 10^{-25}\text{cm}^2$. Note: $r_0 = \alpha^2 a_0 = 2.818 \times 10^{-13}$ cm. (Thomson received the Nobel Prize in 1906.)

2. Rayleigh scattering (sometimes elastic scattering, coherent scattering). The state of the atomic target (usually the ground state) is unchanged. (Lord Rayleigh (John Strutt) was awarded the 1904 Nobel Prize.)

3. Raman scattering (sometimes inelastic scattering, incoherent scattering). The atomic electron makes a transition from the ground state to a bound excited state. There is a photon in the final state. If the energy of the photon is similar to the transition energy, then resonant Raman scattering may occur. (Cf. Heitler pp.189-192.) (Raman's Nobel Prize came in 1930.)

4. Compton scattering (sometimes inelastic scattering or incoherent scattering). The atomic electron makes a transition to a continuum state. There is both an initial and a final photon. (Some people [e.g. Aberg] include bound states and call Raman Scattering one form of Compton Scattering). This is of historical importance because it confirmed the particle nature of photons as suggested by Einstein's Photoelectric Effect (below). (Compton's Nobel Prize was awarded in 1927.) Variations include:

   (a) The classical Thomson scattering cross section from a free electron (no energy threshold). (Cf. Heitler pp.34-45.)

(b) The Thomson cross section multiplied by the total inelastic form factor, $1 - | < i| \exp i\vec{q} \cdot \vec{r}|i > |^2$ (no threshold).

(c) The relativistic Klein-Nishina formula (no threshold). (Cf. Heitler pp. 211-221.)

(d) The cross section arising from $| < f|A^2|i > |^2$ including the correct threshold at $\hbar\omega = I$.

5. Photoionization (photoelectric effect or photo effect, photoabsorption, photoannihilation). Ionization of the target with annihilation of the photon. This effect, explained by Einstein (using data first taken forty years earlier), demonstrated that a photon is particle-like because it is indivisible (i.e quantized). For this Einstein received the Nobel Prize in 1921. For $\hbar\omega\ I = Z/v << 1$, $\sigma_{PE} = \sigma_T 64\alpha^3/Z^2(I/\hbar\omega)^{7/2}$ where K-shell ionization dominates because it is farthest from a free electron where photoannihilation is forbidden by conservation of energy and momentum. (Cf. Heitler pp. 204-11).

6. Photoannihilation. An incident photon is annihilated in the process of exciting an atom (or molecule). If an electron is ionized this process is often referred to as photoionization.

7. Bragg scattering (coherent scattering). This is constructive interference of waves from N centers (e.g. in crystal lattices). It is a generalization of two-slit Young diffraction. It is given by the structure factor listed below. (Cf. Heitler pp. 193-4.) (The Braggs [father and son] received their Nobel Prize in 1915.)

8. Rutherford scattering. This is scattering between two free charged particles. (Rutherford received no Nobel prize in physics, but received the 1908 Nobel Prize in Chemistry.)

9. Photo-emission and absorption. In photo-emission it is usually the case that $A^2 << \vec{p} \cdot \vec{A}$. This is not always the case for photo-absorption (including cases listed above).

10. Bremsstrahlung. This is radiation from an accelerated charge. It contains an infrared divergence.

11. Note: Photon-photon scattering is small ($< 10^{-31}\text{cm}^2$).

2nd Order to 1st Order to 2nd Order. For Compton scattering the transition from 2nd order in $\vec{A}$ at $\hbar\omega/I << 1$ to first order at $\hbar\omega/I >> 1$ is discussed by Heitler pp. 194-6. The subsequent transition to the relativistic 2nd order limit is illustrated in figure 10 on p. 220 of Heitler.

10

Relations between charged particle and photon reactions

Photons and charged particles both interact with matter via the electromagnetic fields they carry. In principle one could describe interactions with both photons and charged particles in terms of electromagnetic wavepackets*†. For charged particles the fields are modulated by the $1/R$ Coulomb potential which is manifested by the Rutherford scattering factor of $1/q^4$ in momentum space. The wavepackets which govern the envelope of the oscillating electromagnetic fields of photons are often well approximated by a constant envelope containing a plane wave, except for short pulsed lasers and other cases where higher harmonics are significant. It is also common [Jackson, 1975, p. 719; Bransden and Joachain, 1983] to regard the interaction of charged particles as an exchange of virtual photons where energy conservation may be violated in intermediate states within the limits of the Uncertainty principle. In this picture the Nth Born approximation is described in terms of an exchange of N virtual photons.

In this chapter the relationship between interactions with charged particles and photons is described within the framework of first order perturbation of the interaction V with the projectile. Here it is useful to differentiate between photoannihilation (PE) (i.e. the photo effect first described by Einstein), where the photon is annihilated, and Compton scattering (C) in which the photon is inelastically scattered (and not an-

* For fast charged particles the electric field is a fast sharp pulse with $t < t_{orbit}$ where t_{orbit} is the time required for an electron to complete one orbit in the target atom. Laser fields, on the other hand, increase and decrease slowly with $t > t_{orbit}$ even for the fastest laser pulses ($\sim$ 30 fs), except possibly for lasers in high Rydberg states. Consequently in strong laser fields the time evolution may depend on the shape of the wavepulse produced by the laser.

† In principle interactions with systems of charged particles may also be described in terms of their electromagnetic fields. A recurring question in physics, chemistry biology, and materials science is the interrelation between electrons and photons. In other words, how are photon waves and matter waves coupled?

nihilated). In lowest order V is $\vec{p} \cdot \vec{A}$ for photoannihilation and A^2 for Compton scattering. For charged particles $V = Z/|\vec{R} - \vec{r}|$. Since charged particle scattering, Compton scattering and photoannihilation are physically distinguishable processes with different interaction operators, it is remarkable that their matrix elements are so similar. To first order in V the matrix element for both charged particle scattering and Compton is proportional to the atomic form factor $< f|e^{i\vec{q}\cdot\vec{r}}|i >$ (itself proportional to the generalized oscillator strength – Cf. (8.7), (9.43) and the appendix to chapter 9). In the dipole limit this matrix element reduces to $< f|i\vec{q}\cdot\vec{r}|i >$ (proportional to the oscillator strength). In this dipole limit charged particle scattering, Compton scattering and photoannihilation cross sections are all proportional to the square of this same matrix element $i < f|\vec{q}\cdot\vec{r}|i >$. Total ionization cross sections for photoannihilation are well described by the dipole approximation in many cases. In Compton scattering, on the other hand, the contributions are mostly non-dipole, except at forward angles (i.e., $qr << 1$).

10.1 Multipoles

Consider a cross section σ for an arbitrary transition. This cross section may be generally expanded in partial waves according to,

$$\sigma = \sum_{\ell=0}^{\infty} \sigma_\ell \,. \tag{10.1}$$

It is assumed at this point that the cross section has been integrated over the scattering angles (θ, ϕ) of the projectile, where each partial wave may be associated with a given spherical harmonic $Y_{\ell,m}(\theta, \phi)$. The $\ell = 1$ contribution is called the dipole term and the remaining $\ell \neq 1$ terms are the non-dipole terms.

If the transition is differential in the energy ΔE transferred to the target by the projectile (i.e., target ionization), then one may write,

$$\begin{aligned} \frac{d\sigma(\Delta E)}{d\Delta E} &= \sum_\ell \frac{d\sigma_\ell(\Delta E)}{d\Delta E} = \frac{d\sigma_1(\Delta E)}{d\Delta E} + \sum_{\ell\neq 1} \frac{d\sigma_\ell(\Delta E)}{d\Delta E} \\ &\equiv \frac{d\sigma_d(\Delta E)}{d\Delta E} + \frac{d\sigma_{nd}(\Delta E)}{d\Delta E} \end{aligned} \tag{10.2}$$

where 'd' and 'nd' represent the dipole and non-dipole contributions to the cross sections.

1 Ratios of cross sections

Now consider a cross section $d\sigma^{++}/d\Delta E$ for a two electron transition (either double ionization as noted or ionization-excitation) singly differential in the energy ΔE transferred by the projectile. This cross section may be expressed in terms of the cross section for single ionization by using the ratio $R(\Delta E)$ of two to one electron cross sections,

$$\begin{aligned}\frac{d\sigma^{++}}{d\Delta E} &= \left(\frac{\frac{d\sigma^{++}}{d\Delta E}}{\frac{d\sigma^{1}}{d\Delta E}}\right)\left(\frac{d\sigma^{+}}{d\Delta E}\right) \\ &= R(\Delta E)\ \frac{d\sigma^{+}}{d\Delta E}\ \frac{\sigma^{+}}{\sigma^{+}} \\ &= R(\Delta E)\ \rho(\Delta E)\ \sigma^{+}\ , \end{aligned} \tag{10.3}$$

where $\rho(\Delta E) \equiv \frac{1}{\sigma^{+}}\frac{d\sigma^{+}}{d\Delta E}$ is the normalized energy spectral distribution for singly ionized states as a function of the energy transfer ΔE. Because $\rho(\Delta E)$ is found from a single electron transition it is relatively easy to calculate. The normalization for $\rho(\Delta E)$ is,

$$\int_{0}^{\infty} \rho(\Delta E)\, d\Delta E = 1\ . \tag{10.4}$$

It is convenient to define,

$$\rho_d(\Delta E) \equiv \frac{1}{\sigma^{+}}\ \frac{d\sigma_d^{+}(\Delta E)}{d\Delta E}$$

and

$$\rho_{nd}(\Delta E) \equiv \frac{1}{\sigma^{+}}\ \frac{d\sigma_{nd}^{+}(\Delta E)}{d\Delta E}\ .$$

Note that $\int \rho_d(\Delta E)$ and $\int \rho_{nd}(\Delta E)$ are both less than one. However,

$$\rho_d(\Delta E) + \rho_{nd}(\Delta E) = \rho(\Delta E) \tag{10.5}$$

is normalized.

Now define the dipole fraction f_d of the energy distribution by,

$$f_d(\Delta E) \equiv \frac{\rho_d(\Delta E)}{\rho_d(\Delta E) + \rho_{nd}(\Delta E)}\ . \tag{10.6}$$

Clearly, the non-dipole fraction is $f_{nd}(\Delta E) = 1 - f_d(\Delta E)$.

The ratio of double to single dipole cross sections is defined as,

$$R_d(\Delta E) \equiv \left[\frac{d\sigma_d^{++}(\Delta E)}{d\Delta E} / \frac{d\sigma_d^{+}(\Delta E)}{d\Delta E} \right] \tag{10.7}$$

The non-dipole ratio R_d is similarly defined (as may be each of the multipole ratios R_ℓ). It is then easily shown that,

$$\begin{aligned} R(\Delta E) &= \sum_{\ell=0}^{\infty} f_\ell(\Delta E) R_\ell(\Delta E) \\ &= f_d(\Delta E) R_d(\Delta E) + f_{nd}(\Delta E) R_{nd}(\Delta E) \end{aligned} \tag{10.8}$$

where the dipole term 'd' corresponds to $\ell = 1$ and the non-dipole 'nd' terms are the $\ell \neq 1$ terms. Relation (10.8) is a simple mathematical identity. This identity holds for total cross sections as well as cross sections differential in ΔE.

2 Compton and charged particle scattering

The Compton scattering cross section doubly differential in the energy transfer (ΔE)-momentum transfer (q) plane for inelastic scattering from an arbitrary initial state $|i>$ to an arbitrary final state $|f>$ is given in (9.42) in first order (in both the fine structure constant α and v^2/c^2 [Wang *et al.*, 1995a; Baym, 1974; Heitler, 1954]) as,

$$\left(\frac{d^2\sigma}{d\Delta E dq^2} \right)_C = \frac{\pi r_0^2}{2k^2} \left(1 + \left(1 - \frac{q^2}{2k^2} \right)^2 \right) F_I(\Delta E, q^2), \tag{10.9}$$

with

$$F_I(\Delta E, q^2) = \int d\Omega_e \sqrt{2(\Delta E + \epsilon_i)} |\langle f| \sum_{j=1}^{N} e^{i\vec{q}\cdot\vec{r}_j} |i\rangle|^2, \tag{10.10}$$

which is the inelastic transition form factor (proportional to the generalized oscillator strength [Aberg and Tulkki, 1985] for the N-electron atom integrated over all emission angles of the emitted electron and weighted with the density of continuum final states. In the case of multiple ionization, F_I is understood to include an integral over emission angles of all electrons and the appropriate density of states. The atomic initial state $|i>$ with binding energy $\epsilon_i = I$ is assumed to be isotropic but arbitrary otherwise. Consequently, F_I depends only on the magnitude of the momentum and energy transfer. The dependence of F_I on the energy transfer ΔE is implicit through the energy of the accessed final state, $\Delta E = \epsilon_f - \epsilon_i$. In Eq(10.9), r_0 denotes the classical electron radius $r_0 = \alpha^2$

(in units of the Bohr radius) and the photon energy and wavenumber are denoted by ω and k. Furthermore, the square of the momentum transfer q^2 is related to the scattering angle θ of the photon by,

$$q^2 = (\vec{k} - \vec{k}\,')^2 = k^2 + k'^2 - 2kk'cos\theta \simeq 2k^2(1 - cos\theta)\,, \qquad (10.11)$$

where primes denote the wavevector of the scattered photon. Eq(10.9) is valid for unpolarized light and for sufficiently high energies $\omega >> |I|$ where the contributions from the $\vec{p}\cdot\vec{A}$ term of the coupling of the atom-photon interaction can be neglected compared to the A^2 term.

The factor of $1 + (1 - \frac{q^2}{2k^2})$ in Eq(10.9) is equal to $1 + cos^2\theta$ and results from the sum over polarization in the Compton formula. For large but non-relativistic photon energies ω,

$$\frac{\Delta E}{\omega} \simeq \frac{q^2}{2\omega} \simeq \frac{q}{c} << 1, \qquad (10.12)$$

this factor becomes independent of ΔE.

For the impact of a high velocity particle of charge Z, the cross section in the plane wave Born approximation [McDowell and Coleman, 1970] is given by,

$$\left(\frac{d^2\sigma}{d\Delta E dq^2}\right)_Z = \frac{4\pi}{\mathrm{v}^2}\frac{Z^2}{q^4}\,F_I(\Delta E, q^2)\,, \qquad (10.13)$$

where $q^2 = (\vec{K} - \vec{K}\,')^2$ is again the momentum transfer of the projectile (here the charged particle) to the atomic system. Note that in this approximation no momentum is transferred by the projectile to the target nucleus, so that the momentum transfer, $\vec{q}$, here is the momentum transferred to the electron(s). All modifications for multiple electron emission are the same for scattering by charged particles as for Compton scattering. From (10.9) and (10.13) it is apparent that both the Compton and the charged particle cross sections are proportional to the atomic form factor (or to the generalized oscillator strength [Aberg and Tulkki, 1985]) $f_{fi}(q) = \Delta E/q^2| < f|e^{i\vec{q}\cdot\vec{r}}|i > |^2)$. Consequently, the cross section for Compton scattering may be expressed in terms of the first Born cross section for scattering by charged particles as,

$$\left(\frac{d^2\sigma}{d\Delta E dq^2}\right)_C = \frac{r_0^2}{8}\frac{\mathrm{v}^2}{k^2}\frac{q^4}{Z^2}\left[1 + \left(1 - \frac{q^2}{2k^2}\right)^2\right]\left(\frac{d^2\sigma}{d\Delta E dq^2}\right)_Z\,. \qquad (10.14)$$

Eq(10.14) holds for each partial wave. This result may be applied to transitions other than ionization.

Since the prefactor in (10.14) is to order $\Delta E/\omega$ independent of the energy transfer ΔE, the direct proportionality extends to the single-differential cross section for fixed momentum transfer, namely,

$$\left(\frac{d\sigma}{dq^2}\right)_C = \frac{r_0^2}{8}\frac{v^2}{k^2}\frac{q^4}{Z^2}\left[1+\left(1-\frac{q^2}{2k^2}\right)^2\right]\left(\frac{d\sigma}{dq^2}\right)_Z, \qquad (10.15)$$

where $d\sigma/dq^2$ has been integrated over ΔE. This relation is valid for arbitrary final states $|f>$ including both single and double ionization. This relation could be tested experimentally by observing differential cross sections for Compton scattering as a function of the momentum transfer and comparing them to existing data for differential cross sections by charged particle impact weighted by the factors in the above relation. This relation may also be used to evaluate cross sections for Compton scattering by modifying calculations for charged particle impact.

Let us now use the above relation to express the ratio R of double to single ionization cross sections by Compton scattering in terms of a corresponding cross section ratio by charged particles. Since the prefactors in (10.15) are independent of the final state, in particular of the ionization stage, one has for the ratio at fixed momentum transfer the identity,

$$R_C(q) = \frac{(d\sigma^{++}/dq^2)_C}{(d\sigma^{+}/dq^2)_C} = \frac{(d\sigma^{++}/dq^2)_Z}{(d\sigma^{+}/dq^2)_Z} = R_Z(q)\,. \qquad (10.16)$$

This relation, valid with the Born approximation, holds for all values of q, where q is chosen to be the same value for both Compton scattering and scattering by charged particles.

One may obtain an equivalence in ratios of cross sections integrated over q at a fixed energy transfer ΔE from (10.14) within the limits of a peaking approximation similar to (10.20) below, namely,

$$R_C(\Delta E) = \frac{(d\sigma^{++}/d\Delta E)_C}{(d\sigma^{+}/d\Delta E)_C} = \frac{(d\sigma^{++}/d\Delta E)_Z}{(d\sigma^{+}/d\Delta E)_Z} \simeq R_Z(\Delta E)\,. \quad (10.17)$$

This holds for each partial wave.

The total cross section for single (double) ionization integrated over all large momentum transfers beyond a threshold value q_0 is given by,

$$\sigma_{Z,q_0}^{+,++} = \int_{q_0^2}^{q_{\max}^2} dq^2 \left(\frac{d\sigma^{+,++}}{dq^2}\right)_Z, \qquad (10.18)$$

where the upper limit $q_{\max}^2$ [McDowell and Coleman, chapter 7] can be set equal to ∞ since the momentum-differential cross section decreases rapidly as $\sim q^{-4}$. The q^{-4} dependence corresponds to Rutherford scattering in close encounters between the target electron and the charged particle. Using now (10.15) one may express this cross section in terms of the Compton cross section as,

$$\sigma_{Z,q_0}^{+,++} = \frac{8k^2Z^2}{\mathrm{v}^2}\int_{q_0^2}^{\infty}\frac{dq^2}{q^4}\left[1+\left(1-\frac{q^2}{2k^2}\right)^2\right]^{-1}\left(\frac{d\sigma^{+,++}}{dq^2}\right)_C . \quad (10.19)$$

Since the singly differential Compton cross section is a slowly varying function of q^2 within the interval $[q_0, 2\omega/c]$, Eq(10.19) can be evaluated in peaking approximation as,

$$\sigma_{Z,q_0}^{+,++} \cong \frac{8k^2Z^2}{\mathrm{v}^2}\left(\frac{d\sigma^{+,++}(q_0)}{dq^2}\right)_C \int_{q_0^2}^{\infty}\frac{dq^2}{q^4}\left(1+\left(1-\frac{q^2}{2k^2}\right)^2\right)^{-1} . \quad (10.20)$$

Consequently, the ratios of double to single ionization total cross sections for large momentum transfers $\geq q_0$ by charged particle and by Compton scattering are related by,

$$R_{Z,q_0} = \frac{\sigma_{Z,q_0}^{++}}{\sigma_{Z,q_0}^{+}} = R_C(q_0)\,, \quad (10.21)$$

provided that $r_{\mathrm{target}}^{-1} << q_0 < 2\omega/c$. Note that the right-hand side is differential in the momentum transfer while the left-hand side includes all $q > q_0$. For sufficiently photon large energies, q_0 may be chosen close to zero so that both ratios are for total cross sections.

10.2 Dipole terms

For soft Coulomb scattering the momentum transfer, q, is small, the atomic form factor then reduces to a dipole matrix element, namely, $< f|e^{i\vec{q}\cdot\vec{r}}|i> \approx i < f|\vec{q}\cdot\vec{r}|i>$, and the generalized oscillator strength reduces to the standard dipole optical oscillator strength. In this limit the matrix elements reduce to the dipole matrix elements used for photoexcitation and ionization‡.

‡ Contributions to photoabsorption originate from a region in the dispersion plane with large $\Delta E \simeq \omega$ and small q distinctly different from the region for either Compton scattering or charged particle ionization with large momentum transfer. Even when retardation is taken into account in the transition matrix element for photoabsorption ($\sim< f|\vec{p}\cdot\hat{e}\, e^{i\vec{q}\cdot\vec{r}}|i>$), the dominant region for photoabsorption lies far above the free particle dispersion curve

If the Born cross section is expanded in inverse powers of the projectile energy, E, then $\sigma_Z(E) \simeq A\, lnE/E + B/E$. The leading $ln\ E/E$ (or $ln\ \mathrm{v}^2/\mathrm{v}^2$) contribution to the cross section for charged particles (which is absent both in $d\sigma_Z/dq$ used above and classically) corresponds to the Bethe-Born limit. It may be expressed in terms of the cross section for excitation or ionization via photoannihilation (i.e. the photo effect) [Byron and Joachain, 1967], namely (using atomic units),

$$\frac{d\sigma_Z}{d\Delta E} = \frac{Z^2}{2\pi\alpha}\frac{ln \mathrm{v}^2}{\mathrm{v}^2}\frac{\sigma_{PE}(\Delta E)}{\Delta E}\,, \qquad (10.22)$$

where v is the velocity of the projectile, ΔE is the energy transferred by the projectile, and α is the fine structure constant. If q is near q_{min}, then $q \approx q_{min} = |\vec{K} - \vec{K}'| = \Delta E/2\mathrm{v}$. This relation applies to both single and double ionization. Eq(10.22) is valid in the dipole limit, i.e. when $qr << 1$ corresponding to small $\Delta E/\mathrm{v}$.

If the energy transfer, ΔE, is fixed then,

$$R_Z(\Delta E) \equiv \frac{d\sigma_Z^{++}(\Delta E)/d\Delta E}{d\sigma_Z^{+}(\Delta E)/d\Delta E} = \frac{\sigma_{PE}^{++}}{\sigma_{PE}^{+}} = R_{PE}\,. \qquad (10.23)$$

In the photo effect, $\Delta E = h\nu$ is uniquely defined by the energy of the incident photon, while for charged particles ΔE must be selected from a range of energy transfers of the projectile for both single and double ionization. Manson [Manson and McGuire, 1994] has pointed out that this relation is the same relativistically as non-relativistically.

The ratios for total cross sections for charged particles and photons may be related using the normalized energy spectral distribution for singly ionized final states. This normalized energy spectral distribution is defined from (10.3) by,

$$\rho_Z^{+}(\Delta E) = \frac{1}{\sigma_Z^{+}}\frac{d\sigma_Z^{+}}{d\Delta E}\,. \qquad (10.24)$$

Then, from (10.22),

for all non-relativistic energies, i.e.,

$$\Delta E \simeq \omega \simeq qc >> q^2/2\,.$$

This difference is also reflected in the final-state angular momentum distribution: unlike for photoabsorption, dipole forbidden transitions dominate both Compton scattering and 'hard' charged particle collisions. In this limit, the photoabsorption cross section can be related to the Bethe- Born limit of the Born approximation for charged particle in which dipole allowed transitions due to soft collisions dominate.

$$\begin{aligned}\sigma_Z^{++} &= \int \frac{d\sigma_Z^{++}(\Delta E)}{d\Delta E} d\Delta E = \frac{Z^2}{2\pi\alpha}\frac{ln v^2}{v^2} \int \frac{\sigma_{PE}^{++}(\Delta E)}{\Delta E} d\Delta E \qquad (10.25)\\ &= \frac{Z^2}{2\pi\alpha}\frac{ln v^2}{v^2} \int \frac{\sigma_{PE}^{++}(\Delta E)}{\Delta E}\frac{\Delta E}{\sigma_{PE}^{+}}\frac{\sigma_{PE}^{+}}{\Delta E} d\Delta E = \int \frac{\sigma_{PE}^{++}}{\sigma_{PE}^{+}}\frac{d\sigma_Z^{+}}{d\Delta E} d\Delta E\\ &= \sigma^{+} \int \frac{\sigma_{PE}^{++}}{\sigma_{PE}^{+}} \rho_Z^{+}(\Delta E)\, d\Delta E\,,\end{aligned}$$

so that,

$$R_Z = \int R_{PE}(\Delta E)\ \rho_Z^{+}(\Delta E)\ d\Delta E\,. \qquad (10.26)$$

The above relation holds only in the dipole limit. A similar relation is expected to hold for Compton scattering. For photoannihilation with photons of incident energy ω one has, $\rho_{PE}(\Delta E) = \delta(\Delta E - \omega)$ and the resulting equivalence in (10.26) is obvious.

The convergence of the first Born cross sections to the high velocity Bethe dipole limit is slow because the Bethe $ln\ v^2/v^2$ dipole term varies slowly compared to the $1/v^2$ term which contains non-dipole terms.

10.3 Dipole and non-dipole terms

The cross section for double ionization by charged particle scattering may be decomposed into dipole and non-dipole contributions according to,

$$\begin{aligned}\frac{d\sigma_Z^{++}(\Delta E)}{d\Delta E} &= \sum_{\ell} \frac{d\sigma_{Z,\ell}^{++}(\Delta E)}{d\Delta E} = \frac{d\sigma_{Z,1}^{++}(\Delta E)}{d\Delta E} + \sum_{\ell\neq 1} \frac{d\sigma_{Z,\ell}^{++}(\Delta E)}{d\Delta E}\\ &\equiv \frac{d\sigma_{Z,d}^{++}(\Delta E)}{d\Delta E} + \frac{d\sigma_{Z,nd}^{++}(\Delta E)}{d\Delta E} \qquad (10.27)\end{aligned}$$

where 'd' and 'nd' represent the dipole and non-dipole contributions to the cross sections.

1 Ratios of cross sections differential in energy transfer

The ratio of double to single ionization in (10.27) at a fixed energy transfer ΔE for a given partial wave is defined by,

$$R_{Z,\ell}(\Delta E) \equiv \frac{d\sigma_{Z,\ell}^{++}(\Delta E)}{d\Delta E} \Big/ \frac{d\sigma_{Z,\ell}^{+}(\Delta E)}{d\Delta E}\,, \qquad (10.28)$$

Here ℓ may denote either a single partial wave or a sum of partial waves.

The dipole term of (10.27) may be expressed as,

$$\begin{aligned} \frac{d\sigma_{Z,d}^{++}(\Delta E)}{d\Delta E} &= \left[\frac{d\sigma_{Z,d}^{++}(\Delta E)}{d\Delta E} \Big/ \frac{d\sigma_{Z,d}^{+}(\Delta E)}{d\Delta E} \right] \frac{d\sigma_{Z,d}^{+}(\Delta E)}{d\Delta E} \qquad (10.29) \\ &= R_{Z,d}(\Delta E) \frac{d\sigma_{Z,d}^{+}(\Delta E)}{d\Delta E} , \end{aligned}$$

where $d\sigma_{Z,d}^{+}(\Delta E)/d\Delta E$ is the dipole component of single ionization cross section for charged particles.

It is evident from (10.29) that the ratio $R_{Z,d}$ involves only dipole terms, which according to (10.23) may be identified as,

$$R_{Z,d}(\Delta E) = R_{PE} = \sigma_{PE}^{++}/\sigma_{PE}^{+} \, . \qquad (10.30)$$

This relationship states that in the dipole limit (Bethe-Born limit), the ratio of double to single ionization for a given energy transfer ΔE by charged particles is the same as the corresponding ratio for photoionization.

The non-dipole term in (10.27) is now given by,

$$\begin{aligned} \frac{d\sigma_{Z,nd}^{++}(\Delta E)}{d\Delta E} &= \left[\frac{d\sigma_{Z,nd}^{++}(\Delta E)}{d\Delta E} \Big/ \frac{d\sigma_{Z,nd}^{+}(\Delta E)}{d\Delta E} \right] \frac{d\sigma_{Z,nd}^{+}(\Delta E)}{d\Delta E} \\ &= R_{Z,nd}(\Delta E) \frac{d\sigma_{Z,nd}^{+}(\Delta E)}{d\Delta E} \, . \qquad (10.31) \end{aligned}$$

From (10.17) the ratio of double to single ionization at fixed ΔE is related to the ratio by Compton scattering, namely,

$$R_{Z,nd}(\Delta E) \simeq R_C(\Delta E) \, . \qquad (10.32)$$

Then,

$$\begin{aligned} \frac{d\sigma_{Z}^{++}(\Delta E)}{d\Delta E} &= R_{Z,d}(\Delta E) \frac{d\sigma_{Z,d}^{+}(\Delta E)}{d\Delta E} + R_{Z,nd}(\Delta E) \frac{d\sigma_{Z,nd}^{+}(\Delta E)}{d\Delta E} \\ &\simeq R_{PE} \frac{d\sigma_{Z,d}^{+}(\Delta E)}{d\Delta E} + R_C(\Delta E) \frac{d\sigma_{Z,nd}^{+}(\Delta E)}{d\Delta E} \, . \qquad (10.33) \end{aligned}$$

Now,

$$R_Z(\Delta E) = \frac{d\sigma_{Z}^{++}(\Delta E)/d\Delta E}{d\sigma_{Z}^{+}(\Delta E)/d\Delta E} \qquad (10.34)$$

$$= \frac{\left(d\sigma^{++}_{Z,d}(\Delta E)/d\Delta E + d\sigma^{++}_{Z,nd}(\Delta E)/d\Delta E\right)}{\sigma^{+}_{Z}\rho_{Z}(\Delta E)}$$

$$= \frac{R_{PE}\ \sigma^{+}_{Z}\ \rho_{Z,d}(\Delta E) + R_{C,nd}(\Delta E)\ \sigma^{+}_{Z}\ \rho_{Z,nd}(\Delta E)}{\sigma^{+}_{Z}(\rho_{Z,d}(\Delta E) + \rho_{Z,nd}(\Delta E))},$$

where $d\sigma^{+}_{Z}/d\Delta E = \sigma^{+}_{Z}\rho_{Z}(\Delta E)$ and $\rho_{Z}(\Delta E) = \rho_{Z,d}(\Delta E) + \rho_{Z,nd}(\Delta E)$.

One may next use the dipole fraction $f_d(\Delta E)$ of the normalized energy spectral distribution $\rho(\Delta E)$,

$$f_d(\Delta E) \equiv \frac{\rho_{Z,d}(\Delta E)}{\rho_{Z,d}(\Delta E) + \rho_{Z,nd}(\Delta E)}. \tag{10.35}$$

Clearly, the non-dipole fraction is $f_{nd}(\Delta E) = 1 - f_d(\Delta E)$. Now (10.35) becomes,

$$\begin{aligned} R_Z(\Delta E) &= R_{PE}\ f_d(\Delta E) + R_{C,nd}(\Delta E)\ f_{nd}(\Delta E) \\ &= R_{PE}\ f_d(\Delta E) + R_{C,nd}(\Delta E)\ (1 - f_d(\Delta E))\,. \end{aligned} \tag{10.36}$$

Finally one has from (10.36),

$$R_Z(\Delta E) - f_d(\Delta E)\ R_{PE} = (1 - f_d(\Delta E))\ R_{C,nd}(\Delta E)\,. \tag{10.37}$$

This corresponds to the relatively obvious result that $R = f_d R_d + f_{nd} R_{nd}$ of (10.8). This relation provides a way to check the self consistency of data for scattering by charged particles (Z), photoannihilation (PE) and Compton (C) scattering. The energy transfer of the projectile ΔE is found differently for charged particles, for photoannihilation and for Compton scattering. For photoannihilation $\Delta E = \omega$, where ω is the energy of the incident photon which transfers all of its energy. For scattering by charged particles $\Delta E = \epsilon_e + I$ where ϵ_e is the energy carried off by the electron(s) and I is the ionization potential. In Compton scattering the backscattered photons transfer an energy $\Delta E = 2E^2_\gamma/c^2$, corresponding to the maximum value of $\Delta\lambda = \lambda_C(1 - cos\theta)$ in the standard formula for Compton scattering.

Noting that $R_C(\Delta E) \neq R_{C,d}(\Delta E) + R_{C,nd}(\Delta E)$, one may nevertheless make the useful approximation $R_C(\Delta E) \simeq R_{C,nd}(\Delta E)$ when the dipole contributions are a small fraction of the total cross section for both $d\sigma^{++}_{C}(\Delta E)/d\Delta E$ and $d\sigma^{1}_{C}(\Delta E)/d\Delta E$. This is valid when either the dipole contributions to Compton scattering are small (i.e. well above the classical threshold) or when $R_{C,d}(\Delta E)$ is similar to $R_C(\Delta E)$. The latter condition is valid if the fractional contribution of the dipole term is the same for double and single electron transitions. This condition that $R_{C,d} \simeq R_C$ holds in helium at incident photon energies near 10 keV since $R_{C,d}(\Delta E) = R_{Z,d}(\Delta E) = R_{PE} \simeq R_C(\Delta E)$ so that $R_{C,d} \simeq R_C$ follows from $R = f_d R_d + f_{nd} R_{nd}$. At energies above 10 keV, it is estimated that the dipole contribution is a sufficiently small fraction of total Compton scattering that the error is 5 - 10 % at most. When this approximation holds, then (10.37) becomes,

$$R_Z(\Delta E) - f_d(\Delta E)\, R_{PE} = (1 - f_d(\Delta E))\, R_C(\Delta E)\,. \qquad (10.38)$$

Eq(10.38) is more easily used than (10.37), which requires separation of the dipole from the non-dipole contributions for Compton scattering.

2 *Ratios of total cross sections*

From (10.33), it is now straight forward to obtain the ratio R_Z of double to single ionization total cross sections by charged particles,

$$\begin{aligned} R_Z &= \sigma_Z^{++}/\sigma_Z^{+} = \int_{I_2}^{\infty} \frac{d\sigma_Z^{++}}{d\Delta E} d\Delta E \Bigg/ \int_{I_1}^{\infty} \frac{d\sigma_Z^{+}}{d\Delta E} d\Delta E \qquad (10.39) \\ &\simeq \int_{I_2}^{\infty} R_{PE}\ \rho_{Z,d}^{+}(\Delta E) d\Delta E + \int_{I_2}^{\infty} R_C\ \rho_{Z,nd}^{+}(\Delta E) d\Delta E\,, \end{aligned}$$

where $\rho_{Z,d}^{+}(\Delta E)$ and $\rho_{Z,nd}^{+}(\Delta E)$ are respectively the dipole and non-dipole normalized energy spectral distribution for single ionization by charged particles.

Eq(10.39) expresses the ratio of double to single ionization by charged particles in terms of ratios for two different photon processes: the dipole term R_{PE} by photoionization, and the non-dipole term R_C by Compton scattering, which is assumed to be mostly non-dipole. The results of calculations [Wang *et al.*, 1995a] using (10.39) are shown in figure 10.1.

10.4 Observations

Observations of the ratio of double to single total ionization cross sections by the impact of charged particles at high energies has been related to observations of the double to single ionization ratio using (10.39), where ρ_Z^{+} was calculated in the first Born approximation for both the dipole and non-dipole contributions and R_γ is taken from figure 9.8 for photoannihilation and from observations of Compton scattering for the Compton contribution. The results are shown in figure 10.1.

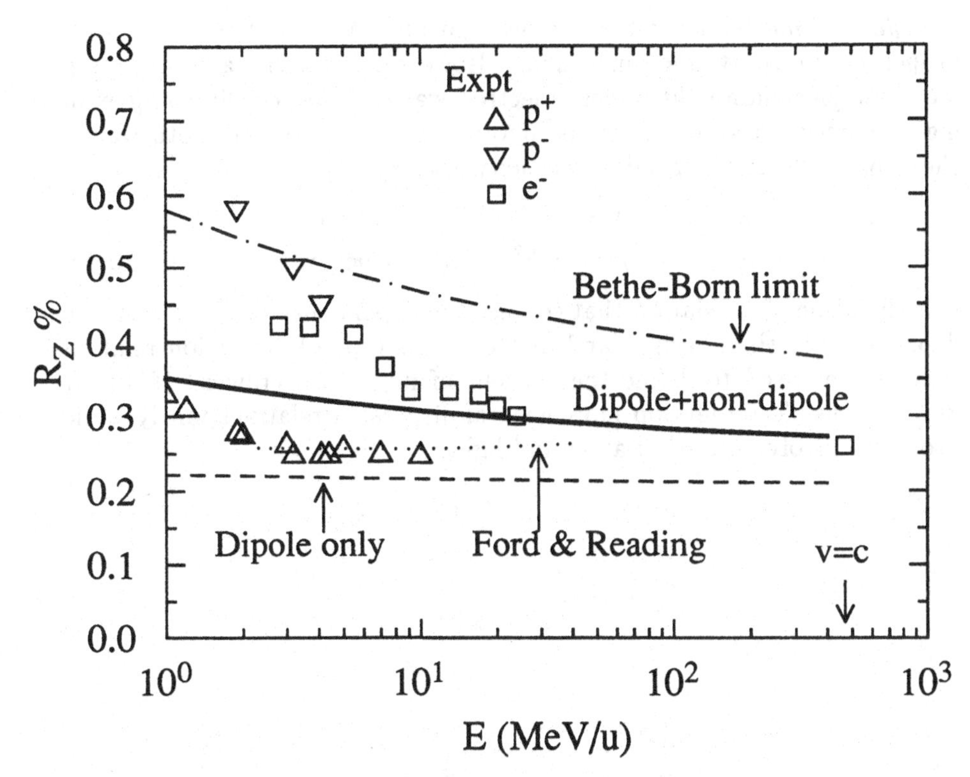

Fig. 10.1. Ratio R_Z of double to single ionization of helium by the impact of charged particles versus collision velocity from Wang *et al.* [1995a]. The solid line gives the sum of separate dipole and non-dipole contributions convoluting observed data for the photon ratio R_γ with a calculated value of the normalized energy distribution ρ for single ionization. The Bethe-Born curve corresponds to the results of Manson and McGuire [1995] who use all partial waves in ρ convoluted with the photoannihilation data alone. The dotted curve represents the Forced Impulse Method of Ford and Reading [1994] which is a direct calculation of both single and double ionization by impact of protons. Data are taken from Levin *et al.* [1993].

Kamber *et al.* [1988] have observed ratios of binary encounter electrons (where there is no internuclear momentum transfer) and obtained a ratio which tends toward a value somewhat less than 2%. This observation partially confirms (10.30), although both non-dipole and Z^3 effects may be non-negligible. The transition from the photoannihilation ratio to the Compton ratio has been observed by Wu *et al.* [1996].

Appendix

Light and matter waves. Electrons and photons may both have wave properties on an atomic dimension[§]. However, electron waves are often more complex than light waves. Electron waves are especially complex in highly correlated systems. While photons and electrons are both waves, they obey different dispersion relations, namely,

$$\begin{aligned} E &= p/\lambda = hc/\lambda \quad &\text{(photon)} \\ K &= p^2/2m = h^2/2m\lambda^2 \quad &\text{(electron)}\ . \end{aligned} \tag{10.40}$$

It is the dispersion relation that couples the dependence in space with the dependence in time in light and matter. This type of dispersion relation may also be used to define the concept of mass [Mysyrowicz, 1996]. It may be sensible to develop more general dispersion relations for dynamic processes involving both matter and light.

[§] Quantum entanglement is how matter and light are tuned.

Appendices

A.1 Hydrogenic wavefunctions

The Schrödinger equation for an electron in the presence of a heavy particle of charge, Z, is,

$$\left(-\nabla^2/2 - Z/r\right)|s>= E|s> \ . \tag{A.1}$$

This may be solved by the method of separation of variables [Messiah, 1961] in both coordinate space and momentum space. The eigenfunctions [Hill, 1996] are of the form,

$$|s>= u_{n\ell}Y_\ell^m(\theta,\phi) \tag{A.2}$$

with $n = 1,2,3,4,....; \ell = 0,...(n-1)$ and $m = -\ell, -\ell+1, ...\ell+1, \ell$ and satisfying the the orthonormality relations,

$$\int_0^\infty u_{n'\ell}u_{n\ell}\ \rho^2 d\rho = \delta_{n'n}$$
$$\int Y_{\ell'}^{*m'}Y_\ell^m d\Omega = \int_0^{2\pi} d\phi \int_0^\pi sin\theta d\theta\ Y_{\ell'}^{*m'}(\theta,\phi)Y_\ell^m(\theta,\phi) = \delta_{m'm}\delta_{\ell'\ell} \tag{A.3}$$

where ρ denotes either r in coordinate space or p in momentum space.

1 Spherical harmonics

The spherical harmonics, $Y_\ell^m(\theta,\phi)$ are defined by,

$$Y_\ell^m(\theta,\phi) = (-1)^m\left[\frac{(2\ell+1)(\ell-m)!}{4\pi(\ell+m)!}\right]^{1/2} P_\ell^m(cos\theta)e^{im\phi} \tag{A.4}$$

where the Associated Legendre function, P_ℓ^m, is defined by

$$P_\ell^m(x) = \frac{(1-x^2)^{m/2}}{2^\ell \ell!} \frac{d^{\ell+m}}{dx^{\ell+m}}(x^2-1)^\ell \tag{A.5}$$

with

$$\int_{-1}^{1} P_{\ell'}^m(x) P_\ell^m(x) dx = \frac{2}{2\ell+1} \frac{(\ell-m)!}{(\ell+m)!} \delta \ell' \ell$$

and $P_\ell = P_\ell^0$.

The first few spherical harmonics are given by,

$$Y_0^0 = \sqrt{\tfrac{1}{4\pi}} \qquad Y_1^0 = \sqrt{\tfrac{3}{4\pi}} cos\theta \qquad Y_2^0 = \sqrt{\tfrac{5}{16\pi}}(3cos^2\theta - 1)$$
$$Y_3^0 = \sqrt{\tfrac{7}{16\pi}}(5cos^3\theta - 3cos\theta)$$
$$Y_1^1 = -\sqrt{\tfrac{3}{8\pi}} sin\theta e^{i\phi} \qquad Y_2^1 = -\sqrt{\tfrac{15}{8\pi}} sin\theta cos\theta e^{i\phi}$$
$$Y_3^1 = -\sqrt{\tfrac{21}{64\pi}} sin\theta(5cos^2\theta - 1)e^{i\phi}$$
$$Y_2^2 = \sqrt{\tfrac{15}{32\pi}} sin^2\theta e^{i2\phi} \qquad Y_3^2 = \sqrt{\tfrac{105}{32\pi}} sin^2\theta cos\theta e^{i2\phi}$$
$$Y_3^3 = -\sqrt{\tfrac{35}{64\pi}} sin^3\theta e^{i3\phi}\ .$$

Closure:

$$\sum_{\ell=0}^{(n-1)} \sum_{m=-\ell}^{+\ell} Y_{\ell'}^{*m'}(\theta', \phi') Y_\ell^m(\theta, \phi) = \delta(\Omega' - \Omega) = \frac{\delta(\theta'-\theta)\delta(\phi'-\phi)}{sin\theta} \tag{A.6}$$

and

$$\sum_{m=-\ell}^{+\ell} |Y_\ell^m(\theta, \phi)|^2 = \frac{2\ell+1}{4\pi}\ .$$

Addition theorem:

$$\frac{2\ell+1}{4\pi} P_\ell(cos\alpha) = \sum_{m=-\ell}^{+\ell} Y_\ell^{*m}(\theta_1, \phi_1) Y_\ell^m(\theta_2, \phi_2) \tag{A.7}$$

where α is the angle between (θ_1, ϕ_1) and (θ_2, ϕ_2).

2 Coordinate space

In coordinate space the radial functions, $u_{n\ell}(r)$, of (A.3) are given by

$$u_{n\ell}(r) = Z^{3/2} N_{n\ell} F_{n\ell}\left(\tfrac{2r}{nZ}\right) \tag{A.8}$$
$$N_{n\ell} = \tfrac{2}{n^2}\sqrt{\tfrac{(n-\ell-1)!}{[(n+\ell)!]^3}}$$
$$F_{n\ell}(x) = x^\ell e^{-x/2} L_{n-\ell-1}^{2\ell+1}(x)$$

where

$$\begin{aligned} L_p^k(x) &= \frac{[(p+k)!)^2]}{p!k!}\,{}_1F_1(-p, k+1; x) \\ &= \sum_{s=0}^{p}(-1)^s \frac{(p+k)!]^2}{(p-s)!(k+s)!s!}x^s \end{aligned} \tag{A.9}$$

is the Laguerre Polynomial and ${}_1F_1(a, c; x)$ is the confluent hypergeometric function [Abramowitz and Stegun, 1964].

The first few of the $u_{n\ell}(r)$ are

$$u_{10} = u_{1s} = Z^{3/2} 2 e^{-Zr}$$
$$u_{20} = u_{2s} = Z^{3/2} 2 \tfrac{\sqrt{2}}{2}\left(1 - \tfrac{1}{2}Zr\right) e^{-Zr/2}$$
$$u_{21} = u_{2p} = Z^{3/2} 2 \tfrac{\sqrt{6}}{12}(Zr)\, e^{-Zr/2}$$
$$u_{30} = u_{3s} = Z^{3/2} 2 \tfrac{2\sqrt{3}}{9}\left(1 - \tfrac{2}{3}Zr + \tfrac{2}{27}(Zr)^2\right) e^{-Zr/3}$$
$$u_{31} = u_{3p} = Z^{3/2} \tfrac{8\sqrt{6}}{27}\left(Zr - \tfrac{1}{6}(Zr)^2\right) e^{-Zr/3}$$
$$u_{32} = u_{3p} = Z^{3/2} \tfrac{2\sqrt{30}}{955}(Zr)^3 e^{-Zr/3}$$

Coulomb waves. The long range $1/r$ behavior of the Coulomb potential influences both the long range phase and the normalization of the the wavefunction for unbound electrons. The Coulomb continuum wavefunction may be generated from (A.9) by analytic continuation of the principal quantum number, n, past infinity by taking $n \to ik$, where k is the magnitude of the momentum of the unbound electron. Using $\eta = Z/k$ the resulting two body Coulomb wavefunction with outgoing boundary conditions may be expressed [McDowell and Coleman, 1970],

$$|\vec{k}> = \psi_{2C}^+(\vec{r}) = e^{-\pi\eta/2}\Gamma(1+i\eta)\,{}_1F_1(-i\eta, 1; i(kr - \vec{k}\cdot\vec{r})\,. \tag{A.10}$$

Here ${}_1F_1$ is the confluent hypergeometric function [Erdelyi *et al.*, 1953].

The conventional normalization is the same as for plane waves, namely,

$$< \vec{k}'|\vec{k} >= (2\pi)^3\delta(\vec{k}' - \vec{k}) \, . \tag{A.11}$$

At $\vec{r} = 0$ the normalization is such that

$$|\psi_{2C}(0)|^2 = |e^{-\pi\eta/2}\Gamma(1+i\eta)|^2 = \frac{2\pi\eta}{e^{2\pi\eta}-1} \, . \tag{A.12}$$

This normalization factor goes to zero as $\eta = Z/k$ goes to infinity.

3 Momentum space

The wavefunction in momentum space may be found from the coordinate space wavefunction by taking a Fourier transform, namely,

$$\psi_{n\ell m}(\vec{p}) = \frac{1}{(2\pi)^{3/2}} \int e^{i\vec{p}\cdot\vec{r}} \psi_{n\ell m}(\vec{r}) d\vec{r} \tag{A.13}$$

In momentum space the $\psi_{n\ell m}(\vec{p})$ obey the Fock sum rule,

$$\sum_{\ell=0}^{n-1} \sum_{m=-\ell}^{+\ell} |\psi_{n\ell m}(\vec{p})|^2 = \frac{8}{\pi^2} n^2 \frac{p_n^5}{(p^2+p_n^2)^4} \tag{A.14}$$

where $p_n = Z/n$ is the velocity of the electron in the nth atomic shell.

The radial wavefunctions defined by (A.2) in momentum space [Bethe and Salpeter, 1957] are,

$$\begin{aligned} u_{n\ell m}(p) = & \left[\frac{2n}{\pi}\frac{(n-\ell-1)!}{(n+\ell)!}\right]^{1/2} 2^{2(\ell+1)} \ell! \frac{p_n^{5/2}}{(p^2+p_n^2)^2} \frac{p^\ell p_n^\ell}{(p^2+p_n^2)^\ell} \\ & \times C_{n-\ell-1}^{\ell+1}\left(\frac{p^2-p_n^2}{p^2+p_n^2}\right) \end{aligned} \tag{A.15}$$

where $C_\mu^\nu(x)$ is the Gegenbauer polynomial [Abramowitz and Stegun, 1964].

The first few momentum space wavefunctions are,

$$u_{10} = u_{1s} = 4\sqrt{\tfrac{2}{\pi}}\tfrac{p_1^{5/2}}{(p^2+p_1^2)^2}$$
$$u_{20} = u_{2s} = 4\sqrt{\tfrac{2}{\pi}}\tfrac{p_2^{5/2}}{(p^2+p_2^2)^2}\ 2\left(\tfrac{p^2-p_2^2}{p^2+p_2^2}\right)$$
$$u_{21} = u_{2p} = 4\sqrt{\tfrac{2}{3\pi}}\ 4\tfrac{p_2^{5/2}}{(p^2+p_2^2)^2}\left(\tfrac{p\,p_2}{p^2+p_2^2}\right)$$
$$u_{30} = u_{3s} = 4\sqrt{\tfrac{2}{\pi}}\tfrac{p_3^{5/2}}{(p^2+p_3^2)^2}\left[4\left(\tfrac{p^2-p_3^2}{p^2+p_3^2}\right)-1\right]$$
$$u_{31} = u_{3p} = \tfrac{8}{\sqrt{\pi}}\tfrac{p_3^{5/2}}{(p^2+p_3^2)^2}\tfrac{p\,p_3}{p^2+p_3^2}\ 4\left(\tfrac{p^2-p_3^2}{p^2+p_3^2}\right)$$
$$u_{32} = u_{3p} = \tfrac{64}{\sqrt{5\pi}}\tfrac{p_3^{5/2}}{(p^2+p_3^2)^2}\left(\tfrac{p\,p_3}{p^2+p_3^2}\right)^2 .$$

It is straightforward to confirm the Fock sum rule of (A.14).

A.2 Expansion of $|\vec{R}-\vec{r}|^{2(i\nu-1)}$

Two useful mathematical identities which are often useful are,

$$\frac{1}{|\vec{R}-\vec{r}|} = \sum_{\ell,m}\frac{4\pi}{2\ell+1}\left(\frac{r_<}{r_>}\right)^\ell \frac{1}{r_>}Y_\ell^{*m}(\hat{R})Y_\ell^m(\hat{r}) \tag{A.16}$$

and the more general identity [Deb *et al.*, 1987],

$$\left[(\vec{R}-\vec{r})^2\right]^{i\nu-1} = \sum_\ell a_\ell(R,r)\sum_m Y_\ell^{*m}(\hat{R})Y_\ell^m(\hat{r}) \tag{A.17}$$

where

$$a_\ell(R.r) = \tfrac{(-1)^\ell 2\pi}{(i\nu)2Rr} \tag{A.18}$$
$$\times\left[(R^2+r^2-2Rr)^{i\nu}\ {}_2F_1\left(-\ell,\ell+1,1+i\nu;\tfrac{2Rr-R^2-r^2}{4Rr}\right)\right.$$
$$\left.+(-1)^{\ell+1}(R^2+r^2+2Rr)^{i\nu}\ {}_2F_1\left(-\ell,\ell+1,1+i\nu;\tfrac{2Rr+R^2+r^2}{4Rr}\right)\right] .$$

Here $i\nu$ is an arbitrary complex number including $i\nu = \frac{1}{2}$ which yields (A.16).

A.3 The Green function

The Green function is the linear machinery that relates a cause $\mathcal{C}$ to an effect $\mathcal{E}$, namely*,

$$\mathcal{E}(t) = \frac{1}{\sqrt{2\pi}} \int_{-\infty}^{\infty} G(t,t')\mathcal{C}(t') \ . \tag{A.19}$$

$G(t,t')$ is the effect (or response) associated with a sharply peaked cause, i.e. $\mathcal{C}(t) = \delta(t-t_0)$. Causality is imposed by the condition that $G(t,t') = 0$ if $t < t'$ since the cause must occur before the effect†. Causality leads to a dispersion relation (often called Kramers-Kronig relations, see (7.18)) between the real and imaginary parts of the Fourier transform of $G(t,t')$. Usually, the principle of reciprocity holds, namely that $G(t,t') = G(t',t)$ so that the effect at t due to a cause at t' is the same as the effect at t' due to the same cause at t. $G(t,t')$ is non-local in time.

If the system is invariant under translation in time‡, then, $G(t,t') = G(t-t')$. Under this condition, one obtains from the Faultung theorem (see (8.16) or Jackson, 1975, section 7.10) for Fourier transforms,

$$\tilde{\mathcal{E}}(\nu) = \tilde{G}(\nu)\tilde{\mathcal{C}}(\nu) \tag{A.20}$$

so that the effect is a local, linear response to the cause in frequency space, where

$$G(t-t') = \frac{1}{\sqrt{2\pi}} \int_{-\infty}^{\infty} d\nu e^{-i\nu(t-t')}\tilde{G}(\nu) \ . \tag{A.21}$$

Alternatively, one may construct[Morse and Feshbach, 1953, chapter 7] a Green function, $\tilde{G}(z,z')$, satisfying,

$$H\tilde{G} - \nu\tilde{G} = \delta(z - z_0) \tag{A.22}$$

from a complete set of eigenfunctions, $f_n(z)$, satisfying,

---

* 'Does a tree falling in a forest make a sound if there is no one to hear it?'. Is an observer necessary for an event to exist? Observation requires a Green function to determine the observation which establishes the existence of the event.

† Causality is applied to both classical and quantum Green functions. However, the Uncertainty principle (which corresponds to the band width theorem classically) leads to effects non-local in time that are inconsistent with causality. This difficulty with quantum causality was addressed by Wigner which suggested that causality might be replaced by the preservation of the normalization of the wavefunction in quantum mechanics [Balescu, 1975].

‡ In the case of a quantum wavefunction, this corresponds to a time independent Hamiltonian H so that $G(t,t') = U(t,t') = e^{iH(t-t')}$.

$$Hf_n - \nu_n f_n = 0 \tag{A.23}$$

subject to $\int f_m^* f_n dz = \delta_{mn}$. Since the f_n forms a complete set, one may set $\tilde{G}(z, z_0) = \sum_n a_n(z) f_n(z_0)$ and quickly find that $a_n(z) = f_n^*(z)/(\nu_n - \nu)$, whence,

$$\tilde{G}(z, z_0) = \sum_n \frac{f_n^*(z) f_n(z_0)}{\nu_n - \nu} . \tag{A.24}$$

This form of the Green function is a sum over terms which are singular at frequencies, ν_n which occur when the system is driven at one of its resonant frequencies. Causality is imposed by moving the poles of these resonances below (or above) the real ν axis corresponding to incoming (or outgoing) boundary conditions [Jackson, section 7.10].

An instructive example is given by an LCR electrical circuit driven at a frequency, ν by an a.c. voltage, $V = V_0 e^{-i2\pi\nu t}$, which is taken as the cause. The effect is the current, $I = I_0 e^{-i2\pi\nu t}$. The equation for this system is,

$$\begin{aligned} V &= L\frac{dI}{dt} + RI + Q/C \qquad \text{(A.25)} \\ &= V_0 e^{-i2\pi\nu t} = -2\pi i L I_0 e^{-i2\pi\nu t} + R I_0 e^{-i2\pi\nu t} + \frac{1}{2\pi\nu} I_0 e^{-i2\pi\nu t} \end{aligned}$$

so that

$$\begin{aligned} \tilde{G}(\nu) &= \frac{\text{effect}}{\text{cause}} = \frac{I(\nu)}{V(\nu)} \qquad \text{(A.26)} \\ &= \frac{1}{-2\pi i\nu L + R + i/(2\pi\nu C)} = \frac{\nu}{R}\left(\frac{1}{\nu - \nu_+} - \frac{1}{\nu - \nu_-}\right) \end{aligned}$$

where $\nu_\pm = \frac{i}{4\pi L}(-R \pm \sqrt{R^2 - 4L/c})$ are the resonant frequencies of the system. Here $\tilde{G}$ is the reciprocal of the impedance of the system (with a factor of i to properly shift the phase by $\pi/2$). A similar example is discussed in detail in section 7.10 of Jackson [1975]. Other examples in both classical and quantum systems are given in standard texts [e.g. Morse and Feshbach, 1953; Jackson, 1975; Goldberger and Watson, 1964].

A.4 Equivalence of observables in the particle and wave pictures

Particle picture. In b space the projectile is regarded as a point particle with a well defined impact parameter, $\vec{b}$. The transition probability is the square of the transition probability amplitude a defined by

$$
\begin{aligned}
a(\vec{b},t) &= e^{iE_f t} < f|\psi_i(t)> e^{-iE_i t} \\
&= e^{-i\Delta E t} < f|U(t,-\infty)|i> \\
&= e^{-i\Delta E t} < f|Te^{-i\int_{-\infty}^{t} V(t')dt'}|i> .
\end{aligned}
\tag{A.27}
$$

This corresponds to the projection of the normalized state vector, $\psi(t)$, onto a particular final state, $|f>$.

Wave picture. In q space the projectile is regarded as a plane wave with a well defined momentum transfer, $\vec{q} = \vec{k}_f - \vec{k}_i$. The differential cross section is the square of the amplitude, $f(\vec{q})$, of the outgoing spherical wave, defined by [McDowell and Coleman, 1970],

$$
\begin{aligned}
f(\vec{q}) &= \frac{\mu}{2\pi}\int e^{i\vec{q}\cdot\vec{R}} < f|V|\psi_i> d^3R \\
&= \frac{\mu}{2\pi}\int e^{i(\vec{q}_\perp\cdot\vec{b}+q_\parallel Z)} < f|V|\psi_i> d^2\vec{b}dz
\end{aligned}
\tag{A.28}
$$

where $\mu = k_i/\mathrm{v}$ is the reduced mass and $\vec{R} = \vec{b} + z$ is the coordinate of the projectile with $z = \mathrm{v}t$.

The probability amplitudes and the scattering amplitude are related by a Fourier transform. To show this we first develop a useful identity from (A.28) , namely,

$$
\begin{aligned}
a(\vec{b},t) &= e^{-i\Delta E t} < f|Te^{-i\int_{-\infty}^{t} V(t')dt'}|i> \\
&= \int_{-\infty}^{t} \frac{d}{dt'} e^{-i\Delta E t'} < f|Te^{i\int_{-\infty}^{t'} V(t'')dt''}|i> \\
&= -i\int_{-\infty}^{t} e^{-i\Delta E t'} < f|V(t')Te^{i\int_{-\infty}^{t'} V(t'')dt''}|i> \\
&= -i\int_{-\infty}^{t} e^{-i\Delta E t'} < f|V|\psi_i(t')> dt' .
\end{aligned}
\tag{A.29}
$$

Using (A.30), $z = \mathrm{v}t$ and $q_\| = \Delta E/\mathrm{v}$, now consider the Fourier transform of $a(\vec{b})$ (where the $\lim_{t\to\infty}$ is now implied), namely,

$$
\begin{aligned}
\tilde{f}(\vec{q}) &= \frac{1}{(2\pi)^2}\int e^{i\vec{q}_\perp\cdot\vec{b}}a(\vec{b})d\vec{b} \qquad (A.30)\\
&= -i\frac{1}{(2\pi)^2}\int\int e^{i\vec{q}_\perp\cdot\vec{b}}e^{-i\Delta Et'} < f|V|\psi_i(t') > d\vec{b}dt'\\
&= -i\frac{1}{(2\pi)^2 v}\int e^{i(\vec{q}_\perp\cdot\vec{b}+q_\parallel Z)} < f|V|\psi_i(t') > d\vec{b}dz\\
&= -i\frac{1}{(2\pi)^2 v}\int e^{\vec{q}\cdot\vec{R}} < f|V|\psi_i > d^3R\\
&= \frac{-i}{2\pi k_i}f(\vec{q})\,.
\end{aligned}
$$

Thus the Fourier transform of $a(\vec{b})$ is equal to the scattering amplitude, $f(\vec{q})$, defined by (A.29), multiplied by an overall factor of $-i/2\pi k_i$.

The total cross section, σ, is the same in both pictures. Using the definition of σ in the wave picture McDowell and Coleman, 1970], and $d\vec{q}_\perp = qdqd\phi_q$ with $-2qdq = k_i k_f d(cos\theta)$ from $q^2 = (\vec{k}_f - \vec{k}_i)^2$, where $\phi_b = \phi_q = \phi$, and $d\Omega = sin\theta d\theta d\phi$, then (in units of a_0^2),

$$
\begin{aligned}
\sigma &\equiv \frac{k_f}{k_i}\int |f(\theta,\phi)|^2 d\Omega \qquad (A.31)\\
&= \frac{1}{k_i^2}\int |f(\vec{q})|^2 d\vec{q}_\perp = \frac{1}{k_i^2}\int |\left(\frac{2\pi k_i}{-i}\right)\tilde{f}(\vec{q})|^2 d\vec{q}_\perp\\
&= (2\pi)^2\int\left|\frac{1}{(2\pi)^2}\int e^{i\vec{q}_\perp\cdot\vec{b}}a(\vec{b})d\vec{b}\right|^2 d\vec{q}_\perp\\
&= \frac{1}{(2\pi)^2}\int\int e^{i\vec{q}_\perp\cdot\vec{b}}a(\vec{b})d\vec{b}\int e^{-i\vec{q}_\perp\cdot\vec{b}'}a^*(\vec{b}')\,d\vec{b}'d\vec{b}d\vec{q}_\perp\\
&= \frac{1}{(2\pi)^2}\int\int\int a(\vec{b})a^*(\vec{b}')e^{i\vec{q}_\perp\cdot(\vec{b}'-\vec{b})}\,d\vec{b}'d\vec{b}d\vec{q}_\perp\\
&= \int\int a(\vec{b})a^*(\vec{b}')(2\pi)^2\delta(\vec{b}'-\vec{b})\,d\vec{b}'d\vec{b}\\
&= \int |a(\vec{b})|^2 d\vec{b}\,.
\end{aligned}
$$

A more thorough discussion of the above results is given by McCarroll and Salin [1966] (see McDowell and Coleman [1970]).

Also, using $k_i \simeq k_f$, the differential cross sections are related by,

$$
|a(\vec{b})|^2 d\vec{b} = |f(\theta,\phi)|^2 d\Omega = |f(\vec{q})|^2 d\vec{q}_\perp \qquad (A.32)
$$

consistent with section 1.2.

Finally it is noted that the above relationships hold in each order in perturbation theory.

A.5 Binomial and multinomial moments

The binomial distribution may be generated by,

$$(1)^N = (P + (1-P))^N = \sum_n \binom{N}{n} P^n (1-P)^{N-n} \tag{A.33}$$

so that

$$\sum_n \binom{N}{n} P^n (1-P)^{N-n} = 1\,. \tag{A.34}$$

The average value of an observable, $\mathcal{O}$ is, by definition, given by

$$< \mathcal{O} >= \sum_n \mathcal{O} \binom{N}{n} P^n (1-P)^{N-n}\,. \tag{A.35}$$

Since P is independent of n, it follows immediately that,

$$< P >= P\,. \tag{A.36}$$

Also, it is useful to note that

$$< n >= \sum_{n=0}^{N} n \binom{N}{n} P^n (1-P)^{N-n} = NP \tag{A.37}$$

and

$$\Delta n^2 =< n^2 > - < n >^2 = NP(1-P)\,. \tag{A.38}$$

The first of the two equations immediately above may be proved as follows. Note that

$$\begin{aligned} n \binom{N}{n} &= nN!/(N-n)!n! \qquad\qquad (A.39) \\ &= N(N-1)!/(N-1-(n-1))!(n-1)! = N \binom{N-1}{n-1}\,. \end{aligned}$$

The sum now runs from $n = 1$ to $n = N + 1$. Then,

$$
\begin{aligned}
<n> &= N \sum_{n=1}^{N+1} \binom{N-1}{n-1} P^n (1-P)^{N-n} && \text{(A.40)} \\
&= N \sum_{n'=0}^{N'} \binom{N'}{n'} P^{n'+1} (1-P)^{N'-n'} = NP \,.
\end{aligned}
$$

The second expression similarly follows from

$$
<n^2> - <n>^2 = <n(n-1)> + <n> - <n>^2 \,.
$$

The general moments may be found from

$$
<n><n-1> \ldots . <n-j> = (N)(N-1)\ldots(N-j)P^{j+1} \,.
$$

It follows that rms width, Δn, of a binomial distribution is constrained by its average value, $<n>$, i.e., $\Delta n^2 = <n> (1 - \frac{<n>}{N})$.

For atoms with two shells denoted by I and V,

$$
\begin{aligned}
P_{nn'}^{N_I N_V} &= P_n^{N_I} P_{n'}^{N_V} && \text{(A.41)} \\
&= \binom{N_I}{n} P_I^n (1-P_I)^{N_I-n} \binom{N_V}{n'} P_V^{n'} (1-P_V)^{N_V-n'} \,.
\end{aligned}
$$

In this case the inclusive probability for having a transition in either shell I or shell V is,

$$
\begin{aligned}
<P> &= <P_I + P_V> \\
&= \sum_n P_I \binom{N_I}{n} P_I^n (1-P_I)^{N_I-n} \sum_{n'} \binom{N_V}{n'} P_V^{n'} (1-P_V)^{N_V-n'} \\
&+ \sum_n \binom{N_I}{n} P_I^n (1-P_I)^{N_I-n} \sum_{n'} P_V \binom{N_V}{n'} P_V^{n'} (1-P_V)^{N_V-n'} \\
&= P_I + P_V \,. && \text{(A.42)}
\end{aligned}
$$

Similarly, the corresponding average energy of these transitions is,

$$
E_j = <n> E_I + <n'> E_V = N_I P_I E_I + N_V P_V E_V \,. \qquad \text{(A.43)}
$$

Historical note. The useful binomial expansion $(P + Q)^N$ may have been writen in stone for $N = 2$ by the Babylonians about 1500 bc. The 3-4-5 right triangle, a special case for $N = 2$, was used before that by the

Egyptians to establish accurate right angles in the construction of pyramids. The binomial coeficients for $N = 2$ were known to Euclid about 300 bc. About 1100 ad Omar Khyyam asserted he could do the binomial expansion for $N = 4, 5, 6$ and higher. Pascal's Triangle which gives the binomial coefficients was given in part by Chu Shï-kié in 1303. The expansion for $N \leq 17$ was done by Stifel [1486/7 - 1567]. The expansion for positive integer N was given by Pascal in 1654 from correspondence with Fermat. The generalization to negative and non-integer numbers was done later by Newton and others [Smith, 1925, Cantor, 1898].

A.6 N body systems

The N body problem in atomic systems may be characterized as follows:

1. Initial states of N electron targets. Effects of more than one electron on the initial state are discussed briefly in section 6.2.1.

2. Final states of N electron targets. Many electron effects on the final state in single electron transitions are discussed briefly in sections 6.2.1, 2.5 and 8.1. Multiple electron excitation, ionization, transfer and combinations thereof, where more than one electron goes to a different final state in an atomic interaction, are discussed throughout this book.

3. Multiple atomic centers in targets. This topic is developed in section 4.2 and by Heitler (pp.192-4).

4. Composite projectiles. Projectiles with internal atomic structure are discussed in chapter 8. There are limits in which the composite particles of the projectile are coherent and limits which are incoherent. Application of an independent particle model for projectiles with multiple atomic centers interacting with targets with multiple atomic centers may be developed using section 4.2.

5. Initial state photons. Absorption of N photons in a strong laser field is discussed briefly in section 9.5.

6. Final state photons. Photon showers are discussed by Heitler (pp. 224-231). Except for the infrared divergence, the probability goes as α^N for an N photon final state.

References

*** A ***

Aberg, T. (1967). *Phys. Rev.* **156**, 35.
Aberg, T. (1976). *Photoionization and Other Probes of Many Electron Interactions* (F. Wuilleumier, ed.), p. 49. Plenum Press, New York.
Aberg, T. and Goscinski, O. (1981). *Phys. Rev.* **A24**, 801.
Aberg, T. and Howat, G. (1982). *Handbuch der Physik* **XXXI**, 469.
Aberg, T., Blomberg, A., Tulkki, J. and Goscinski, O. (1984). *Phys. Rev. Lett.* **52**, 1207.
Aberg, T. and Tulkki, J. (1985). *Atomic Inner Shell Processes* (B. Crasemann, ed.), Chapter 10.
Aberg, T., Blomberg, A. and MacAdam, K.B. (1987). *J. Phys. B.* **20**, 4795.
Aberg, T. (1987). *Nucl. Instr. Meth.* **A262**, 1.
Aberg, T. (1990). *Atlanta Proceedings* (1973), (R.W. Fink *et. al.*, eds.), *U.S. At. Eng. Com. Conf.* No. 720404, 1509.
Aberg, T. (1996). *Raman Emission by X-rays* World Scientific, NY.
Abramowitz, M. and Stegun, I.A. (1964). *Handbook of Mathematical Functions.* National Bureau of Standards, Washington, D.C.
Alston, S. and Macek, J. (1982). *Phys. Rev.* (SPB).
Alston, S. (1990). *Phys. Rev.* **A42**,331.
Alt, E.O. and Mukhamedzhanov, A.M. (1993). *Phys. Rev.* **A47**, 2004.
Altick, P.L. (1983). *J. Phys. B.* **16**, 3543.
Amrien, W.O., Martin, P.A. and Misra, B. (1970). *Helv. Phys. Acta* **17**, 589.
Amusia, M.Y., Drukarev, E.G., Gorshkov, V.G. and Kazachkov, M.D. (1975). *J. Phys. B.***8**, 1248.
Amusia, M.Y. (1996). *Atomic, Molecular and Optical Physics Reference Book* (G.W.F. Drake, ed.), AIP Press, NY, chapter 20.
Andersen, L.H., Hvelplund, P., Knudsen, H., Moller, H.P., Elsner, K., Rensfelt, K.G. and Uggerhoj, E. (1986). *Phys. Rev. Lett.* **57**, 2147.
Andersen, L.H., Hvelplund, P., Knudsen, H., Moller, S.P., Sorensen, A.H., Elsner, K., Rensfeld, K.G. and Uggerhoj, E. (1987). *Phys. Rev.* **A36**, 3612.
Andersen, L.H., Nielsen, L.B. and Sorensen, J. (1988). *J. Phys. B.* **21**, 1587.

Andersen, L.H., Hvelplund, P., Knudsen, H., Moeller, S.P., Pedersen, J.O.P., Uggerhoj, E., Elsner, M. and Morenzoni, E. (1989). *Phys. Rev. Lett.* **62**, 1731.
Andersen, L.H., Hvelplund, P., Knudsen, H., Moller, S.P., Pedersen, J.O.P., Sorensen, A.H., and Uggerhoj, E. (1990). *Phys. Rev. Lett.* **65**, 1687.
Andersson, L.R. and Burgdörfer, J. (1993). *Phys. Rev. Lett.* **71**, 50.
Andersson, L.R. and Burgdörfer, J. (1994). *Phys. Rev.* **A50**, R2810.
Ansersen, N., Gallagher, J.W. and Hertel, I.V. (1988). *Phys. Reports* **165**, 1.
Anholt, R. (1986). *Phys. Lett. A* **114**, 126.
Anholt, R. and Gould, H. (1986). *Adv. At. Mol. and Opt. Phys.* **22**, 315.
Anholt, R., Meyerhof, W.E., Xu, X.Y., Gould, H., Feinberg, B., McDonald, R.J., Wegner, H.E. and Thieberger, P. (1987). *Phys. Rev.* **A36**, 1586.
Anton, J. (1994). *Diplom thesis*, University of Kassel.
Anton, J., Detelffsen, D. and Schartner, K.H. (1993). *J. Phys. B.* **26**, 1.
Arcuni, P.W. and Schneider, D. (1987). *Phys. Rev.* **A36**, 3059.
Au, C.K. and Drachman, Richard, J. (1988). *Phys. Rev.* **A37**, 6333.
Auger, P. (1926). *Ann. Phys.* (Paris) 6, 183.
Augst, S., Strickland, D., Meyerhofer, D., Chin, S.L. and Eberly, J.H. (1989). *Phys. Rev. Lett.* **63**, 2212.
Augst, S., Talebpour, A., Chin, S.L., Beaudoin, Y. and Chaker, M. (1995). *Phys. Rev.* **A52**, R917.
Azuma, Y., Berry, H.G., Gemmel, D.S., Suielman, J., Westerlind, M., Sellin, I.A. and Kirkland, J. (1995). *Phys. Rev.* **A51**, 447.

*** B ***

Balashov, V.V., Lipovetsky, S.S. and Senashenko, V.S. (1973). *Sov. Phys. - JEPT* **36**, 858.
Balescu, R. (1975). *Equilibrium and Non-Equilibrium Statistical Mechanics* John Wiley, NY.
Barat, M. (1988). *Comment At. and Mol. Phys.* **21**, 307.
Barany, A. (1990). *The Physics of Electronic and Atomic Collisions* (A. Dalgarno, R.S. Freund, P.M. Koch, M.S. Lubell and T.B. Lucatorto, eds.), p. 246. AIP Press, NY.
Barkus, W.H., Birnbaum, W. and Smith, F.H. (1956). *Phys. Rev.* **10**, 778.
Barnett, C.F. (1990). *Atomic Data for Fusion*, Controlled Fusion Atomic Data Center, Oak Ridge, TN.
Bartschat, K. (1993). *The Physics of Electronic and Atomic Collisions*, (ed., T. Andersen, B. Fastrup, F. Folkmann, H. Knudsen and N. Andersen, *AIP Conf. Proc.* **295**, 251.
Basbas, G. (1982). *Nucl. Instr. Meth.* **B4**, 227.
Bassel, R.H. and Gerjuoy, E. (1960). *Phys. Rev.* **117**, 749.
Bates, D.R. and Griffing, G. (1953). *Proc. Phys. Soc. (London)* **A66**, 961.
Bates, D.R. (1958). *Proc. Royal Soc.* **A247**, 294.
Batka, J.J. and Berry, R. Stephen (1993). *J. Phys. Chem.* **97**, 2435.
Baym, G. (1974). *Lectures on Quantum Mechanics*, Addison Wesley, NY.
Becker, A. and Faisal, F.H.M. (1996). *J. Phys. B.* **29**, L197.
Becker, R.L., Ford, A.L. and Reading, J.F. (1980). *J. Phys. B.* **13**, 4059.
Becker, R.L., Ford, A.L. and Reading, J.F. (1981). *Phys. Rev.* **A23**, 510.
Becker, R.L. (1988). *High Energy Atomic Collisions* (D. Berenyi and G. Hock, eds.), p. 447. Springer Verlag, Berlin.

Beigmann, I. (1995). *Phys. Reports* **250**, 95.
Belkic, Dz., Gayet, R. and Salin, A. (1979). *Phys. Reports* **56**, 279.
Bell, K.L. and Kingston, A.E. (1994). *Adv. At. Mol. and Opt. Phys.* **32**, 1
Ben-Itzhak, I. and McGuire, J.H. (1988). *Phys. Rev.* **A38**, 6422.
Ben-Itzhak, I., Gray, T.J., Legg, J.C. and McGuire, J.H. (1988). *Phys. Rev.* **A37**, 3685.
Berakdar, J. and Briggs, J.S. (1994). *Phys. Rev. Lett.* **72**, 3799.
Berg, H., Dörner, R., Kelbch, C., Kelbch, S., Ullrich, J., Hagmann, S., Richard,P., Schmidt-Böcking, H., Schlachter, A.S., Prior, M., Crawford, H.J., Engelange, J.M., Flores, J., Loyd, D.H., Pedersen, J. and Olson, R.E. (1988). *J. Phys. B.***21**, 3929.
Bergou, J., Varro, S., and Fedorov, M.V. (1981). *J. Phys. A.* **14**, 2305.
Bergstrom, P.M., Suric, T., Pisk, K. and Pratt, R.H. (1993). *Phys. Rev.* **A48**, 1134.
Bergstrom, P.M., Hino, K. and Macek, J.H. (1995). *Phys. Rev.* **A51**, 3044.
Berry, R. Stephen, and Krause, J.L. (1986). *Phys. Rev.* **A33**, 2865.
Berry, R. Stephen (1989). *Contemporary Physics* **30**, 1.
Berry, R. Stephen (1995). *Structure and Dynamics of Atoms and Molecules* Kluwer Academic Publ., Netherlands, 155.
Bethe, H.A. and Salpeter, E.E. (1957) *Quantum mechanics of one- and two-electron atoms*, (Academic Press, NY).
Bhalla, C.P. and Karim, K.R. (1989). *Phys. Rev.* **A39**, 6060.
Bhalla, C.P. (1990). *Phys. Rev. Lett.* **64**, 1103.
Blomberg, A., Aberg, T. and Goscinski, O. (1986). *J. Phys. B.***19**, 1063.
Bohr, N. (1913). *Phil. Mag.* **26**, 857.
Bohr, N. (1948). *Danske. K. Vidensk. Selsk. Mat. Fys. Meddr.* **18**, No. 8.
Bordenave-Montesquieu, A. and Beniot-Cattin, P. (1971). *Phys. Lett. A.* **37**, 243.
Bordenave-Montesquieu, A., Gliezes, A. and Benoit-Cattin, P. (1982). *Phys. Rev.* **A25**, 245.
Bordenaev-Montesquieu *et al.* (1995). *J. Phys. B.* **28**, 653.
Bottcher, C. (1982). *Phys. Rev. Lett.* **48**, 85.
Bottcher, C. (1985). *Adv. At. Mol. and Opt. Phys.* **20**, 241.
Bottcher, C. (1988). *Adv. At. Mol. and Opt. Phys.* **25**, 303.
Bottcher, C., Bottrell, G.J. and Strayer, M. (1991). *Comput. Phys. Commun.* **63**, 63.
Bottcher, C., Schultz, D.R. and Madison, D.H. (1994). *Phys. Rev.* **A49**, 1714.
Boyle, J.J. and Pindzola, M.S. (1995). *Many-body Atomic Physics*, Cambridge Univ. Press, Cambridge, England.
Braier, P.A. and Berry, R. Stephen (1994). *J. Phys. Chem.* **98**, 3506.
Bransden, B.H. and Joachain, C.J. (1983). *Physics of Molecules and Atoms*, Longman, NY, p. 162.
Bransden, B.H. and Dewangan, D.P. (1988). *Adv. At. Mol. and Opt. Phys.* **25**, 343.
Bransden, B.H. and McDowell, M.R.C. (1992). *Charge Exchange and the Theory of Ion-Atom Collisions*, Clardon Press, Oxford.
Brandt, W. (1983). *Phys. Rev.* **A27**, 1314.
Brauner, M., Briggs, J.S. and Klar, H. (1989). *J. Phys. B.***22**, 2265.
Bray, I., Konovalov, D.A., McCarthy, I.E. and Stelbovics, A.T. (1994). *Phys. Rev.* **A50**, R2818.
Breinig, M. and Sellin, I. (1981). *Bull. Am. Phys. Soc.* **76**, 601.
Briggs, J.S. and Dettman, K. (1974). *Phys. Rev. Lett.* **33**, 1123.

Briggs, J.S. and Roberts (1974). *J. Phys. B.*7, 1370.
Briggs, J.S. (1977). *Phys. Rev.* **A10**, 3075.
Briggs, J.S. and Taulbjerg, K. (1978). in *Topics in Current Physics* (I. Sellin, ed.), Vol. 5, p. 105. Springer Verlag, NY.
Briggs, J.S. and Taulbjerg, K. (1979). *J. Phys. B.* **12**, 2565.
Briggs, J.S., Macek, J.H. and Taulbjerg, K. (1982). *Comm. in At. and Mol. Phys.* **12**,1.
Briggs, J.S. (1988). in *Electronic and Atomic Collisions* (H.B. Gilbody, W.R. Newell, F.H. Read and A.C.H. Smith, eds.), p. 13. Elsevier Science Publishers, NY.
Briggs, J.S. (1989). *Comments At. Mol. Phys.* **23**, 155.
Briggs, J.S. and Macek, J. (1990). *Adv. in At. Mol. and Opt. Phys.* **29**, 1.
Briggs, J.S. (1993). *The Physics of Electronic and Atomic Collisions*, (ed., T. Andersen, B. Fastrup, F. Folkmann, H. Knudsen and N. Andersen, *AIP Conf. Proc.* **295**, 221.
Brinkman, H.C. and Kramers, H.A. (1930). *Proc. Acad. Sci. Amsterdam* **33**, 973.
Brion, C. (1995). private communication.
Bruch, R., Schneider, D., Schwartz, W.H.E., Meinhart, M., Johnson, B.M., and Taulbjerg, K. (1979). *Phys. Rev.* **A19**, 587.
Bruch, R., Altick, P.L., Träbert, E. and Heckmann, P.H. (1984). *J. Phys. B.* **17**, L655.
Bruch, R., Kocbach, L., Träbert, E., Heckmann, P.H., Raith, B. and Will, U. (1985). *Nucl. Instr. Meth.* **B9**, 438.
Bruch, R., Biegman, I.L., Rauscher, E.A., Fuelling, S., McGuire, J.H., Träbert, A.E., Heckmann, P.H. (1993). *J. Phys. B.* **26**, L413.
Bruckner, K.A. (1955). *Phys. Rev.* **100**, 36.
Bruckner, K.A. (1959). *The many body problem, les houches session.* John Wiley, NY, 47.
Burgdörfer, J., Andersson, L.R., McGuire, J.H. and Ishihara, T. (1994). *Phys. Rev.* **A50**, 349.
Burgdörfer, J. (1984). in *Forward Electron Ejection in Ion Collisions (Lecture Notes in Physics)* Springer, Berlin, 32.
Burke, K., Perdew, J.P. and Langreth, D.C. (1994). *Phys. Rev. Lett.* **73**, 1283.
Burke, K., Angulo, J.C. and Perdew (1994a). *Phys. Rev.* **A50**, 297.
Burke, K., Perdew, J.P. and Levy, M. (1995). in *Modern Density Functional Theory: A Tool for Chemistry*, ed. Seminario and Politzer, P. Eslevier, Amsterdam.
Burke, P.G. and Robb, W.P. (1975). *Adv. in At. Mol. and Opt. Phys.* **11**, 144.
Burke, P.G. (1996). *Atomic, Molecular and Optical Physics Reference Book* (G.W.F. Drake, ed.), AIP Press, NY, chapter 32.
Byron, F.W. and Joachain, C.J. (1967). *Phys. Rev.* **164**, 1.

*** C ***

Caldwell, C.D. and Krause, M.O. (1996). *Atomic, Molecular and Optical Physics Reference Book* (G.W.F. Drake, ed.), AIP Press, NY, chapter 51.
Cantor, M. (1898). *Geschichte der Mathematik*, B.G.Teubner, Leipzig.
Cara ,P. (1996). *Raman Emission by X-rays*, World Scientific, NY.
Carlson, T.A. (1967). *Phys. Rev.* **156**, 142.

Carlson, T.A., Nestor, C.W., Tucker, T.C. and Malik, F.B. (1968). *Phys. Rev.* **169**, 127.
Carter, S.L. and Kelly, H.P. (1981). *Phys. Rev.* **24**, 170.
Cavagnero, M.J. (1995). *Phys. Rev.* **A52**, 2865.
Ceraulo, Sandra C., Stehman, R.M., and Berry, R. Stephan (1994). *Phys. Rev.* **A49**, 1730.
Chakravorty, S.J., Gwaltney, S.R., Davidson, E.R., Parpia, F.A., and Froese Fischer, C. (1993). *Phys. Rev.* **A47**, 3649.
Charlton, M., Andersen, L.H., Brun-Nielsen, L., Deutch, B.I., Hvelplund, P., Jacobsen, F.M., Knudsen, H., Lariccha, G., Poulsen, M.R. and Pedersen, J.O. (1988). *J. Phys. B.* **21**, L545.
Chase, L.L., Peyghambarian, N., and Mysyrowicz, A. (1983). *Phys. Rev.* **B27**, 2325.
Chen, J.C.Y. and Chen, A.C. (1972). *Adv. in At. and Mol. Phys.* **8**, 71.
Cheshire, I.M. (1964). *Proc. Phys. Soc. (London)* **84**, 89.
Cheshire, I.M. (1965). *Phys. Rev.* **138**, A 992.
Cheshire, I.M. (1967). *Proc. Phys. Soc. (London)* **92**, 862.
Christensen-Dalsgaard, B. (1988). *J. Phys. B.* **21**, 2539.
Chu, S.I. and Cooper, J. (1985). *Phys. Rev.* **A32**, 2769.
Cocke, C.L. (1979). *Phys. Rev.* **A20**, 749.
Cocke, C.L. (1992). *Invited talks of ICPEAC XVII*, (Brisbane, Australia), 49.
Cocke, C.L. (1996). *Atomic, Molecular and Optical Physics Reference Book* (G.W.F. Drake, ed.), AIP Press, NY, chapter 55.
Coleman, J.P. (1968). *J. Phys. B.* **2**, 567.
Collins, L.A. and Schneider, B.I. (1988). in *Electronic and Atomic Collisions* (H.B. Gilbody, W.R. Newell, F.H. Read, A.C.H. Smith, eds.) 57. Elsevier Science Publishers BV.
Cook, C.J., Smith, N.R.A. and Heinz, O. (1975). *J. Chem. Phys.* **63**, 1218.
Cooper, J.W. (1993). *Phys. Rev.* **A47**, 1841.
Corkum, P.B. (1993). *Phys. Rev. Lett.* **71**, 1994.
Cowan, P.L. (1994). in *Resonant Anomalous X-Ray Scattering: Theory and Applications* (G.Materlick, C.J.Sparks and K. Fischer, eds.) North-Holland, Amsterdam.
Cowan, R.D. (1981). *Theory of Atomic Structure and Spectra* University of California Press, Berkeley.
Crasemann, B. (1975). *Inner-Shell Processes.* Academic Press, NY.
Crasemann, B. (1992). *Invited Talks of ICPEAC XVII*, (Brisbane, Australia), 69.
Crasemann, B. (1996). *Atomic, Molecular and Optical Physics Reference Book* (G.W.F. Drake, ed.), AIP Press, NY, chapter 52.
Cromer, D.T. and Mann, J.B. (1967). *J. Chem.* Phys. **27**, 1892.
Crooks, G.B. and Rudd, E. (1970). *Phys. Rev. Lett.* **25**, 1599.
Crothers, D.S.F. and Dunseath, K.M. (1990). *J. Phys. B.* **23**, L365.
Crothers, D.S.F. (1991). *Invited talk at the Intl. Sympos. on Ion Atom Collisions*, Gold Coast, Australia
Crothers, D.S.F. and Dube, L.J. (1992). *Adv. in At. Mol. and Opt. Phys.* **30**, 287.
Crothers, D.S.F. (1996). *Atomic, Molecular and Optical Physics Reference Book* (G.W.F. Drake, ed.), AIP Press, NY, chapter 43.
Crowe, A. (1987). *Adv. in At. Mol. and Opt. Phys.* **24**, 269.
Cvejanovic, S. and Read, F.H. (1974). *J. Phys. B.* **10**, 1841.

*** D ***

Dalgarno, A., and Sadeghpour, H., (1992). *Phys. Rev.* **A46**, 3591.
Dalgarno, A. (1994). *Abstracts of the Department of Energy Annual Workshop.*
Datz, S., Hippler, R., Anderson, L.H., Dittner, P.F., Knudsen, H., Krause, H.F., Miller, P.D., Pepmiller, P.L., Rassel, T., Stolterfoht, N., Yamazaki, Y. and Vane, C.R. (1987). *Nucl. Instr. Meth.* **A262**, 62.
Datz, S., Hippler, R., Anderson, L.H., Dittner, P.F., Knudsen, H., Krause, H.F., Miller, P.D., Pepmiller, P.L., Rassel, T., Schuch, R., Stolterfoht, N., Yamazaki, Y. and Vane, C.R. (1990). *Phys. Rev.* **A41**, 3559.
Day, M.H. (1981). *J. Phys. B.* **14**, 231.
Davidson, E.R. (1976). *Reduced Density Matricies in Quantum Chemistry* Academic Press, NY.
Deb, N.C., Sil, N.C. and McGuire, J.H. (1987). *J. Phys. B.* **20**, 443.
Deb, N.C.and Sil, N.C. (1996). *J. Phys. B.*, **29**, 741.
Deitrich, P., Burnett, N.H., Ivanov, M. and Corkum, P.B. (1994). *Phys. Rev.* **A50**, R3585.
Delos, J.B. and Thorson, W.R. (1972). *Phys. Rev.* **A6**, 709.
DePaola, B.D. (1991). *Nucl. Instr. Meth.* **B56/57**, 154.
Derrida, J. (1976). *Of Grammatology* Johns Hopkins University Press.
Derrida, J. (1988). *Limited, Inc.* Northwestern University Press.
Desjardins, S.J., Bawagan, A.D.O., Liu, Z.F., Tan, K.H., Wang, Y and Davidson, E.R. (1995). *J. Chem. Phys.* **102**, 6385, and references therein.
Detleffsen, D., Anton, M., Werner, A. and Schartner, K.H. (1995). *J. Phys. B.*, in press.
Dettmann, K., Harrison, K.G. and Lucas, M.W. (1974). *J. Phys. B.* **7**, 269.
Devi, K.R. and Garcia, J.D. (1984). *J. Phys. B.* **18**, 4589.
Dewangen, D.P. and Walters, H.R.J. (1979). *J. Phys. B.* **11**, 3983.
Dollard, J.D. (1964). *J. Math Phys.* **5**, 729.
Dörner, R., Vogt, T., Mergel, V., Khemliche, H., Kravis, S., Cocke, C.L., Ullrich. J., Unverzagt, M., Spielberger, L., Damrau, M., Jagutzki, O., Ali, I., Weaver, B., Ullman, K., Hsu. C.C., Jung, M., Kanter, E.P., Sonntag, B., Prior, M., Rotenberg, E., Denlinger, J., Warwick, T., Manson, S.T. and Schmidt-Böcking, H. B. (1996). *Phys. Rev. Lett.* **76**, 2654.
Drachman, R.J., Bhatia, A.K. and Shabazz, A.A. (1990). *Phys. Rev.* **A42**, 6333.
Dreizler, R.M. and Gross, E.K.U. (1990). *Density Functional Theory.* Springer-Verlag, Berlin.
Dreizler, R.M., Ast, H.J., Henne, A., Lüdde, H.J. and Starg, C. (1990). in *The Physics of Electronic and Atomic Collisions* (A. Dalgarno, R.S. Freund, P.M. Koch, M.S. Lubell and T.B. Lucatorto, eds.), p. 258. AIP, NY.
DuBois, R.D. and Manson, S.T. (1987). *Phys. Rev.* **A35**, 2007.
DuBois, R.D. and Kover, A. (1989). *Phys. Rev.* **A40**, 3605.
Duguet, A ., Duprë, C. and Lahmam-Bennani, A. (1991). *J. Phys. B.* **24**, 675.
Duke, D.W. and Owens, R. (1984). *Phys. Rev.* **D30**, 49.
Dunn, G.H. and Kieffer, L.J. (1963). *Phys. Rev.* **132**, 2109.
Duncan, M.M. and Menendez, M.G. (1977). *Phys. Rev.* **A16**, 1799.

*** E ***

Eadie, W.T., Drijard, D., James, F.E., Roos, M., and Soudoulet, B. (1971). *Statistical Methods in Experimental Physics.* North Holland, Amsterdam.
Eberly, J.H., Javanainen, J. and Rzazewski,K. (1991). *Phys. Reports* **204**, 5.
Eckhardt, M. and Schartner, K.H. (1983). *Z. Physik* **A312**, 321.
Ederer, D.L. and McGuire, J.H. (1996). *Raman Emission by X-rays,* World Scientific, NY.
Edwards, A.K., Wood, R.M. and Ezell, R.L. (1986). *Phys. Rev.* **A34**, 4411.
Edwards, A.K., Wood, R.M., Beard, A.S., Ezell, R.L. (1988). *Phys. Rev.* **A37**, 3697.
Eichler, J. (1977). *Phys. Rev.* **A15**, 1856.
Eichler, J. and Chan, F.T. (1979). *Phys. Rev.* **A20**, 104.
Erdelyi, A., Magnus, F., Oberhettinger, F. and Tricomi, F.G. (1953). *Higher Transcendental Functions.* McGraw Hill, NY.
Ericson, T. (1963). *Annals of Physics* **23**, 390.
Esry, B. and Greene, C.H. (1994). *Contributed Papers of ICAP XIV,* (Boulder, CO), 1C-5.

*** F ***

Fano, U. (1961). *Phys. Rev.* **A124**, 1866.
Faisal, F.H.M. (1973). *J. Phys. B.* **6**, L89.
Faisal, F.H.M. (1987). *Theory of multiphoton processes.* Plenum Press, NY.
Faisal, F.H.M. (1994). *Phys. Let. A* **187**, 180.
Faisal, F.H.M., Dimou, L., Stiemke, H.J. and Nurhuda, M. (1995). *Jour. Nonlinear Opt. Phys. and Materials,* **4**, No. 3, 701.
Feagin, J.M. (1984). *J. Phys. B.* **17**, 2433.
Feagin, J.M., Briggs, J. and Reeves, T.M. (1984). *J. Phys. B.* **17**, 1057.
Fetter, A.L. and Walecka, J.D. (1971). *Quantum Theory of Many-Particles Systems.* McGraw Hill, San Francisco, CA.
Fink, R.W., Manson, S.T., Palms, J.M. and Rao, P.V. (1973). *Proc. Intl. Conf. Inner Shell Ionz. Phenomena and Future Appl.*, Altanta, U.S. Atomic Energy Comm. Rep. No. CONF-720404, Oak Ridge, TN.
Filippi, C., Umrizer, C.J. and Taut, M. (1994). *J. Chem. Phys.* **100**, 1290.
Fish, S. (1982). *Is there a text in this class?* Harvard University Press.
Fisher, V., Ralchenko, Yu, Goldgirsh, A., Fisher, D., and Maron, Y. (1995). *J. Phys. B.* **28**, 3027.
Fittinghoff, D.N., Bolton, P.R., Chang, B. and Kulander, K.C. (1992). *Phys. Rev. Lett.* **69**, 2642.
Fogel, Ya.M., Mitin, R.V., Kozlov, V.F. and Romashko, N.D. (1959). *Sov. Phys.-JEPT* **8**, 390.
Ford, A.L., Reading, J.F. and Becker, R.L. (1979). *J. Phys. B.* **12**, 2905.
Ford, A.L., Reading, J.F. and Becker, R.L. (1982). *J. Phys. B.* **15**, 3257.
Ford, A.L. and Reading, J.F. (1989). *Nucl. Instr. Meth.* **B40/41**, 362.
Ford, A.L. and Reading, J.F. (1994). *J. Phys. B.* **27**, 4215.
Ford, A.L. and Reading, J.F. (1996). *Atomic, Molecular and Optical Physics Reference Book,* (G.W.F. Drake, ed.), AIP Press, NY, chapter 42.
Forrey, R.C., Sadeghpour, H.R., Baker, J.D., Morgan, J.D. and Dalgarno, A. (1995). *Phys. Rev.* A **51**, 2112.

Fortin, E., Fafared, S. and Mysyrowicz (1993). *Phys. Rev. Lett.* **70**, 3951.
Franz, A. and Altick, P.L. (1995). *J. Phys. B.* **28**, 4639.
Fritsch, W. and Lin, C.D. (1983). *Phys. Rev.* **A27**, 3361.
Fritsch, W. and Lin, C.D. (1986). *J. Phys. B.* **19**, 2683.
Fritsch, W. and Lin, C.D. (1988). *Phys. Rev. Lett.* **61**, 690.
Fritsch, W. and Lin, C.D. (1990). *Phys. Rev.* **41** 4776.
Fritsch, W. and Lin, C.D. (1991). *Rep. Prog. Phys.* 202, 1 (1991).
Froese Fischer, C. (1977). *The Hartree-Fock Method for Atoms.* John Wiley and Sons, NY.
Froese Fischer, C. (1996). *Atomic, Molecular and Optical Physics Reference Book* (G.W.F. Drake, ed.), AIP Press, NY, chapter 21.
Fulde, P. (1995). *Electron Correlation in Molecules and Solids.* Springer, Berlin.
Fülling, S., Bruch, R., Rauscher, E.A., Neill, P.A., Träbert, E., Heckmann, P.H., and McGuire, J.H. (1992). *Phys. Rev. Lett.* **68**, 3152.

*** G ***

Gale, G. M. and Mysyrowicz, A. (1974). *Phys. Rev. Lett.* **32**, 727.
Gardiner, C.W. (1985). *Handbook of Stochastic Methods.* Springer Verlag, Berlin.
Gasiorowicz, S. (1996). *Quantum Physics.* John Wiley, NY.
Gau, J.N. and Macek, J. (1975). *Phys. Rev.* **A12**, 1760.
Gavrila, M. (1992). *Adv. in At. Mol. Opt. Phys.*, supplement 1.
Gayet, R. and Salin, A. (1990). private communication.
Georgiades, N. Ph., Polzik, E.S., Edamatsu, K., Kimble, H.J. and Parkins, A.S. (1995). *Phys. Rev. Lett.* **75**.
Gerjuoy, E. (1966). *Phys. Rev.* **148**, 54.
Gersdorf, P., John, W., Perdew, J.P. and Ziesche, P. (1996). *Intl. J. Quantum Chem.*
Giese, J.P. and Horsdal, E. (1988). *Phys. Rev. Lett.* **57**, 1414.
Giese, J.P., Schulz, M., Swensen, J.K., Schöne, H., Benhennu, M., Varghese, S.L., Vane, C.R., Dittner, P.F., Shafroth, S.M. and Datz, S. (1990). *Phys. Rev.* **A42**, 1231.
Gilbody, H.B. (1994). *Adv. in At. Mol. Opt. Phys.* **33**, 149.
Gilbody, H.B. (1995). *Invited Talks of ICPEAC XIX*, (Whistler, Canada).
Giusti-Suzor, A. and Zoller, P. (1987). *Phys. Rev.* **36**, 5178.
Glauber, R.J. (1959). in *Lectures in Theoretical Physics.* (W.E. Brittin and L.G.Dunham, eds.), Interscience, NY, **1**, 315.
Godunov, A.L., Mileev, V.N. and Senashenko, V.S. (1983). *Sov. Phys. J. Tech. Phys.* **28**, 236.
Godunov, A.L., Kunikeev, Sh.D., Novikov, N.V. and Senashenko, V.S. (1989). In *Contributed Papers of ICPEAC XVI*, (New York), (A. Dalgarno, R.S. Freund, M.S. Lubell and T.B. Lucatoro, eds.), 502.
Godunov, A.L., Kunikeev, Sh.D., Novikov, N.V. and Senashenko, V.S. (1989a). *Sov. Phys. - JEPT* **69**, 927.
Godunov, A.L., Novikov, N.V. and Shenashenko, V.S. (1992). *J. Phys. B.***25** L43.
Godunov, A.L. and Schipakov, V.A. (1993). *J. Phys. B.* **26**, L811.
Godunov, A.L. and Schipakov, V.A. (1995). *Nucl. Instru. Meth.* **B98**, 354.
Godunov, A.L., Schipakov, V.A. and McGuire, J.H. (1996). preprint.

Goldberger, M.L. and Watson, K.M. (1964). *Collision Theory*, Chapter 11. Wiley, NY.
Golden, J.E. (1975). *Ph.D. thesis*, Kansas State University, Manhattan, KS.
Goldstein, H. (1950). *Classical Mechanics*. Addison-Wesley, Reading, MA.
Goldstone, J. (1957). *Proc. Royal Soc.* **A239**, 237.
Gradshteyn, I.S. and Ryzhik, I.M. (1980). *Table of Integrals, Series and Products*. Academic Press, NY.
Graham, W.G. (1990). in *The Physics of Electronic and Atomic Collisions* (A. Dalgarno, R.S. Freund, P.M. Koch, M.S. Lubell and T.B. Lucatorto, eds.), p. 544. AIP, NY.
Gramlich, K., Grün, N. and Scheid, W. (1986). *J. Phys. B.* **19**, 1457.
Grant, I.P. (1974). *J. Phys. B.* bf 7, 1458.
Gray, L.G. and MacAdam, K.B. (1994). *J. Phys. B.* **14**, 3055.
Gravielle, M.S. and Miraglia, J.E. (1988). *Phys. Rev.* **A38**, 504.
Greene, C.H. and Rau, A.R.P. (1982). *Phys. Rev. Lett.* **26**, 533.
Greene, C.H. and Rau, A.R.P. (1982). *J. Phys. B.* **16**, 99.
Greene, C.H. and Zare, R.N. (1982). *Annu. Rev. Phys Chem.* **33**, 119.
Greene, C.H. and Zare, R.N. (1983). *J. Chem. Phys.* **78**, 6741.
Greene, C.H. and Aymar, M. (1991). *Phys. Rev.* **A44**, 1773.
Greene, C.H. (1995). *Invited Talks of ICPEAC XIX*, (Whistler, Canada).
Grobe, R. and Eberly, J.H. (1993). *Phys. Rev.* **A47**, RC1605.
Grobe, R., Rzazewski, K. and Eberly, J.H. (1994). *J. Phys. B.* **27**, L503.
Gross, E.K.U., Dobson, J,F. and Petersilka, M. (1966). *Density Functional Theory* ed. Nalewajski, R.F., *Topics in Chemistry* Springer Verlag, NY.
Gryzinski, M. (1965). *Phys. Rev.* **138**, A349.
Gryzinski, M. (1987a). *International Journal of Theoretical Physics*, **26**, No. 10, 967.
Gryzinski, M. (1987b). *Journal of Magnetism and Magnetic Materials*, **71**, 53.

*** H ***

Hägg, L. (1993). *PhD. thesis*, Uppsala University.
Hägg, L. and Goscinski, O. (1993). *J. Phys. B.* **26**, 2345.
Hägg, L. and Goscinski, O. (1996). *Intl. J. Quantum Chem.* **58**, 689.
Hahn, Y. (1990). *The Physics of Electronic and Atomic Collisions*. (A. Dalgarno, R.S. Freund, P.M. Koch, M.S. Lubell and T.B. Lucatorto, eds.), 550. AIP, NY.
Hahn, Y., Z. (1995). Phys. D.
Hall, J., Richard, P., Gray, T.J., Newcomb, J., Pepmiller, B., Lin, C.D., Jones, K., Johnson, B. and Gregory, D. (1983). *Phys. Rev.* **A28**, 99.
Hall, R.I., Avaldi, L., Dawber, G., Zubeck, M., Ellis, K. and King, G.C. (1991). *J. Phys. B.* **24**, 115.
Hall, R.I., Avaldi, L., Dawber, G., Ellis, K., King, G.C., McConkey, A.G., MacDonald, M.A. and Zubeck, M. (1993). *The Physics of Electronic and Atomic Collisions*. (ed., T. Andersen, B. Fastrup, F. Folkmann, H. Knudsen and N. Andersen), *AIP Conf. Proc.* **295**, 61.
Halpern, A. and Thomas, B.K. (1979). *Phys. Rev. Lett.* **43**, 33.
Hanamura, E. (1973). *Solid State Commun.* **12**, 951.
Handke, G., Tarantelli, F., and Cederbaum, L.S. (1996). *Phys. Rev. Lett.* **76**, 896.
Hansteen, J.H. and Mosebekk, O.P. (1972). *Phys. Rev. Lett.* **29**, 1361.

Hansteen, J.M., Johansen, A.M. and Kocbach, L. (1975). *At. Data Nucl. Data Tables* **15**, 305.
Hatano, Y. (1995). *Invited Talks of ICPEAC XIX*, (Whistler, Canada).
Haugen, H.K., Anderson, L.P., Hvelplund, P. and Knudsen, H. (1982). *Phys. Rev.* **A26**, 1950.
Hazi, A. (1981). *Phys. Rev.* **A23**, 2232.
Heber, O., Bandson, B.B., Sampoll, G. and Watson, R.L. (1990). *Phys. Rev. Lett.* **64**, 851.
Heitler, W. (1954). *Quantum Theory of Radiation.* Oxford University Press, London.
Heller, E.J., Sundberg, R.L., Tanor,D. (1982). *J. Phys. Chem.* **86**, 1822.
Heller, E.J. (1995). *Invited Talks of ICPEAC XIX*, (Whistler, Canada).
Helm, H. (1992). *Invited Talks of ICPEAC XVII*, (Brisbane, Australia), 169.
Henry, R.W.J. and Kingston, A.E. (1988). *Adv. in At. Mol. and Opt. Phys.* **25**, 267.
Herrick, D.R., Kellman, M.E. and Poliak, R.D. (1980). *Phys. Rev.* **A22**, 1517.
Hertal, C. and Salin, A. (1980). *J. Phys. B.* **13**, 785.
Hill, R.N. (1996). *Atomic, Molecular and Optical Physics Reference Book.* (G.W.F. Drake, ed.), AIP Press, NY, chapter 10.
Hino, K.I., Ishihara, T., Shimizu, F., Toshima, N. and McGuire, J.H. (1993). *Phys. Rev.* **A48**, 1271.
Hino, K., Bergstrom, P. and Macek, J. (1994). *Phys. Rev. Lett.* **72**, 1620.
Hippler, R. (1988). *Coherence in Atomic Collision Physics.* (H.J. Beyer, K. Blum and R. Hipper, eds.) 137. Plenum Publishing.
Hitchcock, A.P. (1995). *Invited Talks of ICPEAC XIX*, (Whistler, Canada).
Hofmann, G., Neumann. J., Pracht, U., Tinschert, K., Stenke. M., Vöpel, R., M Müller, A. and Salzborn E. (1993). *VIth Intl. Conf. on the Phys. of Highly Charged Ions*, (ed., P. Richard, M. Stöckli, C.L.Cocke and C.D. Lin), *AIP Conf. Proc.* **279**, 485.
Holland, D.M.P., Codling, K., West, J.B. and Marr, G.V. (1979). J. Phys. B. **12**, 2465.
Horbatsch, M. (1986). *J. Phys. B.* **19**, L193.
Horbatsch, M. and Dreizler, R.M. (1986). *Z. Physik* **D2**, 183.
Horsdal-Pedersen, E., Cocke, C.L. and Stöckli, M. (1983). *Phys. Rev. Lett.* **50**, 1910.
Horsdal, E., Jenson, B. and Nielson, K.O. (1986). *Phys. Rev. Lett.* **57**, 1414.
Huang, K. (1987) in *Statistical Mechanics.* John Wiley, NY.
Huelskötter, H.P., Meyerhof, W.E., Dillard, E.D. and Guardola, N. (1989). *Phys. Rev. Lett.* **63**, 1938.
Huelskötter, H.P., Feinberg, B., Meyerhof, W.E., Belkacem, A., Alonso, J.R., Blumenfeld, L., Dillard, E.D., Gould, H., Guardala, N., Krebs, G.F., McMahan, M.A., Rohades-Brown, M.E., Rude, B.S., Schweppe, J., Spooner, D.W., Street, K., Thieberger, P. and Wegner, H.E. (1991). *Phys. Rev.* **A44**, 1712.
Huetz, A. (1996). *Invited Talks of ICPEAC XIX*, (Whistler, Canada).
Hvelplund, P., Haugen, H.K. and Knudsen, H. (1980). *Phys. Rev.* **A22**, 1930.
Hvelplund, P., Knudsen, H., Mikkelsen, U., Morenzoni, E., Moeller, S.P., Uggerhoj, E. and Worm, T. (1994). *J. Phys. B.* **27**, 925.

*** I ***

Inokuti, M. (1971). *Rev. Mod. Phys.* **43**, 297.
Inokuti, M. (1979). *Argonne National Laboratory Report* **ANL-76-88**, 177. unpublished.
Irby, V.D., Rolfes, R.G , Makarov, O.P. and MacAdam K.B. (1994). *Nucl. Instr. Meth.* **B99**, 174.
Ishihara, T. and McGuire, J.H. (1988). *Phys. Rev.* **A38**, 3310.
Ishihara, T., Hino, K. and McGuire, J.H. (1991). *Phys. Rev.* **A44**, R6980.
Itoh, A., Zouros, T.J., Schneider, D., Stettner, U., Zeitz, W. and Stolterfoht, N. (1985). *J. Phys. B.* **18**, 4581.

*** J ***

Jackson, J.D. and Schiff, H. (1953). *Phys. Rev.* **89**, 359.
Jackson, J.D. (1975). *Classical Electrodynamics.* John Wiley and Sons, NY.
Jacobs, V.L. and Burke, P.G. (1972). *J. Phys. B.* **5**, L67.
Jain, A., Lin, C.D. and Fritsch, W. (1989). *Phys. Rev.* **A39**, 1741.
Jakubassa-Amundsen, D.H. (1989). *Intl. Jour. Mod. Phys. A* **4**, No. 4, 769.
Jakubassa-Amundsen, D.H. (1992). *Z. Physik* **D22**, 701.
Janev, R.K., Presnyakov, L.P. and Shevelko, V.P. (1985). *Physics of Highly Charged Ions.* Springer Verlag, Berlin.
Jaynes, E.T. (1957). *Phys. Rev.* **108**, 171 and Phys. Rev. **106**, 620.
Jaynes, E.T. (1979). *The Maximum Entropy Formalism.* (R.D. Levine and M. Tribus, eds.) 15-118. MIT, Cambridge, MA.
Joachain, C.J. (1983). *Quantum Collision Theory.* North Holland, Amsterdam.
Jones, S., Madison, D.H., Franz, A. and Altick, P. (1994). *Phys. Rev.* **A48**, R22.
Jones, S. and Madison, D.H. (1994). *J. Phys. B.* **27**, 1423.
Jones, S. and Madison, D.H. (1995). *Invited Talks of ICPEAC XIX*, (Whistler, Canada).

*** K ***

Kabachnik, N.M. (1993). *J. Phys. B.* **26**, 3803.
Kais, S., Herschbach, D.R. and Levine, R.D. (1989). *J. Chem. Phys.* **91**, 7791.
Kalman, R. (1995). *Mathematica Japonica*, **41**, No. 1.
Kamber, E.Y., Cocke, C.L., Cheng, S. and Varghese, S.L. (1988). *Phys. Rev. Lett.* **60**, 2026.
Karule, E. (1990). *Adv. in At. Mol. and Opt. Phys.* **27**, 265.
Kauffman, R.L., McGuire, J.H., Richard, P. and Moore, C.F. (1973). *Phys. Rev.* **A8**, 1233.
Keldysh, L.V. (1965). *Sov. Phys. - JEPT* **20**, 1307.
Kelly, H.P. (1969). *Adv. Theor. Phys.* **2**, 75.
Kimura, M., Sato, H. and Olson, R.E. (1985). *Phys. Rev.* **A28**, 2085.
Kittel, C. (1956). *Introduction to Solid State Physics*, John Wiley, N.Y.
Knudsen, H., Andersen, L.H., Hvelplund, P., Astner, G., Cederquist, H., Darnared, H., Liljeby, L. and Rensfelt, K.G. (1984). *J. Phys. B.* **17**, 3545.
Knudsen, H. and Reading. J.F. (1992). *Phys. Reports* **212**, 107.

Kocbach, L. (1976). *J. Phys. B.* **9**, 2269.
Kohland, H. and Dreizler, R.M. (1986). *Phys. Rev. Lett.* **56**, 1993.
Kossmann,H., Schmidt, V., and Andersen, T. (1988). *Phys. Rev. Lett.* **60**, 1266.
Krishnakumar, E. and Rajgara, F.A. (1993). *J. Phys. B.* **26**, 4155.
Kristensen, F.G. and Horsdal, P.E. (1990). *J. Phys. B.* **23**, 4129.
Kuerpick, P., Lüdde, H.J., Sepp, W.D. and Fricke, B. (1994). *Nucl. Instru. Meth.* **B94**, 183.
Kuerpick, P., Sepp, W.D. and Fricke, B. (1995). *Phys. Rev.* **A51**, 3693.
Kulander, K.C. (1987a). *Phys. Rev.* **A35**, 445.
Kulander, K.C. (1987b). *Phys. Rev.* **A36**, 2726.
Kulander, K.C. (1996). *Atomic, Molecular and Optical Physics Reference Book.* (G.W.F. Drake, ed.), AIP Press, NY, chapter 67.
Kunikeev, Sh.D. and Senashkenko, V.S. (1989). *Contributed Papers of ICPEAC XVI,* (New York), (A. Dalgarno, R.S. Freund, M.S. Lubell and T.B. Lucatorto, ed.), 503.
Kutzelnigg, W., Del Re, G. and Berthier, G. (1968). *Phys. Rev.* bf 172, 49.

*** L ***

Lablanquie, P. (1992). *Invited Talks of ICPEAC XVII,* (Brisbane, Australia), 507.
Lablanquie, P., Ito, K., Morin, P., Nenner, I., and Eland, J.H.D. (1990). *Z. Physik* **D16**, 77.
Lahmam-Bennani, A., Dupré, C. and Duguet, A. (1989). *Phys. Rev. Lett.* **63**, 1582.
Lahmam-Bennani, A. and Duguet, A. (1992). Invited Talks of ICPEAC XVII (Brisbane, Australia), 209.
Lahmam-Bennani, A. and Duguet, A. (1996). *Can. J. Phys.*
Lambropoulos, P. (1976). *Adv. in At. and Mol. Phys.* **12**, 87.
Landau, L.D. and Liftshitz, E.M. (1965). *Quantum Mechanics.* 2nd ed. Oxford, Permagon Press.
Landsberg, P.T. (1990). *Thermodynamics and Statistical Mechanics.* Dover Publications, NY. Sec. 9.1.
Langhoff, P. (1996). *Raman Emission by X-rays* World Scientific, NY.
Laufer, P.M. and Krieger, J.B. (1986). *Phys. Rev.* **A33**, 1480.
Lee, S.Y. and Heller, E.J. (1979). *J. Chem. Phys.* **71**, 4777.
Levin, J.C., Lindle, D.W., Keller, N., Miller, R.D., Azuma. Y., Berrah Monsour, N., Berry, H.G. and Sellin,I. (1991). *Phys. Rev. Lett.* **67**, 968.
Levin, J.C., Lindle, D.W., Keller, N., Miller, R.D., Azuma, Y., Berrha, N., Berry, H.G. and Sellin, I.A., (1992). *Phys. Rev. Lett.* **67**, 968.
Levin, J.C., Sellin, I.A., Johnson, B.M., Lindle, D.W., Miller, R.D., Mansour, N.B., Azuma, Y., Berry, H.G. and Lee, D.H. (1993). *Phys. Rev.* **A47**, R16.
Levin, J.C., Armen, G.B. and Sellin, I.A. (1995). *Contributed Papers of ICPEAC XIX,* (Whistler, Canada), 147.
Levin, J.C., Armen, G.B. and Sellin, I.A. (1996). *Phys. Rev. Lett.* **76**, 1220.
Levine, R.D. and Bernstein, R.B. (1974). *Molecular Reaction Dynamics.* Oxford University Press, NY. pp. 21, 37, 46, 108.
Lin, C.D. and Richard, P. (1981). *Adv. in At. Mol. and Opt. Phys.* **17**, 275.
Lin, C.D. (1986). *Adv. in At. Mol. and Opt. Phys.* **22**, 77.

Lindle, D.W., Ferrett, T.A., Beckerll, Korbin, P.H., Truesdale, C.M., Kerkhoff, H.G. and Shirley, D.A. (1985). *Phys. Rev.* **A31**, 714.
Liu, C.R. and Starace, A.F. (1990). *Phys. Rev.* **A42**, 2689.
Löwdin, P.O. (1959). *Adv. Chem. Phys.* **2**, 207.
Löwdin, P.O. (1995). *Intl. J. Quantum Chem.* **55**, 77.
Lucatorto, T., Phillips, W.D. and Rolston, S. (1996). *Atomic, Molecular and Optical Physics Reference Book.* (G.W.F. Drake, ed.), AIP Press, NY, chapter 65.
Lüdde, H.J. and Dreizler, R. (1985). *J. Phys. B.* **18**, 107.

*** M ***

Ma, Y. (1996). *Raman Emission by X-rays*, World Scientific, NY.
McDaniel, E.W., Mitchell, J.B.A. and Rudd, M.E. (1993). *Atomic Collisions: Heavy Particle Projectiles*, John Wiley, NY.
Macek, J. (1968). *J. Phys. B.* **2**, 831.
Macek, J. (1970). *Phys. Rev.* **A1**, 235.
Macek, J. and Alston. S. (1982). *Phys. Rev.* **A26**, 250.
Madison, D.H. and Merzbacher, E. (1975). Atomic Inner Shell Processes. (B. Crasemann, ed.) Academic Press, NY.
Madison, D.H. (1990). *The Physics of Electronic and Atomic Collisions.* (A. Dalgarno, R.S. Freund, P.M. Koch, M.S. Lubell and T.B. Lucatorto, eds.), p. 149. AIP, NY.
Madison, D.H. (1991). *Comments At. Mol. Phys.* **26**, 59.
Madison, D.H. (1996). *Many-body Atomic Physics* Cambridge Univ. Press, Cambridge, England.
Magnus, W. (1954). *Commun. Pure Appl. Math.* **7**, 649.
Maidagan, J.M. and Rivarola, R.D. (1984). *J. Phys. B.* **17**, 2477.
Malhi, N.B., Ben-Itzhak, I., Gray, T.J., Legg, J.C., Needham, V., Carnes, K. and McGuire, J.H. (1987). *J. Chem. Phys.* **87**, 6502.
Manivannan, K., Ramanan, S.V., Mathias, R.T. and Brink, P.R. (1992). *Biophys. J.* **61**, 216.
Manson, S. and McGuire, J.H. (1995). *Phys. Rev.* **A 51**, 400.
Marcus, C. (1993). *Invited talk at the DAMOP Meeting of the Am. Phy. Soc.*
Marji, B.E., Duguet, A., Lahmam-Bennani, A, Lecas, M. and Wellenstein, H.F. (1995). *J. Phys. B.* **28**, L733.
Martin, F. and Salin, A. (1995). *J. Phys. B.* **28**, 639.
Martin, F. and Salin, A. (1996). *Phys. Rev.* **A**, in press.
Massey, H.S.W. *et al.* (1969). *Electronic and Ionic Impact Phenomena.* Oxford University Press.
Matsuzawa, M. (1980). *J. Phys. B.* **13**, 3201.
McCarroll, R. and Salin, A. (1966). *Compt. Rend Acad. Sci.* **263**, 329.
McCarroll, R. and Salin, A. (1968). *Proc. Phys. Soc.* **19**, 63.
McCarthy, I.E. and Weigold, E. (1990). *Adv. in At. Mol. and Opt. Phys.* **27**, 165.
McCarthy, I.E. and Weigold, E. (1995). *Electron-Atom Collisions*, Cambridge Univ. Press.
McConkey, J.W., Trajmar, S. and King, G.C.M. (1988). *Comments At. Mol. Phys.*, **17**.

McDaniel, E.W., Mitchell, J.B.A. and Rudd, M.E. (1993). *Atomic Collisions: Heavy Particle Projectiles*, John Wiley, NY.
McDaniel, E.W. and Mansky, E.J. (1994). *Adv. in At. Mol. and Opt. Phys.* **33**, 390.
McDowell, M.R.C. and Coleman, J.P. (1970). *Introduction to the Theory of Ion-Atom Collisions.* North-Holland, NY.
McGuire, E.J. (1975). *Atomic Inner Shell Processes.* (B. Crasemann, ed.) Academic Press, NY, Chap. 7.
McGuire, J.H. and Richard, P. (1973). *Phys. Rev.* **A8**, 1374.
McGuire, J.H. and Macdonald, J.R. (1975). *Phys. Rev.* **A11**, 146.
McGuire, J.H. and Weaver, O.L. (1977). *Phys. Rev.* **A16**, 41.
McGuire, J.H., Land, D. J., Brennan, J.G. and Basbas, G. (1970). *Phys. Rev.* **A 19**, 2180.
McGuire, J.H., Stolterfoht, N. and Simony, P.R. (1981). *Phys. Rev.* **A24**, 97.
McGuire, J.H. (1982a). *Phys. Rev. Lett.* **49**, 1153.
McGuire, J.H. (1982b). *Phys. Rev.* **A26**, 143.
McGuire, J.H. (1982c). unpublished.
McGuire, J.H., Simony, P.R, Weaver, O.L. and Macek, J. (1982). *Phys. Rev.* **A26**, 1109.
McGuire, J.H. (1983). *J. Phys. B.* **16**, 3805.
McGuire, J.H. (1984). *J. Phys. B.* **17**, 1779.
McGuire, J.H. and Weaver, O.L. (1984). *J. Phys. B.* **17**, L583.
McGuire, J.H. (1985). *J. Phys. B.* **18**, L75.
McGuire, J.H. and Weaver, O.L. (1986). *Phys. Rev.* **A34**, 2473.
McGuire, J.H. and Sil, N.C. (1986). *Phys. Rev.* **A33**, 725.
McGuire, J.H. (1987). *Phys. Rev.* **A36**, 1114.
McGuire, J.H. and Heil, T.G. (1987). *Nucl. Instr. and Meth.* **B24/25**, 243.
McGuire, J.H., Müller, A. and Salzborn, E. (1987). *Phys. Rev.* **A35**, 3265.
McGuire, J.H. and Burgdörfer, J. (1987). *Phys. Rev.* **A36** 4089.
McGuire, J.H., Deb, N.C., Aktas, Y. and Sil, N.C. (1988). *Phys. Rev.* **A38**, 3333.
McGuire, J.H. (1988). *High Energy Ion-Atom Collisions.* (D. Berenyi and G. Hock, eds.), 415. Springer Verlag, Berlin.
McGuire, J.H., Straton, J.C., Axmann, W.J., Ishihara, T. and Horsdal, E. (1989). *Phys. Rev. Lett.* **62**, 2933.
McGuire, J.H. and Straton, J.C. (1990a). *The Physics of Electronic and Atomic Collisions.* (A. Dalgarno, R.S. Freund, P.M. Koch, M.S. Lubell and T.B. Lucatorto, eds.), p. 280. AIP, NY.
McGuire, J.H. and Straton, J.C. (1990b). *Nucl. Instru. Meth.* **B56/57**, 192.
McGuire, J.H. (1991). *Adv. in At. Mol. and Opt. Phys.* **29**, 217.
McGuire, J.H., Straton, J.C. and Ishihara, T. (1993). *Many Body Atomic Physics.* Cambridge University Press.
McGuire, J.H. (1995). *Physics Today*, November, 15.
McGuire, J.H., Berrah, N., Bartlett, R.J., Samson, J.A.R., Tanis, J.A., Cocke, C.L., and Schlachter, A.S. (1995). *J. Phys. B.* **28**, 913.
McGuire, J.H., Straton, J.C. and Ishihara, T. (1996a). *Atomic, Molecular and Optical Physics Reference Book.* (G.W.F. Drake, ed.), AIP Press, NY, chapter 40.
McGuire, J.H., Wang, J., Straton, J.C., Wang, Y.D., Weaver, O.L., Corchs, S.E., and Rivarola, R.D. (1996b). *J. Chem. Phys.* **105**, 1846.

Mehlhorn, W. (1978). *Summer lecture notes*, University of Aarhus.
Mehlhorn, W. (1985). *Atomic Innter-Shell Processes.* (B. Crasemann, ed.). Plenum Publishing.
Melchert, F., Debus, W., Krüdeneer, S., Schulze, R., Olson, R.E. and Salzborn, E. (1991). *Atomic Physics of Highly Charged Ions.* eds., Salzborn, E., Mokler, P.H. and Müller, A. (Springer Verlag, Berlin), 249.
Menendez, M.G. and Duncan, M.M. (1989). *Phys. Rev.* **A39**, 1534.
Merkuriev, S.P. (1976). *Yad. Fiz.* **24**, 289.
Merkuriev, S.P. (1980). *Ann. Phys.* **130**, 395.
Merkuriev, S.P. and Fadeev, L.D. (1985). *Quantum theory of scattering for the few-body system.* *'Nauka.'* (in Russian).
Merzbacher, E. (1984). *Electronic and Atomic Collisions* (J. Eichler, I.V. Hertel and N. Stolterfoht, eds.), p. 1. North Holland, Amsterdam.
Messiah, A. (1961). *Quantum Mechanics.* Chapter XIX, Sec. 24. John Wiley.
Meyerhof, W.E. and Huelskötter, H.P. (1990). U.S.-Japan Seminar, Anchorage, AL.
Milonni, P.W. and Singh, S. (1990). *Adv. in At. Mol. and Opt. Phys.* **28**, 76.
Miraglia, J.E., Piacentini, R.D., Rivarola, R.D. and Salin, A. (1981). *J. Phys. B* **14**, 197. (Also, Reunion Nacional de Fisica (Argentina) in 1978.)
Miraglia, J.E. (1982). *J. Phys. B.* **15**, 4205.
Miraglia, J.E. (1983). *J. Phys. B.* **16**, 1029.
Miraglia, J.E. and Gravielle, M.S. (1987). *Contributed Papers of ICPEAC XVI,* (Brighton, England), (J. Geddes, H.B. Gilbody, A.E. Kingston and C.J. Latimer, eds.) 517.
Mittleman, M.H. (1966). *Phys. Rev. Lett.* **16**, 498.
Moiseiwitsch, B.L. (1994). *Adv. in At. Mol, and Opt. Phys.* **33**, 279.
Mokler, P.H., Reusch, S., Stöhlker, T., Schuch, R., Schulz, M., Wintermeyer, G., Stachura, Z., Warczak, A., Müller, A., Awaya, T. and Kambara, T. (1989). *Radiation Effects and Defects in Solids* Gordon and Breach, 39.
Montenegro, E.C. and Meyerhof, W.E. (1991). *Phys. Rev.* **A43**, 2289.
Montenegro, E.C., Melo, W.S., Meyerhof, W.E. and de Phino, A.G. (1992). *Phys. Rev. Lett.* **69**, 3033.
Montenegro, E.C., Belkacem, A., Spooner, D.W., Meyerhof, W.E. and Shah, M.B. (1993). *Phys. Rev.* **A47**, 1045.
Montenegro, E.C., Meyerhof, W.E. and McGuire, J.H. (1994). *Adv. in At. Mol. and Opt. Phys.* **34**, 250.
Montenegro, E.C. and Zouros, T.M.J. (1994). *Phys. Rev.* **A50**, 3186.
Montenegro, E.C. (1995). *Invited Talks of ICPEAC XIX,* (Whistler, Canada).
Morse, P.M. and Feshbach, H. (1953). *Methods of Theoretical Physics.* McGraw-Hill, NY.
Morishita, T, Hino, K., Watanabe, S. and Matsuzawa, M. (1994). *J. Phys. B.* **27**, L287.
Morishita, T., Hino, K., Watanabe, S. and Matsuzawa. M. (1995).
Mott N.F. and Masey, H.S.W. (1965). *The Theory of Atomic Collisions* Oxford University Press.
Müller, A., Groh, W., Kneissel, U., Heil, R., Stroher, H. and Salzborn, E. (1983a). *J. Phys. B.* **16**, 2039.
Müller, A., Groh, W. and Salzborn, E. (1983b). *Phys. Rev. Lett.* **51**, 107.
Müller, A., Tinschert, K., Achenbach, Ch. and Salzborn, E. (1985). *Phys. Rev. Lett.* **54**, 414.

Müller, A., Schuch, B., Groh, W. and Salzborn, E. (1987). *Z. Physik* **D7**, 251.
Mukoyama, T. (1986). *Bull. Chem. Res. Koyoto Univ.* **64**, 1.
Mysyrowicz, A. (1980). *J. Phys. (Paris)* **41**, Suppl. 7, 281.
Mysyrowicz, A., Snoke, D., and Wolfe, J. (1990). *Phys. Stat. Sol.* **159**, 387.
Mysyrowicz, A., Benson, E. and Fortin, E. (1996). *Phys. Rev. Lett.* **77**, 896.
Mysyrowicz, A. (1996). *Invited Papers of the 30th Anniversary Meeting of the Physical Society of Brazil*, (Aguas de Lindoia).

*** N ***

Nagy, L. and Vegh, L. (1994). *Nucl. Instru. Meth.* **B86**, 165.
Nagy, L., Wang, J., Straton, J.C. and McGuire, J.H. (1995). *Phys. Rev.* **A52**, R902.
Nagy, L., McGuire, J.H., Vegh, L., Sulik, B. and Stolterfoht, N. (1996), preprint.
Newell, W.R. (1992). *Comments At. Mol. Phys.* **28**, 59.
Niehaus, A. (1986). *J. Phys. B.* **19**, 2925.
Nikolaev, N.S., Fateeva, L.N., Dmitriev, I.S., and Teplova, Yu. A. (1961). *Sov. Phys. - JEPT* **12**, 627.

*** O ***

Ohya, M. and Petz, D. (1993). *Quantum Entropy and Its Use.* Springer, Berlin.
Olson, R.E. (1987). *Phys. Rev.* **A36**, 1519.
Olson, R.E. (1988). in *Electronic and Atomic Collisions.* (H.B. Gilbody, W.R. Newell, F.H. Read and A.C.H. Smith, eds.) p. 271. Elsevier Science Publishers, BV, London.
Olson, R.E., Ullrich, J. and Schmidt-Böcking, H. (1989). *Phys. Rev.* **A39**, 5572.
Olson, R.E., Ullrich, J. and Schmidt-Böcking, H. (1989a). *Phys. Rev.* **A40**, 2843.
Olson, R.E., Ullrich, J. and Schmidt-Böcking, H. (1989b). *Phys. Rev.* **A39**, 5572.
Olson, R.E. (1993). *VIth Intl. Conf. on the Phys. of Highly Charged Ions,* (ed., P. Richard, M. Stöckli, C.L.Cocke and C.D. Lin), *AIP Conf. Proc.* **279**, 229.
Oppenheimer, J.R. (1928). *Phys. Rev.* **31**, 349.

*** P ***

Palinkas, J., Schuch, R., Cederquist, H. and Gustafsson (1989). *Phys. Rev. Lett.* **22**, 2464.
Peach, G., Willis, S.L. and McPowell, M.R.C. (1985). *J. Phys. B.* **18**, 3921.
Pedersen, J.O.P. and Folkmann, F. (1990). *J. Phys. B.* **23**, 441.
Pedersen, J.O.P. and Hvelplund, P. (1989). *Phys. Rev. Lett.* **62**, 2373.
Peng, C.K., Buldyrev, S.V., Goldberger, A.L., Havlin, S. Mantegna, R.N., and Stanley, H.E.
(1995a). *Physica* **A 221**, 180.
Peng, C.K., Havlin, S., Stanley, H.E. and Goldberger, A.L. (1995b). *Chaos* **5**, 82.
Pepmiller, P.L., Richard, P., Newcomb, J., Dillingham, R., Hall, J.M., Gray, T.J. and Stöckli, M. (1983). *IEEE Trans. Nucl. Sci.* **NS30**, 1002.

Perdew, J.P., Chevary, J.A., Vosko, S.H., Jackson, K.A., Pederson, M.R., Singh, D.J. and Fiolhais, C. (1992). *Phys. Rev.* B. **46**, 6671.
Phaneuf, R. (1996). *Atomic, Molecular and Optical Physics Reference Book.* (G.W.F. Drake, ed.), AIP Press, NY, chapter 54.
Pireaux, B., L'Huillier, A. and Rzazewski, J. (1993). *Super Intense Laser Atom Physics.* Plenum, NY.
Placzek, G., Nijbor, B.R.A. and Van Hove, L. (1951). *Phys. Rev.* **82**, 392.
Potvliege, R.M. and Shakeshaft, R. (1992). *Invited Talks of ICPEAC XVII,* (Brisbane, Australia), 497.
Potvliege, R.M. and Shakeshaft, R. (1990). *Phys. Rev.* **A41**, 1609.
Press, W.H., Flannery, B.P., Teukolsky, S.A. and Vetterling, W.T. (1986). *Numerical Recipies.* Cambridge University Press, Sec. 12.8.
Presnyakov, L.P. and Uskov, D.B. (1984). *Sov. Phys. - JEPT* **59**, 515.
Presnyakov, L.P., Tawara, H., Tolstikhina, I. Yu and Uskov, D.B. (1995). *J. Phys. B.* **28**, 785.
Prost, M. Morgenstern, R., Schneider, D. and Stolterfoht, N. (1977). *Contributed Papers of ICPEAC X,* (Paris), (M. Barat and J. Reinhardt, eds.).
Proulx, D., and Shakeshaft, R. (1993). *Phys. Rev.* **A48**, R875.

*** R ***

Rau, A.R.P. (1970). *Phys. Rev.* **A4**, 207.
Rau, A.R.P. (1976). *J. Phys. B.* **9**, L283.
Rau, A.R.P. (1983). *J. Phys. B.* **16**, L699.
Rau, A.R.P. (1984). *Phys. Reports* **110**, 369.
Rau, A.R.P. (1984). *Atomic Physics* **9**, ed. N. Fortson (World Scientific, Singapore), 491.
Rau, A.R.P. and Molina, Q. (1989). *J. Phys. B.* **22**, 189.
Rau, A.R.P. (1990). *Rep. Prog. Phys.* **53**, 181.
Rau, A.R.P. (1990). *Aspects of Electron-Molecule Scattering and Photoionization. AIP Conf. Proc.* **204**, ed.A. Herzenberg, 24.
Rau, A.R.P. (1992). *Science* **258**, 1444.
Reaker, A., Bartschat, K. and Reid, R.H.G. (1994). *J. Phys. B.* **14**, 3129.
Read, F.H. (1992). *Invited Talks of ICPEAC XVII,* (Brisbane, Australia), 35.
Reading, J.F. (1973). *Phys. Rev.* **A8**, 3262.
Reading, J.F. and Ford, A.L. (1980). *Phys. Rev.* **A21**, 124.
Reading, J.F., Ford, A.L., Smith, J.S. and Becker, R.L. (1984). *Electronic and Atomic Collisions.* (J. Eichler, I.V. Hertel and N. Stolterfoht, eds.). Elsevier Science Publishing BV, Berlin.
Reading, J.F. and Ford, A.L. (1987a). *Phys. Rev. Lett.* **58**, 543.
Reading, J.F. and Ford, A.L. (1987b). *J. Phys. B.* **20**, 3747.
Reading, J.F., Bronk, T. and Ford, A.L. (1996). *J. Phys. B.*
Redmond, P.J. unpublished.
Reinhardt, J., Müller, B. and Griener, W. (1981). *Phys. Rev.* **A24**, 103.
Reinhold, C.O. and Falcon, C.A. (1986). *J. Phys. A.* **33**, 3859.
Reiss, H.R. (1980). *Phys. Rev.* **A22**, 1786.
Rhoades, R. (1986). *The Making of the Atomic Bomb.* Simon and Schuster, NY).
Richard, P. (1975). *Inner-Shell Processes.* (B. Craseman, ed.), p. 74.

Academic Press, NY.
Richard, P. (1990). *Invited Talks of X-90, 15th Intl. Conf. on X-ray and Inner-Shell Processes.* Knoxville, TN.
Rivarola, R.D. (1981). *Ph.D. thesis* University of Bordeaux.
Rivarola, R.D. (1993). *VIth Intl. Conf. on the Phys. of Highly Charged Ions,* (ed., P. Richard, M. Stöckli, C.L.Cocke and C.D. Lin), *AIP Conf. Proc.* **279**, 237.
Roberts, M.J. (1985). *J. Phys. B.* **18**, L707.
Rodbro, M. and Andersen, P.D. (1979). *J. Phys. B.* **12**, 2883.
Rocin, P., Barat, M., Gaborland, M.N., Guillemot, L., Laurant, H. (1989). *J. Phys. B.* **22**, 509.
Rosenstock, H.M., Wallenstein, M.B., Wahrhaftig, A.L. and Eyring, H. (1952). *Proc. Natl. Acad. Sci. USA* **38**, 667.
Rottke, H., Wolff, B., Brickwedde, M., Feldmann, D. and Welge, K.H. (1990). *Phys. Rev.* **A64**, 404.
Rudd, M.E. (1956). *undergraduate thesis.*
Rudd, M.E., Kim, Y.K., Madison, D.H. and Gallagher, J.W. (1985). *Rev. Mod. Phys.* **57**, 967.
Russek, A. (1963). *Phys. Rev.* **A132**, 246.
Ryufuku, H. and Watanabe, T. (1979). *Phys. Rev.* **A19**, 1538.
Ryufuku, H., Sasaki, K., and Watanabe, T. (1980). *Phys. Rev.* **A21**, 745.

*** S ***

Sadeghpour, H. and Greene, C.H. (1990). *Phys. Rev. Lett.* **65**, 312.
Salin. A. (1969). *J. Phys. B.* **2**, 631.
Salin, A. (1987). *Phys. Rev.* **A36**, 5471.
Salzborn, E. (1990). *The Physics of Electronic and Atomic Collisions.* (A. Dalgarno, R.S. Freund, P.M. Koch, M.S. Lubell and T.B. Lucatorto, eds.) p. 290, AIP, NY.
Samson, J.A.R., Greene, C.H., and Bartlett, R.J. (1993). *Phys. Rev. Lett.* **71**, 201.
Samson, J.A.R., He, Z.X., Bartlett, R.J., and Sagurton, (1994). *Phys. Rev. Lett.* **72**, 3329.
Sant'Ana, M., Montenegro, E.C. and McGuire, J.H. (1997). unpublished.
Schafer, K.J., Yang, B., DiMauro, L.F. and Kulander, K.C. (1993). *Phys. Rev. Lett.* **70**, 1599.
Schafer, K. (1995). *Invited Talks of ICPEAC XIX*, (Whistler, Canada).
Schartner, K.H. (1990). *The Physics of Electronic and Atomic Collisions.* (A. Dalgarno, R.S. Freund, P.M. Koch, M.S. Lubell and T.B. Lucatorto, eds.), p. 215. AIP, NY.
Schaut, K.J., Kwong, N.H. and Garcia, J.D. (1991). *Phys. Rev.* **A43**, 2294.
Schiff, L.I. (1968). *Quantum Mechanics.* McGraw-Hill, NY, Chap. 11.
Schlachter, A.S., Stearns, J.W., Berkner, K.H., Bernstein, L.M., Clark, M.W., DuBois, R.D., Graham, W.G., Morgan, T.J., Müller, D.W., Stöckli, M.W., Tanis, J.A. and Woodland, W.T. (1990). *The Physics of Electronic and Atomic Collisions.* (A. Dalgarno, R.S. Freund, P.M. Koch, M.S. Lubell and T.B. Lucatorto, eds.) p. 366, AIP, NY.
Schlemmer, P., Röstel, T., Jung, K. and Ehrhard, H. (1989). *J. Phys. B.* **22**. 2179.

Schmidt, V., Sandner, N., Kuntzemüller, H., Dhez, P., Wuilleumier, F., and Källne, E. (1976). *Phys. Rev.* **A13**, 1748.
Schramm, U., Berger, J., Grieser, M., Habs, D., Kilgus, G., Schüssler, T., Schwain, D., Wolf, A., Neumann, R. and Schuch, R. (1992). *Invited Talks of ICPEAC XVII*, (Brisbane, Australia), 647.
Schryber, U. (1967). *Helv. Phys. Acta* **40**, 1023.
Schuch, R, Justiniano, E., Schulz, M., Mokler, P.M., Reusch, S., Datz, S., Dittner, P.F., Giese, J., Miller, P.D., Schöne, H., Kambara, T., Müller, A., Stachura, Z., Vane, R., Warzcak, A. and Wintermeyer, G. (1990). *The Physics of Electronic and Atomic Collisions.* (A. Dalgarno, R.S. Freund, P.M. Koch, M.S. Lubell and T.B. Lucatorto, eds.), AIP, NY, 562.
Schulz, M., Giese, J.P., Swenson, J.K., Datz, S., Dittner, P.F., Krause, H.F., Schöne, H. Vane, C.R., Benhenni, M. and Shafroth, S.M. (1989). *Phys. Rev. Lett.* **62**, 1738.
Schulz, M., Htwe, W.T., Gaus, A.D., Peacher, J.L. and Vajnai, T. (1995). *Phys. Rev.* **A51**, 2140.
Schulz, M. (1995). *Intl. J. Mod. Phys.* **B9**, 3269.
Schultz, D.R., Bottcher, C., Madison. D.H., Peacher, J.L., Buffington, G., Pindzola, M.S., Gorczyca, T.W., Gavras, P. and Griffin, D.C. (1994). *Phys. Rev.* **A50**, 1348.
Schwarzkopf, O., Krässig, B., Elmiger, J. and Schmidt, V. (1993). *Phys. Rev. Lett.* **70**, 3008.
Shah, M.B. and Gilbody, M.B. (1982). *J. Phys. B.* **15**, 413.
Shah, M.B. and Gilbody, M.B. (1985). *J. Phys. B.* **18**, 899.
Shah, M.B., Patton, C.J., Geddes, J. and Gilbody, H.B. (1995). *Nucl. Instru. Meth.* **B98**, 280.
Shakeshaft, R. and Spruch, L. (1979). *Rev. Mod. Phys.* **51**, 369.
Shakeshaft, R. (1980). *Phys. Rev.* **A18**, 1930.
Shakeshaft, R. (1995). *Workshop of the U.S.Department of Energy*, (San Antonio), Nov. 3-4.
Shimamura, I. and Takayanagi, K. (1984). *Electron-Molecule Collisions.* Phenum Press, NY.
Shingal, R., Bransden, B.H., Ermalov, A.M., Flower, D.R., Newby, C.W. and Noble, C.J. (1986). *J. Phys. B.* **19**, 309.
Shingal, R. and Lin, C.D. (1989). *Phys. Rev.* **A40**, 1302.
Shore, B. (1967). *Rev. Mod. Phys.* **39**, 439.
Shore, B.W. and Menzel, D.H. (1968). *Principles of Atomic Spectra.* Wiley, NY.
Sidorovitch, V.A. and Nikolaev, V.S. (1983). *J. Phys. B.* **16**, 3743.
Sidorovitch, V.A., Nikolaev, V.S. and McGuire, J.H. (1985). *Phys. Rev.* **A31**, 2139.
Sidorovitch, V.A. (1994). *Physica Scripta* **50**, 119.
Sil, N.C. and McGuire, J.H. (1985). *J. Math. Phys.* **26**, 845.
Skogvall, B. and Schweitz, G. (1990). Submitted to *Phys. Rev. Lett.*
Smith, D.E. (1925). *History of Mathematics*, Ginn and Co,, NY.
Sokolnikoff, I.S. and Redheffer,R.M. (1958). *Mathematics of Physics and Modern Engineering.* McGraw Hill, NY.
Spielberger, L., Jagutzki, O., Dörner, R., Ullricj, J., Meyer, U., Mergel, V., Unverzagt, M., Damrau, M., Vogt, T., Ali, I., Khayyat, Kh., Bahr, D., Schmidt, H.G., Frahm, R. and Schmidt-Böcking. H. (1995). *Contributed Papers of ICPEAC XIX*, (Whistler, Canada), 148.

Stanley, H.E. (1996). *Invited Papers of the 30th Anniversary Meeting of the Physical Society of Brazil,* (Aguas de Lindoia).
Stanley, M.H.R., Amaral, L.A., Buldyrev, L.A., Havlin, S., Leschhorn, H., Maass, P., Salinger,
M.A. and Stanley, H.E. (1996). *Nature* **379**, 804.
Starace, A.F. (1971). *Phys. Rev.* **A3**, 1242.
Starace, A.F. (1982). *Handbuch der Physik,* **XXXI**, 1.
Starace, A.F. (1996). (1996). *Atomic, Molecular and Optical Physics Reference Book.* (G.W.F. Drake, ed.), AIP Press, NY, chapter 35.
Stelbovics, A.T. (1992). *Invited Talks of ICPEAC XVII,* (Brisbane, Australia), 21.
Stich, W., Ludde, H.J. and Dreizler, R.M. (1985). *J. Phys. B.* **18**, 1195.
Stolterfoht, N. (1971). *Phys. Lett. A* **37**, 117.
Stolterfoht, N., Ridder, D. and Ziem, P. (1972). *Phys. Let. A* **42**, 240.
Stolterfoht, N., Miller, P.D., Krause, H.F., Yamazaki, Y., Swenson, J.K., Bruch, R., Dittner, P.F., Pepmiller, P.L. and Datz, S. (1987a). *Nucl. Inst. Meth.* **B24/25**, 168.
Stolterfoht, N., Schneider, D., Tannis, J., Altevogt, H., Salin, A., Fainstein, P.D., Rivarola, R., Grandin, J.P., Scheurer, J.N., Andriamonje, S., Bertault, D. and Chemin, J.F. (1987b). *Europhys. Lett.* **4**, 899.
Stolterfoht, N. (1988). private communication.
Stolterfoht, N. (1989). *Spectroscopy and Collisions of Few Electron Ions.* (M. Ivaseu, V. Floreseu and V. Zoran, eds.), p. 342, Bucharest. World Scientific, Singapore.
Stolterfoht, N. (1990). *Physica Scripta* **42**, 192.
Stolterfoht, N. (1991). *Nucl. Instr. Meth. Meth.* **B53**, 477.
Stolterfoht, N., Mattis, A., Schneider, D., Schiweitz, G. and Skovall, B. (1995). *Phys. Rev.* **A51**, 350.
Straton, Jack C. (1990). *Phys. Rev.* **A42**, 307.
Straton, Jack C. (1991). *Phys. Rev.* **A43**, 1381.
Straton, Jack C., McGuire, J.H. and Chen, Zheng (1992). *Phys. Rev.* **A46**, 5514.
Strayer, M.R., Bottcher, C., Oberacker, V.E. and Umar, A.S. (1990). *Phys. Rev.* **A41**, 1399.
Suric. T., Pisk, K., Logan, B.A. and Pratt, R.H. (1994). *Phys. Rev. Lett.* **73**, 790.
Swenson, J.K., Yamazaki, Y., Miller, P.D., Krause, H.F., Dittner, P.F., Pepmiller, P.D., Datz, S. and Stolterfoht, N. (1986). *Phys. Rev. Lett.* **57**, 3042.
Szabo, A. and Ostlund, N.S. (1982). *Modern Quantum Chemistry, Introduction to Advanced Electronic Structure Theory* (Macmillan Publishing, NY).

*** T ***

Tang, Z. and Shakeshaft, R. (1993). *Phys. Rev.* **A47**, R3487.
Tanis, J.A., Bernstein, E.M., Graham, W.G., Clark, M., Shafroth, M., Johnson, B.M., Jones, K.M. and Meron, M. (1982). *Phys. Rev. Lett.* **49**, 1325.
Tanis, J.A., Bernstein, E.M., Clark, M.W., McFarland, R.H., Morgan, T.J., Muller, A., Stöckli, M.P., Berkner, K.H., Gohil, P., Schlachter, A.S., Stearns, J.W., Johnson, B.M., Jones, K.W., Meron, M. and Nason, J. (1985). *Electronic and Atomic Collisions.* (D.C. Lorents, W.E. Meyerhof and J.R. Peterson, eds.) p. 425. North Holland, Amsterdam.

Tanis, J.A. (1990). *The Physics of Electronic and Atomic Collisions.* (A. Dalgarno, R.S. Freund, P.M. Koch, M.S. Lubell and T.B. Lucatorto, eds.), p. 538. AIP, NY.
Tanis, J.A. and Woodland, W.T. (1990). *The Physics of Electronic and Atomic Collisions.* (A. Dalgarno, R.S. Freund, P.M. Koch, M.S. Lubell and T.B. Lucatorto, eds.), p. 366. AIP, NY.
Tanis, J.A. (1993). *VIth Intl. Conf. on the Phys. of Highly Charged Ions,* (ed., P. Richard, M. St"ockli, C.L.Cocke and C.D. Lin), *AIP Conf. Proc.* **279**, 262.
Taulbjerg, K., Barrachina, R.O. and Macek, J.H. (1990). *Phys. Rev.* **A41**, 207.
Taulbjerg, K. (1990). *Phys. Ser.* **42**, 205.
Taut, M. (1993). *Phys. Rev.* **A48**, 3561.
Taut, M. (1994). *J. Phys. A.* **27**, 1045.
Taut, M. (1994a). *Solid State Commun.* **89**, 189.
Taylor, J.R. (1972). *Scattering Theory: The Quantum Theory of Nonrelativistic Collisions* (John Wiley and Sons, NY).
Temkin, A. (1985). *Autoionization. Recent developments and applications,* Pleunm Press, NY.
Theisen, T.C. and McGuire, J.H. (1979). *Phys. Rev.* **A20**, 1406.
Thomas, L.H. (1927). *Proc. Royal Soc.* **A114**, 561.
Thompson, J.J. (1913). *Phil. Mag.* **26**, 792.
Toburen, L.H., Nakai, M.Y. and Langly, R.A. (1968). *Oak Ridge National Laboratory Report,* ORNL -TM-1988.
Tokesi, K. and Hock, G. (1996). *J. Phys. B.* **29**, L119.
Tolstikhin, O. (1995). private communication.
Tolstikhin, O., Watanabe, S. and Matsuzawa, M. (1995). *Phys. Rev. Lett.* **74**, 3573.
Trajmar, S. and McConkey, J.W. (1996). *Atomic, Molecular and Optical Physics Reference Book.* (G.W.F. Drake, ed.), AIP Press, NY, chapter 50.
Tuan, T.F. and Gerjuoy, E. (1960). *Phys. Rev.* **117**, 756.
Tulkki, J., Armen, G.B., Aberg, T., Cransmenn, B., and Chen, M.H. (1987). *Z. Physik* **D5**, 241.
Tweed, R.J. (1992). *Z. Physik* **D23**, 309.

*** U ***

Ullrich, J. Berg, H., Cocke, C.L., Euler, J., Lencinas, S., Froschauer, K., Spielberger, L., Ullmann, K., Schmidt-Böcking, H., Olson, R.E., and Schluz, M. (1993a). *VIth Intl. Conf. on the Phys. of Highly Charged Ions,* (ed., P. Richard, M. Stöckli, C.L.Cocke and C.D. Lin), *AIP Conf. Proc.* **279**, 251.
Ullrich, J. Moshammer, P., Berg, H., Mann, R., Tawara. H., Dörner, R., Euler, J., Schmidt-Böcking, H., Hagmann, S. Cocke, C.L., Unvergzagt, M., Lencinas, S., and Mergel, V. (1993). *Phys. Rev. Lett.* **71**, 1697.

*** V ***

Vana, M., Aumayr, F., Lemell, C. and Winter, HP. (1995). *Intl. J. Mass Spectr. and Ion Processes* **149 - 50**, 45.
Van den Bos, J. and de Heer, F.J. (1967). *Physica* **33**, 333.

Van den Bos, J. and de Heer, F.J. (1968). *Cerreta Physica* **40**, 161.
VanKampen, N.G. (1981). *Stochastic Processes in Physics and Chemistry.* North Holland, Amsterdam.
van Hemmen, L. (1996). *Invited papers of the 30th Anniversary Meeting of the Physical Society of Brazil,* (Aguas de Lindoia).
Vegh, L. (1988). *Phys. Rev.* **A37**, 992.
Vegh, L. (1989). *J. Phys. B.* **22**, L35.
Vegh. L., and Burgdörfer, J. (1990). *Phys. Rev.* **A42**, 665.
Vilenkin, N.Ya. (1971). *Combinatorics.* p. 38. Academic Press, NY.
Viswanathan, G.M., Afanasyev, V., Buldyrev, S.V., Murphy, E.J., Prince, P.A. and Stanley, H.E. (1996). *Nature* **381**, 418.
Vogt, H., Schuch, R., Justiniano, E., Schulz, M. and Schwab, W. (1986). *Phys. Rev. Lett.* **57**, 2256.
Vriens, L. (1966). *Proc. Royal Soc. London* **90**, 935.

*** W ***

Walker, B., Sheehey, B., DiMauro, L.F., Agostini, P., Schafer, K.J. and Kulander, K.C. (1994). *Phys. Rev. Lett.* **73**, 1227.
Walker, B. (1995). *Ph.D. thesis,* S.U.N.Y, Stony Brook.
Waller, I and Hartree, D.R. (1929). *Proc. Royal Soc. (London)* **A124**, 119.
Walters, H.R.J., (1988). *Electronic and Atomic Collisions.* (H.B. Gilbody, W.R. Newell, F.H. Read and A.C.H. Smith, eds.), p. 147. North Holland, Amsterdam.
Wang, J., Reinhold, C.O., and Burgdörfer, J. (1991). *Phys. Rev.* **A44**, 7243.
Wang, J., Reinhold, C.O., and Burgdörfer, J. (1992). *Phys. Rev.* **A45**, 4507.
Wang, J., Olson, R.E., Tökési, Burgdörfer, J., and Reinhold, C. (1993). *VIth Intl. Conf. on the Phys. of Highly Charged Ions,* (ed., P. Richard, M. Stöckli, C.L.Cocke and C.D. Lin), *AIP Conf. Proc.* **279**,
Wang, J., McGuire, J.H., and Burgdörfer, J. (1995a). *Phys. Rev.* **A51**, 4687.
Wang, J., McGuire, J.H. and Montenegro, E.C. (1995b). *Phys. Rev.* **A51**, 504.
Wang, J., Amusia, M. Ya. and McGuire, J.H. (1994). unpublished.
Wang, L.J., King, M. and Morgan, T.J. (1986). *J. Phys. B.* **19**, 623.
Wang, Y.D., McGuire, J.H. and Rivarola, R.D. (1989). *Phys. Rev.* **A40**, 3673.
Wang, Y.D., Straton, J.C., McGuire, J.H., and DuBois, R.D. (1990). *J. Phys. B.* **23**, L133.
Wang, Y.D. (1992). *Ph.D. thesis,* Tulane University.
Wang, Y.D. and Callaway, J. (1994). *Phys. Rev.* **A50**, 2327.
Wannier, G.H. (1953). *Phys. Rev.* **90**, 817.
Weaver, O.L. and McGuire, J.H. (1985). *Phys. Rev.* **A32**, 1435.
Webster, B.C., Jamieson, M.J. and Stewart, R.F. (1978). *Adv. in At. Mol. Phys.* **14**, 87.
Weigold, E., Frost,l. and Nygaard, K.J. (1981). *Phys. Rev.* **A21**, 1950.
Wentzel, G. (1927). *Z. Physik* **43**, 524.
Wells, J.C., Oberacker, V.E., Umar, A.S., Bottcher, C., Strayer, M.R., Wu, J.S. and Plunien, G. (1992). *Phys. Rev.* **A45**, 6296.
Whelitz, R., Heiser, F., Hemmers, O., Langer, B., Menzel, A. and Becker, U., (1991). *Phys. Rev. Lett.* **67**, 3864.

White, M.J. (1996). *Atomic, Molecular and Optical Physics Reference Book.* (G.W.F. Drake, ed.), AIP Press, NY, chapter 66.
Wigner, E.P. and Seitz, F. (1933). *Phys. Rev.* **43**, 804.
Williams, J.F. (1981). *J. Phys. B.* **14**, 1197.
Williams, J.F. (1986). *Aust. J. Phys.* **39**, 621.
Wight, G.R. and Van der Weil, M.J. (1976). *J. Phys. B.* **9**, 1319.
Wuilleumier, F.J., (1982). *Ann. Phys. Fr.* **4**, 231.
Wu, W., Wong, K.L., Montenegro, E.C., Ali, R., Chen, C.Y., Cocke, C.L., Dörner, R., Frohne, V., Giese, J.P., Mergel, V., Meyerhof, W.E., Raphaelian, R., Schmidt-Böcking, H. and Walch, B. (1995). *Phys. Rev.* A.
Wu, W., Datz, S., Jones, N.L., Krause, H.F., Sorge, K.A., and Vane, C.R. (1996). *Phys. Rev. Lett.* **76**, 4324.

*** Z ***

Zajfman, D. and Maor, D. (1986). *Phys. Rev. Lett.* **56**, 320.
Zare, R.N. (1967). *J. Chem. Phys.* **47**, 204.
Zare, R.N. (1988). *Angular Momentum: Understanding Spatial Aspects in Chemistry and Physics.* John Wiley, NY.
Zhang, L. and Rau, A.R.P. (1993). *Phys. Rev.* **A42**, 6342.
Zhang, X, Whelan, C.T. and Walters, H.R.J. (1990). *J. Phys. B.* **23**, L173.
Ziesche, P. (1995). *Int. J. Quantum Chem.* **56**, 363.
Ziman, J.M. (1970). *Elements of Advanced Quantum Theory.* Cambridge University Press.
Zouros, T.M., Lee, D.H., Sanders, J.M., Shinpaugh, J.L., Tipping, T.N., Richard, P. and Varghese, S.L. (1990). *Phys. Rev.* **A42**, 678.
Zouros, T.J.M., Schneider, D. and Stolterfoht, N. (1987). *Phys. Rev.* **A35**, 1963.
Zouros, T.J.M., Lee, D.H. and Richard, P. (1989). *Phys. Rev. Lett.* **62**, 2261.
Zouros, T.J.M., Lee, D.H., Sanders, J.M. and Richard, P. (1993). *Nucl. Instr. Meth.* **B97**, 166.

Index